The Physiology of Flowering

Volume III
The Development of Flowers

Authors

Jean-Marie Kinet, Ph.D.
Assistant
Centre de Physiologie Végétale Appliquée (I.R.S.I.A.)
University of Liège
Liège, Belgium

Roy M. Sachs, Ph.D.
Professor and Plant Physiologist
Department of Environmental Horticulture
University of California
Davis, California

Georges Bernier, Ph.D.
Professor of Plant Physiology
Director, Centre de Physiologie Végétale Appliquée (I.R.S.I.A.)
University of Liège
Liège, Belgium

CRC Press
Taylor & Francis Group
Boca Raton London New York

CRC Press is an imprint of the
Taylor & Francis Group, an **informa** business

First published 1985 by CRC Press
Taylor & Francis Group
6000 Broken Sound Parkway NW, Suite 300
Boca Raton, FL 33487-2742

Reissued 2018 by CRC Press

Library of Congress Cataloging in Publication Data
(Revised for volume 3)

Bernier, Georges, 1934-
The physiology of flowering.

Includes bibliographical references and indexes.
Partial contents: v. 3. The development of flowers.
1. Plants, Flowering of--Collected works.
I. Kinet, Jean-Marie. II. Sachs, Roy M. III. Title.
QK830.B47 1981 582.13'04166 82-159289
ISBN 0-8493-5709-8 (v. 1)
ISBN 0-8493-5710-3 (v. 2)
ISBN 0-8493-5711-X (v. 3)

A Library of Congress record exists under LC control number: 82159289

ISBN 13: 978-1-315-89656-4 (hbk)
ISBN 13: 978-1-351-07566-4 (ebk)

Visit the Taylor & Francis Web site at http://www.taylorandfrancis.com and the
CRC Press Web site at http://www.crcpress.com

OBJECTIVES AND ORGANIZATION OF THE WORK

> It is to this concentration of studies upon the problem of the *initiation* of flowering that we owe the tendency in so much reported experimental work to accept the formation of barely differentiated primordia as synonymous with flowering, and even the naive concept of the single flower-'forming' substance itself.
>
> J. Heslop-Harrison*

The arbitrary choices, often required for clarity of exposition, have driven us to consider in separate volumes floral evocation and initiation, on one hand, and floral development to anthesis, on another hand. This division should not lead one to forget, however, the essential continuity and unity of the various stages constituting the flowering process. Flowering is, after all, about sexual reproduction in higher plants and we must expect that the entire physiological process is designed to expedite recombination of genetic characters and reproduction of the organism from seed.

Our division of the flowering process into two major stages may in fact be justified *a posteriori* because it is clear that evocation (together with initiation of flower primordia) and development of inflorescence or flower structures are two aspects that have concerned different research workers and even entirely different scientific institutions. Evocation and initiation, which are characterized essentially by a morphogenetic event triggered by cellular and molecular changes, have occupied primarily the interest of scientists engaged in fundamental research. By contrast floral development has been largely the object of research programs with economic applications in mind. The economic value of the flower is undeniable in floriculture and also generally in horticulture and agronomy since it is the prerequisite step for fruit and grain production. This difference in research groups and ultimate interest has resulted in a much more profound separation than one expected at first sight; namely, often entirely different plant species have been investigated by physiologists interested in evocation and by those involved in the study of development.

The first two volumes of this series are devoted to the floral transition, i.e., the shift-over from leaf production to flower or inflorescence initiation by shoot meristems. This volume is concerned with the next steps of the flowering process from the earliest stages of initiation through differentiation, growth, and maturation of flower and inflorescence parts to anthesis.

This development phase is extremely complex; in fact it is much more complex than the stage of floral evocation or transition. It is the result of a large number of parallel reaction sequences that are coordinated in time and space to produce in harmonious fashion numerous different reproductive structures. Among the different component processes of floral development we have specified certain qualities for detailed analyses, namely, the rate of development, flower number and size, axis elongation, sex expression, and flower bud dormancy. Intentionally certain stages dealing with differentiation of specific tissues of floral organs, most notably sporogenesis, have not been covered because numerous synthetic works on this material, above all on microsporogenesis, have already been published. Other aspects not developed in this volume, like flower opening or perianth pigmentation, have received relatively less attention from physiologists.

This volume begins with a detailed description of the reproductive structures and of their morphogenesis and development. This is essential information for everything that follows because we make frequent references to stages of development and specific structures affected.

* From Heslop-Harrison, J., *Biol. Rev.*, 32, 38, 1957. With permission.

The organization of the remaining part of this volume is guided largely by that of the first two volumes and, insofar as possible, we have retained the same outline. That is to say, we have considered separately the influences of environmental factors, each class of growth regulators and photosynthesis on the different processes of floral development. Such organization of work is not without problems. Information on each of the important stages appears in a discontinuous fashion. This confers a degree of repetitiveness to the entire work, and perhaps interferes with deeper insights into the morphogenetic sequences and physiological processes that control development.

Another ''philosophy'' of organization would have been subdivision of the volume into chapters on stages of development and underlying physiological processes. But it is clear to us that even with this latter organization it would not have been possible to avoid considering the role of different environmental and chemical factors in a somewhat repetitive fashion.

Finally, the present volume ends with a chapter on the applied aspects of the research on flower initiation and development.

The treatment of the different chapters is neither simplistic nor exhaustive. Our general philosophy has been to avoid extreme positions, either abusive generalizations that mask the real complexity of the problem and the diversity of plant behaviors or complete descriptions of all possible types of plant responses that create confusion and discourage the readers. Evidently, when one attempts to cover such an extensive subject in a limited number of pages there is inevitably a problem of topic selection. While our aim was to provide a balanced account of the most important and recent contributions in all aspects of the subject, some topics have wittingly received special treatment. Their selection reflects essentially our personal interest; other writers would have certainly made other choices and presented a differently balanced book.

It is important to underline that constant reference to source material and use of a rich illustration should assist the unspecialized reader to obtain a full understanding of the discussed topics. Concluding sections are also inserted in many places and hopefully will be considered as resting spots. The busy reader may begin with these sections and the short Chapter 10, and return to the main text for examination of important details. A glossary is also included for the reader who is unfamiliar with the scientific jargon of the field. In a work like this, there is some unavoidable repetition of material, but this has been reduced by frequent use of cross references. The species most commonly used in flowering studies will be usually referred to by their generic or common names alone.

In this work we deal mainly with angiosperms, although gymnosperms are occasionally considered.

We hope that these three volumes will convey some of the excitement that we have felt during their preparation as well as during our investigations on flowering.

THE AUTHORS

Jean-Marie Kinet, Ph.D., is a member of the Centre de Physiologie Végétale Appliquée (I.R.S.I.A.), Liège, Belgium.

He graduated in 1964 from the University of Liège, with a "licence" degree in botany and obtained his doctoral degree in 1972 from the same university. He worked in the laboratory of Professor Wareing, University College of Wales, Aberystwyth, U. K. (1975).

Dr. Kinet is a member of the American Society for Horticultural Science, the Belgian Society of Cell Biology, the Federation of European Societies of Plant Physiology, and the International Society for Horticultural Science.

He has published more than 40 research papers. His current major research interests include the physiological and cellular aspects of flower initiation, inflorescence and flower development, and the inference that fundamental studies have on research concerning species of economic interest.

Roy M. Sachs, Ph.D., is Professor and Plant Physiologist, Department of Environmental Horticulture, University of California, Davis.

Dr. Sachs graduated in 1951 from the Massachusetts Institute of Technology, Cambridge, with a B.Sc. in Biology, and obtained the Ph.D. degree in Plant Physiology in 1955 from the California Institute of Technology, Pasadena, under the guidance of Dr. Went. He was a Fulbright Fellow at the Università di Parma, Instituto di Botanica, Italy, in 1955 and 1956, working with Dr. Lona. From 1956 to 1958 he was a Research Botanist at the University of California, Los Angeles with Dr. Lang. In 1958 he began his academic career with the University of California, Los Angeles (1958 to 1961) and was with the Department of Floriculture and Ornamental Horticulture, then moved to Davis to join the Department of Landscape Horticulture (later renamed Environmental Horticulture). In 1969 to 1970 he was a visiting research scientist at the Vegetation Control Laboratories, U.S. Department of Agriculture, Fort Detrick, Frederick, Md., in 1975 a North Atlantic Treaty Organization (NATO) visiting Professor at the Facoltà di Agraria, Universita di Padova, Italy, and in 1977 a visiting Research Scientist at the Département de Botanique, University of Liège, Belgium.

Dr. Sachs is a member of the American Society of Plant Physiologists. He has been the recipient of numerous grants for research on control of vegetative and reproductive development in higher plants. He has presented over 20 invited lectures on the control of vegetative and reproductive development.

Dr. Sachs is the author of more than 100 publications in scientific and technical journals, including review articles on reproductive development in higher plants. His current major research interests are in the areas of reproductive development, chemical control of growth, biomass production, and fuel conversion systems.

Georges Bernier, Ph.D., is Professor of Plant Physiology, University of Liège and Director of the Centre de Physiologie Végétale Appliquée (I.R.S.I.A.), Liège, Belgium.

Dr. Bernier graduated in 1956 from the University of Liège, with a "licence" degree in botany and obtained his doctoral degree in 1963 from the same university. He performed an in-depth investigation on the uses of plants in nutrition, medicine, and other aspects of life in semiprimitive human societies in Zaire (1957 to 1958). He worked in the laboratories of Professors Buvat and Nougarède, Ecole Normale Supérieure, Paris, France (1960 to 1961), and of Professor Jensen, University of California, Berkeley (1964 to 1965).

Dr. Bernier is a member of the American Society of Plant Physiologists, the French Society of Plant Physiology, the Society for Experimental Biology (U.K.), the Federation of European Societies of Plant Physiology, the Belgian Society of Biochemistry, and the

Belgian Society of Cell Biology. He was elected as a member of the Académie royale des Sciences, des Lettres et des Beaux-Arts of Belgium in 1982 and was president of the Belgian Association for Plant Physiology (1983 to 1984).

Dr. Bernier has presented many inivited lectures at international meetings, universities, and institutes. He has published more than 70 research papers and book chapters, and a book on *Cellular and Molecular Aspects of Floral Induction* (Longman, London, 1970) including the proceedings of a symposium that he organized at the University of Liège. His current major research interests include the molecular and cellular changes occurring in meristems at floral evocation and subsequent flower initiation, and the nature of the factors that control these developmental steps.

ACKNOWLEDGMENTS

In preparing this work we acknowledge the support of our colleagues, first in our own departments (M. Bodson, W. P. Hackett, A. Havelange, and A. Jacqmard), and secondly, those who have contributed so much to the original data base on flowering. But in a particular way we have appreciated the efforts of A. H. Blaauw, E. Galun, A. H. Halevy, J. Heslop-Harrison, J. P. Nitsch, and C. W. Wardlaw who have "paid their dues" in writing provocative and sometimes encyclopedic reviews of the subject.

We are greatly indebted to L. Simon and J. Brulfert for supplying unpublished materials, and to many copyright owners for permission to reproduce illustrations and data from their publications. These sources are fully acknowledged in the captions of the figures and tables. One of us (Jean-Marie Kinet) would especially like to acknowledge the help and patience of his wife, Cécile, during the preparation of the manuscript. Our particular thanks are due to Michelyne Dejace for her patient cooperation in typing the successive stages of the manuscript. The assistance of Lucie Debaisieux, Maggy Grossart and André Parmentier in collecting and classifying the bibliography and preparing the graphs is also gratefully acknowledged.

THE PHYSIOLOGY OF FLOWERING

Volume I
THE INITIATION OF FLOWERS

Experimental Systems
Control by Nutrition and Water Stress
Control By Daylength
Light Perception and Time Measurement in Photoperiodism
Control by Low Temperature
Classical Theories of Induction
Age and Flower Initiation

Volume II
TRANSITION TO REPRODUCTIVE GROWTH

The Basic Role of Correlative Influences
Thge Flowering Process at the Shoot Apex: Macromorphological Events
The Flowering Process at the Meristem: Anatomical and Histocytological Events
The Flowering Process at the Meristem: Cell Proliferation and Molecular Events
The Nature of Floral Evocation
Exogenous Substances
Levels, Distribution, and Metabolism of Endogenous Substances
Photosynthesis, Energetics, and Assimilate Supply
A Model of Control of Floral Evocation

Volume III
THE DEVELOPMENT OF FLOWERS

Morphology of Inflorescence and Flower Development
Experimental Systems
Control by Nutrition
Control by Light
Control by Temperature and Water Stress
Basic Role of Correlative Influences
Exogenous Substances
Levels, Distribution, and Metabolism of Endogenous Substances
Photosynthesis, Assimilate Supply, and Utilization
Concluding Analysis of the Flowering Problem
Applied Aspects of Floral Initiation and Development

Each volume contains a bibliography, glossary, abbreviation list, and species and subject indexes.

TABLE OF CONTENTS

Volume III

Chapter 1

MORPHOLOGY OF INFLORESCENCE AND FLOWER DEVELOPMENT

TABLE OF CONTENTS

> Some of the controversial matters, which figure so largely in the morphological literature, either become unimportant, or can be harmonized, if the concepts of organogenesis are transferred from the morphological to the physiological plane.
>
> C. W. Wardlaw*

I. INTRODUCTION

The various changes that occur in shoot apices during floral evocation were described in Volume II, Chapters 2 to 5. Evoked apices normally develop into either a solitary flower or more often a cluster of flowers. Although we concluded in Volume II that, as far as the available evidence goes, the features of evocation seem fairly universal, there is on the contrary an extraordinary diversity of structural changes occurring during inflorescence and flower morphogenesis. All of these possible variations result in an enormous array of forms that makes inflorescences and flowers the beautiful objects they are.[1,2] As will become clear in the following chapters, the many stages that can be distinguished in the ontogeny of these reproductive structures do not all react similarly to environmental variables and growth regulators, and thus are not all alike. In physiological investigations they should be considered independently and in relationship to one another. These investigations should therefore have a firm morphological basis and the frequent reluctance of physiologists to enter into the details of structural changes should be overcome. Following this view we shall describe in Section II of this chapter the reproductive structures of a few species among those which are most preferentially used in the studies reviewed in this volume. Emphasis will be given to early ontogeny of these structures since macroscopic characteristics of mature inflorescences and flowers are determined by their early morphogenetic pattern. Then, starting from the diversity of morphological processes, we shall try to see in Sections III and IV if there are rules governing the morphogenesis of these varied forms and their further growth and development. A small Section V on cellular and molecular changes observed during flower development terminates this chapter.

Some of the terms, e.g., carpel, we shall use in our descriptions are unfortunately not simply descriptive but interpretive since they have been created or used by morphologists within the framework of conflicting theories on the interpretation of the nature of inflorescences and flowers. There is little need and no gain to the physiologist in entering these controversies. The unavoidable use of morphological terms in this volume is purely descriptive and does not imply acceptance of the corresponding theories.

II. SOME CASE DESCRIPTIONS

A. Solitary Flowers

Tulip (*Tulipa gesneriana*) plants produce a solitary flower at the top of their main axis. The apical meristem, from which next year's flower will develop, produces three to five foliage leaves, then increases in size and height and initiates in rapid sequence: (1) the three outer perianth members or tepals (Figure 1A), (2) the three inner tepals, (3) the outer whorl of three stamens, each located in the axil of an outer tepal (Figure 1B), (4) the inner whorl of three stamens, each located in the axil of an inner tepal, and (5) the three carpels which

* From Wardlaw, C. W., *Proc. R. Soc. Edinburgh Sect. B*, 66, 394-408, 1957. With permission.

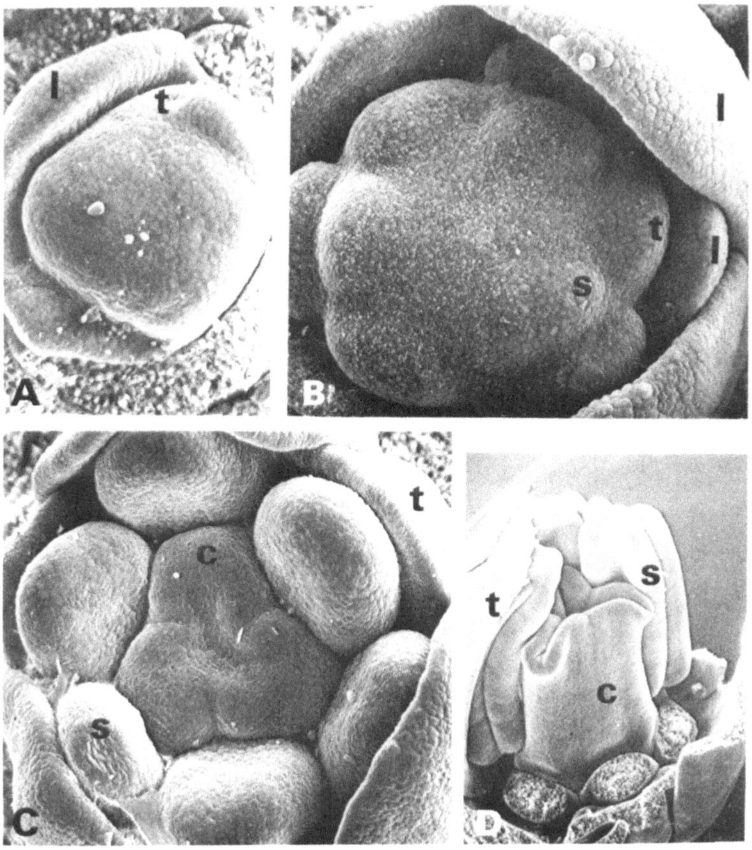

FIGURE 1. SEM micrographs of apical meristem of *Tulipa gesneriana* (tulip). (A) At the stage of initiation of the three outer perianth members or tepals (t). The last leaf (l) primordium is visible. (B) At the stage of initiation of the outer whorl of three stamens (s). Leaf primordia and the two sets of tepals are visible. (C) At the stage of initiation of the three carpels (c). (D) Young developing flower bud. Magnification in D is four times lower than in A, B, and C. (From Shoub, J. and de Hertogh, A. A., *J. Am. Soc. Hortic. Sci.*, 100, 32, 1975. With permission.)

alternate with the three inner stamens (Figure 1C). Several workers claimed that each tepal and associated stamen, in other Liliaceae species related to the tulip, arise together from a unique primordium and become distinct only during later growth.[1,3,4] Work by Shoub and de Hertogh with the more refined technique of scanning electron microscopy (SEM) has shown, however, that tepals and stamens are initiated independently (Figure 1), even though they may have the appearance of a united primordium with a dissecting light microscope.[5] The meristem, which is dome shaped until stamen initiation, becomes flat topped later on. Carpels appear as three crescent-like ridges which are initially free. The carpellary ridges grow towards the center of the flower where their margins meet and fuse. The compound ovary then grows upward and produces a three-lobed stigma at its top (Figure 1D). Placentae develop at the junctions of the carpellary ridges and each bears numerous ovules.

Solitary flowers may also be found exclusively in axillary positions, e.g., in *Anagallis arvensis* and *Impatiens balsamina*. In conditions continuously favorable to flowering, these plants bear one flower in each leaf axil. This kind of floriferous stem is very similar to a raceme without a terminal flower (see below), except for the presence of leaves instead of bracts. In other plants, like *Pharbitis nil*, both a terminal and several axillary flowers are formed.

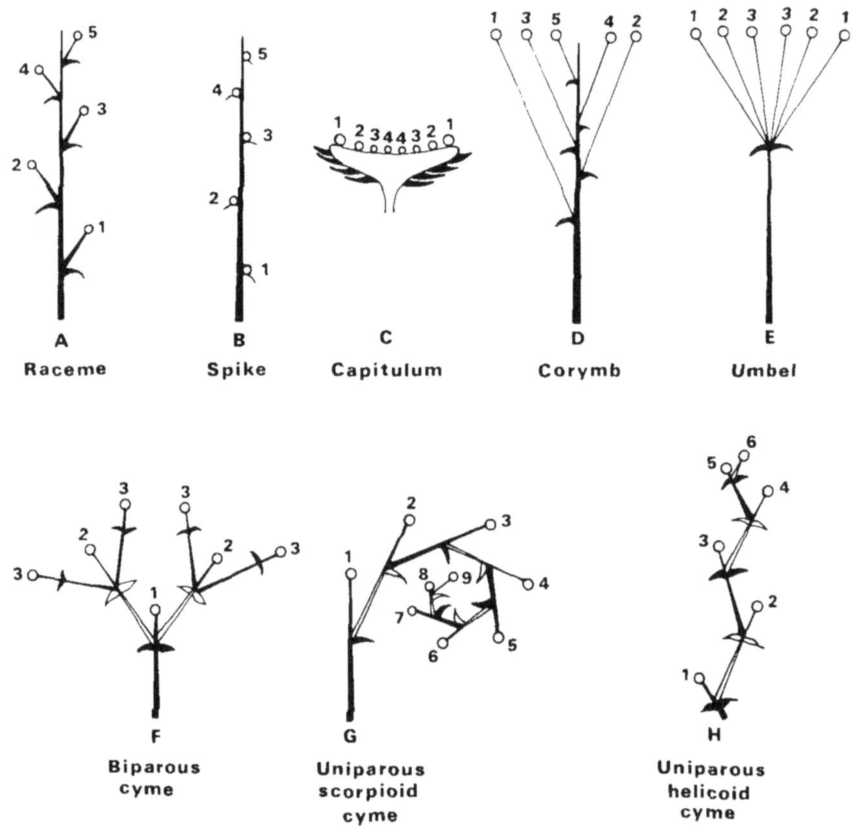

FIGURE 2. Diagrams of some of the common types of inflorescences. Flowers are represented by circles and their order of initiation indicated in numerical sequence with flower 1 being the oldest. (From Camefort, H. and Boué, H., *Reproduction et Biologie des Végétaux Supérieurs*, Doin, Paris, 1980. With permission.)

B. Racemose Type of Flower Groupings

In the racemose inflorescences, the apical reproductive meristem produces flowers typically in an acropetal direction. The oldest flowers are thus at the base of these inflorescences and the youngest near their apex. A terminal flower may or may not be formed. Growth is considered to be indeterminate because it is theoretically possible for such inflorescences to continue developing new flowers at their apex for an extended period. Actually the number of flowers produced may be small and growth may stop rather soon. The classical type of racemose inflorescence is the "raceme", which consists of a unbranched, elongated axis with pedicellate lateral flowers subtended or not by bracts (Figure 2A). All other racemose inflorescences can be derived from this prototype.

1. Raceme

Cruciferae, like *Arabidopsis, Brassica,* and *Sinapis*, typically produce racemes. In *Sinapis* prior to flower initiation, a "buttress" is formed on one flank of the reproductive meristem, essentially by repeated anticlinal divisions in the four to five most superficial cell layers. Initiation then occurs by periclinal divisions in the third and/or fourth layers of the buttress. This is followed by other variously oriented divisions in the surrounding cells, and the flower primordium, uniformly meristematic, emerges above the meristem surface (see Volume II, Chapter 3, Figure 6).[6] With initiation of a flower primordium, a sector of the flank is largely consumed in the process and this sector is usually restored by divisions — mostly anticlinal — in the remaining cells in the axils of the primordium, not in the central cells of the meristem. In this manner the reproductive meristem of *Sinapis* and other Cruciferae functions very similarly to that of vegetative meristems described in Volume II, Chapter 3, Section III.A.2.[6,8] Likewise, these reproductive meristems exhibit a distinct zonation just as vegetative meristems, but they lack the zone of flattened cells dividing transversely, called the "pith-rib meristem", which in vegetative meristems lies typically just below the central zone.

A great number of flower primordia are successively laid down by the flanks of the meristem, and reproductive development ends without formation of a terminal flower. Typically, flowers are not subtended by bracts in *Sinapis* and many other Cruciferae, although a close examination of flower primordia indicates that there is a vestigial bract at their base.[9] There is increased phyllotactic complexity in flower arrangement along the raceme axis as compared to leaf arrangement in the basal vegetative part of the shoot.[10] Vertical growth of the raceme in these plants is due to cell proliferation and elongation in the subapical tissues, or "primary elongating meristem",[11] which in turn derive from the activity of the apical reproductive meristem.[7,12]

Flower morphogenesis in Cruciferae follows a very characteristic pattern. The first part to differentiate from the nascent uniformly meristematic floral primordium is the pedicel.[10,12] This occurs by cell vacuolation and elongation in the basal region of the protuberance. Then, the flower meristem initiates four sepals in such a rapid sequence that different authors disagree on their order of initiation.[1] The four petals appear simultaneously in alternate positions relative to the sepals. Photographs by Sattler clearly show that they are much smaller than the sepal and stamen primordia at initiation and occupy a much smaller area of the meristem periphery.[1] Petals have a retarded development compared to other floral organs: they grow very little until the late stages of flower development during which they reach their final large size by very rapid growth. After petal inception the six stamens are initiated, first the two outer ones opposite to two sepals, and then the four inner ones in two pairs each opposite to one of the remaining sepals. Each stamen is initiated as an individual primordium. Finally, the remainder of the meristem, except for a few centrally located cells, is given over to the formation of a terminal gynoecial ridge of elliptical shape.[1,10] This ridge, as well as primordia of sepals and stamens, are initiated by periclinal divisions occurring in the second cell layer of the meristem.[10,12] As the gynoecial ridge grows upward, it forms the ovary wall surmounted by two stigmatic lobes. Ovules then develop from two pairs of placental bands arising from the ovary wall. Further ontogenetic changes include the formation of a false septum within the ovary.

Labiatae, such as *Perilla* and *Coleus*, also produce racemes, but these differ from those of the Cruciferae in a number of ways.[13,14] First, the flowers in Labiatae are subtended by bracts which are initiated by periclinal divisions of the second layer of the meristem flanks. Further development of bracts is limited and they are ultimately smaller than normal foliage leaves. Flowers arise as precocious axillary buds of the bracts. Second, the bracts are decussate as are the foliage leaves of the basal vegetative portion of the plant axis. Thus, contrary to Cruciferae, there is no phyllotactic change during the floral transformation of the apical meristem in these Labiatae. Third, there is no zonation in the reproductive meristems of *Perilla* and *Coleus* and no buttress is formed in preparation for bract initiation. On the other hand, the racemes of Labiatae, like those of Cruciferae, are indeterminate since their apical meristem finally vacuolates and dies without having formed a terminal flower.

2. Spike

The "spike", typical of some orchids, is similar to a raceme except that the flowers are sessile (Figure 2B). In some grasses, like wheat (*Triticum aestivum*), rye (*Secale cereale*), barley (*Hordeum vulgare*), and *Lolium* species, the lateral primordia from the reproductive meristem are not flowers but secondary spikes or "spikelets".

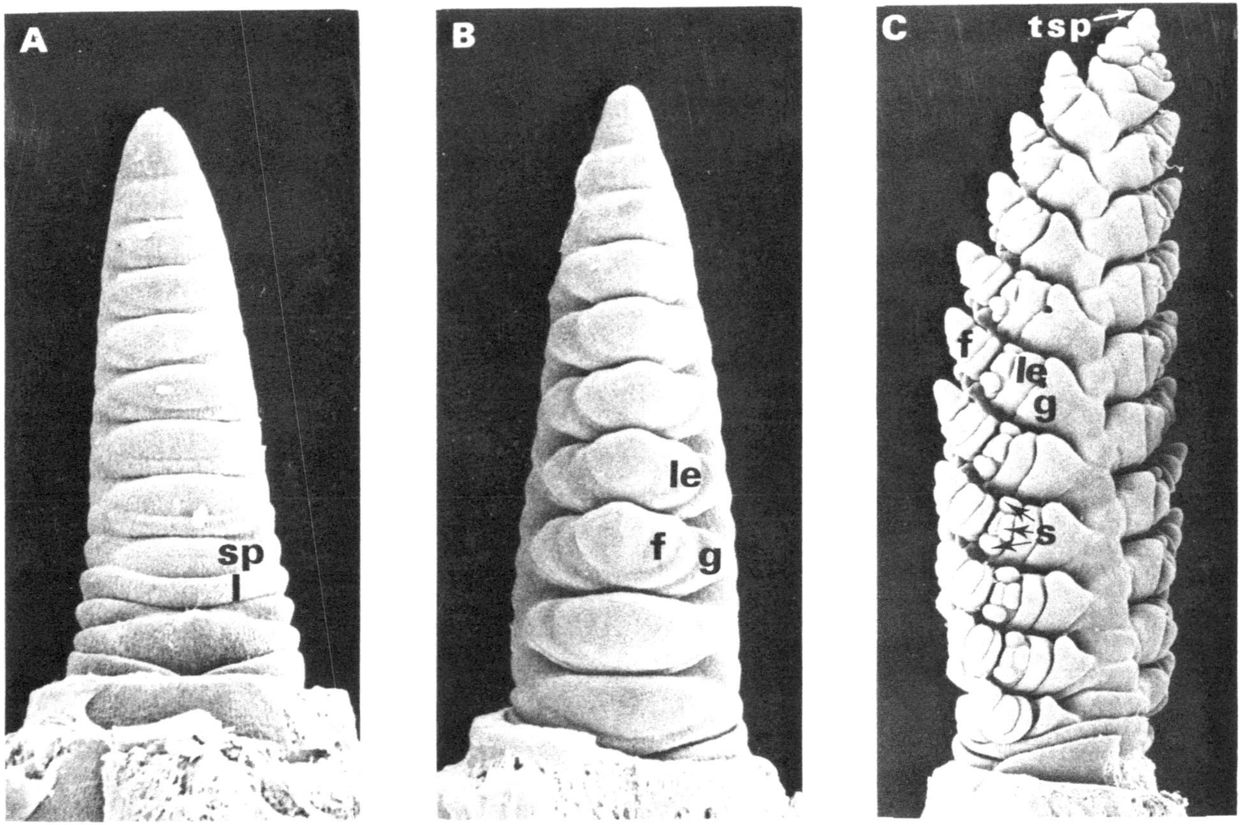

FIGURE 3. SEM micrographs of reproductive meristem of wheat (*Triticum aestivum*). (A) At the double-ridge stage. The lower ridge is the leaf (l) primordium and the upper ridge a spikelet (sp) primordium. Note that double ridges are most advanced about the middle of the elongated apex. (B) At the start of floret (f) initiation. Note the *glumes* (g), *the first spikelet parts to differentiate*, and the lemmas (le). At this stage the leaf primordia are no longer visible. (C) At the stage of formation of the terminal spikelet (tsp). The three stamens (s) appear as swellings beneath the meristem of the floret. Magnification in C is half that in A and B. (From Moncur, M. W., *Floral Initiation in Field Crops. An Atlas of Scanning Electron Micrographs*, CSIRO, Canberra, 1981. With permission.)

In wheat, the first indication of spike (ear) formation is the appearance of "double ridges" at the apex, i.e., the formation of axillary meristems in the axils of the distichously arranged leaf primordia (Figure 3A).[15,16] These axillary meristems develop into spikelets. After a short period the growth of the foliar primordia completely ceases, whereas that of the axillary structures becomes dominant and soon obscures the presence of the leaf initials (Figure 3B). Spikelet primordia form initially in the middle of the elongated apex or prospective spike and then extend acro- but also basipetally along the main axis or "rachis" (Figure 3A). Finally, the apical meristem becomes the primordium of the terminal spikelet (Figure 3C). Between 15 and 30 spikelets are produced per spike depending on the cultivar (cv.) and growing conditions.[17,19] Spikelet number, one of the most important attributes of yield in wheat, was shown by Rawson to depend on the durations of (1) the period from sowing to double-ridge initiation when leaf primordia, which are potential spikelet sites, are accumulated acropetally along the elongating meristem, and (2) the period from double-ridge initiation to terminal spikelet formation during which all the spikelet primordia appear.[17] In cold-requiring cvs., e.g., 'Late Mexico 120', the principal determinant is the length of the first period (Figure 4), whereas it is the length of the second in cvs. with no vernalization response.

The first organs formed by a spikelet meristem are two transverse overlapping bracts with no axillary structure, the so-called "empty glumes" (Figure 3B). Then other bracts, the "lemmas", each with one floret in its axil are laid down in acropetal succession along the spikelet axis or "rachilla". Glumes and lemmas, as foliage leaves, are arranged in an alternate sequence in two rows, i.e., distichously. Each spikelet produces six to eight lemmas and floret primordia, but some of these abort leaving only two to four floret primordia developing to recognizable complete florets (Figure 5).[15,19] Under special growing conditions, spikelets may differentiate many more florets.[16] The floret meristem is an elongated hemispherical structure which initiates in close succession, on a very short axis, first a bract called the "palea", then two scales or "lodicules", three stamens, and one unilocular ovary.

According to Barnard,[16,20] both the apical meristem of a spikelet and the floret meristem have the same histological organization as the apical meristem of the main axis, i.e., the meristem of the spike (illustrated in Volume II, Chapter 3, Figure 8 in the case of *Lolium temulentum*). Glume, lemma, palea, and lodicule primordia are all derived from periclinal divisions in the two most superficial cell layers, as are leaf primordia, and they also develop in the same way as leaves. Stamens, on the contrary, have a deeper origin since periclinal divisions associated with their initiation occur in the second and third cell layers but never in the epidermis. Barnard stressed the fact that their origin is similar to that of the lateral spikelet and floret primordia as well as to that of vegetative axillary buds.

A crescent-shaped carpel emerges from the remaining floret meristem by periclinal divisions in the two superficial layers and grows rapidly as a cowl-shaped structure. Development from the edges eventually closes the base of the cowl and forms the loculus of the ovary. The apex of the floret meristem then develops directly into the ovule.

At maturity, the large anthers are exserted from the floret by active growth of the filaments and the ovary is surmounted by two long feathery styles. The awn (in awned cvs.) arises from a vigorous outgrowth of the abaxial side of the lemma.

Contrary to the cases of wheat and other grasses mentioned above, in which inflorescences are only once branched, the inflorescences of oat *(Avena sativa)*, *Sorghum*, sugarcane (*Saccharum officinarum*), timothy (*Phleum pratense*), orchard grass (*Dactylis glomerata*), etc. are multiple branched. In such species Evans and Grover showed that each secondary protuberance (axillary meristem of the double ridge), instead of developing into a spikelet, elongates and tertiary protuberances are initiated along its length distichously and acropetally in a plane at right angles to the plane of the secondary protuberances.[21] The tertiary protuberances, as they in turn elongate, may develop quaternary protuberances, and so on to

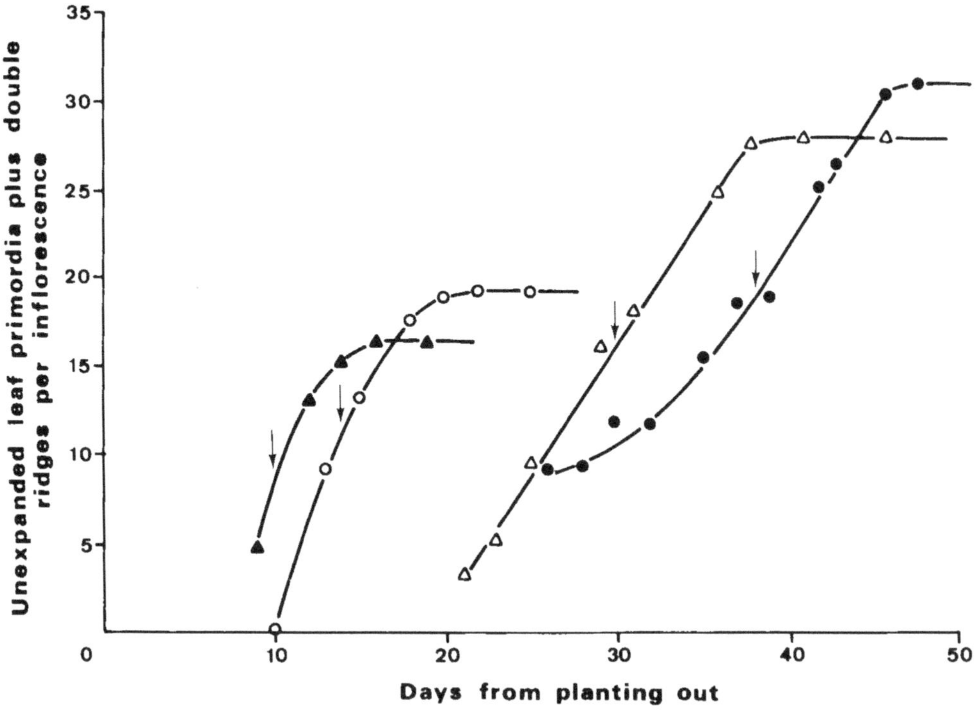

FIGURE 4. Spikelet primordia appearance on the main spike (ear) of the 'Late Mexico 120', a cold-requiring cv. of wheat (*Triticum aestivum*) grown at 21/16°C in 16 hr LD after imbibed seeds had been exposed to cold (2 to 4°C) for 0 (●), 4 (△), 8 (○), or 12 (▲) weeks before planting. Note that 12 weeks of cold reduce spikelet number from 31 to 16 and that this decrease parallels that in number of unexpanded leaf primordia (potential spikelet sites) present at the time double ridges first appear (indicated by arrows). The difference between final spikelet number and that present when spikelet initiation begins (double-ridge initiation) is due to continued development of the inflorescence apical meristem for several days until a terminal spikelet is formed. Thus, 10 to 12 days may elapse for the treatments lasting 0 and 4 weeks vs. 7 days for other regimes, between first double-ridge formation and terminal spikelet initiation. In this time 12 more spikelets are initiated in plants receiving 0 and 4 weeks of cold, whereas this number is only 8 in plants submitted to 8 and 12 weeks of cold. (From Rawson, H. M., *Aust. J. Biol. Sci.*, 23, 1, 1970. With permission.)

even higher orders. Each successive order of branches is at right angles to those of the next lower order. All protuberance primordia arise in principle in the axils of leaf primordia, i.e., derive from the upper bulges of double ridges, but the growth of leaf primordia may become so restricted that they may pass unnoticed in branches of high order. The extent of branching and internodal elongation in the inflorescence differs greatly in different species. Elongation is so negligible in some cases that heads or spikes of a compact character result, as in timothy. In other cases, like oat and orchard grass, some internodes of the rachis and its branches elongate to such an extent as to form loose and relatively large inflorescences called "panicles". Finally, spikelets differentiate on all branches. In oat, they form first at the tip of the main axis and proceed basipetally at the tips of the first-order branches. At the nodes, the sequence of spikelet differentiation is from branches of first order to branches of higher order.[22]

The inflorescences in different species of Gramineae also differ in a number of other details.[1,20,22-28] Thus, while wheat, rye, and *Lolium* have determinate spikes ending in a terminal spikelet, barley and corn (*Zea mays*) have indeterminate inflorescences without any terminal spikelet. Whereas wheat, rye, and *Lolium* have only one spikelet at each node of the rachis, corn and sugarcane have two and barley three. The "triple-mound" stage, typical of barley, corresponds to the initiation of two lateral spikelets surrounding the central initial one at each node.

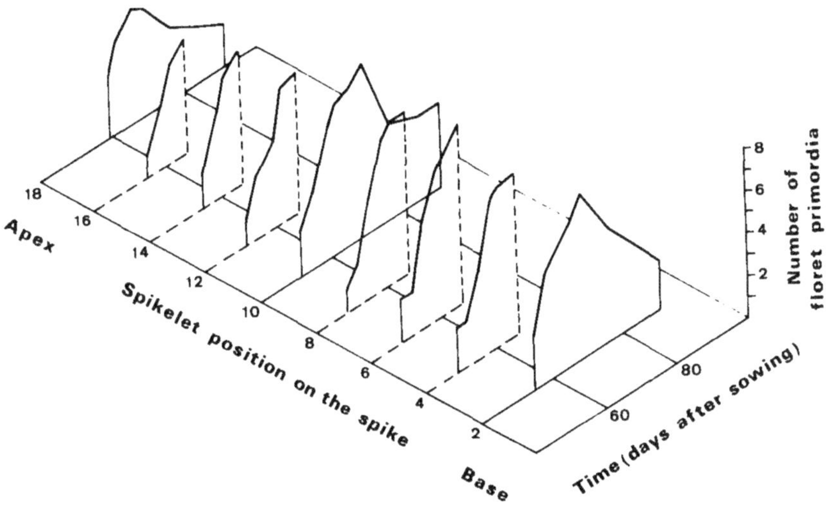

FIGURE 5. Number of floret primordia per spikelet at various positions on the spike (ear) of wheat (*Triticum aestivum* cv. 'Kolibri', spring variety) at various times after sowing in the field in March. Note that there is little difference in the rate of floret initiation between spikelets, but initiation starts first in spikelet 8 and becomes progressively later in more apical or basal spikelets. On the 70th day after sowing and later floret number was counted only in the spikelet positions 2, 10, and 18 (terminal spikelet) but no more florets are produced. The terminal 3 to 5 floret primordia die from 70 to 90 days leaving only 2 to 4 developing florets per spikelet. (From Kirby, E. J. M., *J. Agric. Sci. Camb.*, 82, 437, 1974. By permission of the Cambridge University Press.)

Barley spikelets have one floret, whereas rye, oat, and corn have up to three and wheat, bamboos (*Bambusa*), and *Lolium* have many. Where more than one floret is produced the number that develops to maturity is dependent upon the environmental conditions and is usually less than the number initiated. Other differential characters are found in the number of bracts and floral appendages produced. For example, wheat, rye, barley, and sugarcane all possess two glumes, two lodicules, and three stamens; *Lolium* has only one glume; *Anthoxanthum* has no lodicules and only two stamens; and bamboos have three lodicules and six stamens arranged in two whorls.

Florets in Gramineae are usually hermaphroditic, but they are unisexual in corn. Male and female florets in corn are in separate inflorescences which differ in shape and location in the plant. The male inflorescence or "tassel" is a multiple-branched panicle and is terminal, i.e., it derives from the activity of the apical shoot meristem, whereas the female inflorescences or "cobs" are once-branched spikes and are in the axils of one or more upper leaves of the stem.[25] According to Bonnett, primordia that become lateral branches of the tassel do not differ externally from those which differentiate into spikelets in the cobs.[25]

Despite the great diversity of forms exhibited by mature inflorescences in Gramineae, close examination reveals that the different species have much in common in the early morphogenesis of their reproductive structures.

3. Capitulum

Compositae, like *Chrysanthemum, Aster, Dahlia, Cosmos,* and *Xanthium,* typically produce "capitula" or "heads", in which all florets are inserted on the more or less flattened axis (receptacle) and are surrounded by bracts (involucre). The capitulum is a racemose inflorescence because the florets are initiated acropetally on this broadened axis (Figure 2C).

The early steps of floral transition in *Chrysanthemum,* up to the stage of receptacle formation and attainment of the mantle-core type of organization by the apical meristem,

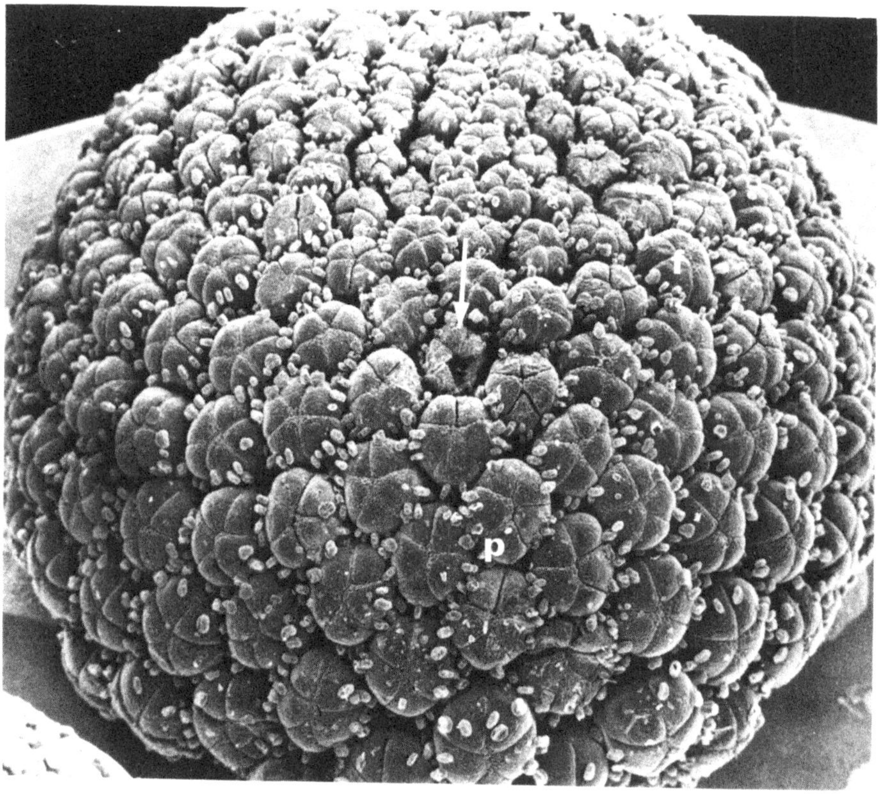

FIGURE 6. SEM micrograph of a developing capitulum of *Chrysanthemum morifolium* cv. "Dramatic". At this stage the receptacle is entirely covered with well-developed florets (f). All visible florets are of the disc type. Marginal ray florets are invisible in this picture. The five petal lobes (p) of the corolla tube are evident on disc florets. Note that the most apical floret (arrow) is abnormal. (Reproduced by permission of the National Research Council of Canada, from Vermeer, J. and Peterson, R. L., *Can. J. Bot.*, 57, 705, 1979.)

are described in Volume II, Chapter 3, Section III.C.2. Later changes in *C. segetum* and *C. morifolium* include the further enlargement of the inner parenchymatous core due to divisions in the internal cell layers of the superficial meristematic mantle, and the initiation of floret primordia by this mantle. For Popham, the first indication of floret initiation may be elongation and anticlinal divisions occurring in a small group of cells of the epidermis.[29] This is followed rapidly by periclinal divisions in subjacent cell layers, essentially in the third one, according to Lance.[30] This contrasts with leaf or bract initiation which results, as a rule, from periclinal divisions in the subepidermal or second layer. Derivatives of these periclinally dividing cells elongate and undergo additional periclinal and anticlinal divisions, producing a little protuberance — the floret primordium — in which the cells of the two superficial tunica layers divide mostly anticlinally.

Differentiation inside the floret primordium starts by formation of a ring-shaped corona which is formed as a result of enlargement and periclinal divisions of lateral subepidermal cells.[29,30] This corona eventually grows into the corolla tube. Five petal lobes arise, then, by periclinal divisions in the subepidermis of the upper margin of the corona (Figure 6). As a result of upward growth of the corolla tube, the central part of the floret becomes cup-shaped. Shortly afterwards stamen initiation is detectable at five places on the inner side of the nascent corolla tube, due to cell elongation in the epidermis and periclinal divisions in the subepidermis.[29,30] The stamens, seemingly inserted below the petal lobes, are actually initiated closer to the meristem center than the corolla. As the stamen primordia develop, growth occurs in a ring zone below their insertion and they are carried upward on the elongating corolla tube.

At the same time, the pappus primordia are formed low on the outer side of the floret by periclinal mitoses in the subepidermis. Initiation of the two carpels and the ovule follows the same cellular pattern as that of other floral members. Carpel primordia arise at the base of the corolla tube closer to the meristem center than stamens; the ovule is "dead" center at the bottom of the now cup-like floret primordium. Upon elongation the two carpel primordia develop into a compound style with a two-parted stigma. The ovarian cavity originates due to active upward growth of a ring zone below the insertion of carpel primordia.

The individual florets arise first as one or two rows around the basal margin of the broadly expanded reproductive meristem (see Volume II, Chapter 3, Figure 5), then form acropetally and rapidly occupy every available position on the receptacle (Figure 6). A total of 245 to 282 florets is formed in *C. morifolium* depending on cultural conditions.[31] The central or disc florets have an actinomorphic tubular corolla due to the uniform growth of the five-toothed corolla tube, but the marginal or ray florets develop a zygomorphic ligulate corolla due to the late asymmetrical growth of the corolla tube. The corolla tube, prolongate along one side only, is a strap-shaped "ligule". In *Chrysanthemum* ray florets are female, while disc florets are hermaphroditic.

Capitulum formation in other Compositae follows the same general developmental pattern, with many variations in details.[30,32,33] The involucre may include only one or more rows of bracts which may vary greatly in number, shape, and arrangement. Receptacle may be flat as in *Helianthus,* convex as in *Chrysanthemum, Cosmos,* and *Rudbeckia,* or elongated-conical as in male *Xanthium* heads. Receptacular bracts, each subtending a floret, are present in *Rudbeckia, Cosmos,* and *Xanthium,* but absent in *Chrysanthemum* and *Aster.* All florets are ligulate and hermaphroditic in lettuce (*Lactuca*), whereas in *Rudbeckia, Cosmos,* and *Aster* marginal florets differ from central ones as in *Chrysanthemum.*

In *Xanthium* both male and female florets are tubular, but they are in separate heads: the male head is a typical capitulum with 150 to 170 florets, whereas a female head has only 2 florets.[34] The distribution of male and female heads in the flowering shoot of plants induced to flower by a single long night is very complex. As shown in Figure 7, Léonard and colleagues have distinguished three zones of flowering branches: the basal zone A in which all terminal heads are male; the intermediate zone B in which they are all female; and the apical zone C in which they are all male again.[35] The apical meristem of the main axis, which always transforms into a male capitulum, is included in zone C.

4. *Other Racemose Inflorescences*

The "corymb" is characteristically a racemose inflorescence in which the length of the various flower pedicels decreases acropetally in such a way that the flowers are finally all at about the same level (Figure 2D).

In the "umbel", typical of Umbelliferae like carrot (*Daucus carota*), all flower pedicels arise from a common point, which as a rule is the top of the stem at the center of a whorl of bracts (Figure 2E). Thus, there is no elongation of internodes of the inflorescence axis in this case. The receptacle grows radially instead of vertically and becomes a flattened disc. An umbel is typically a racemose inflorescence since the outer flowers are initiated before the inner ones.[36,37] It is sometimes described as a capitulum of pedicellate florets. Structural changes during the floral transition at the meristem in carrot are similar to those occurring in species producing capitula, and the reproductive meristem has a typical mantle-core organization[37] (Volume II, Chapter 3, Section III.C.2).

Catkins, found in some monoecious and dioecious trees, are inflorescences related ontogenetically to a raceme.

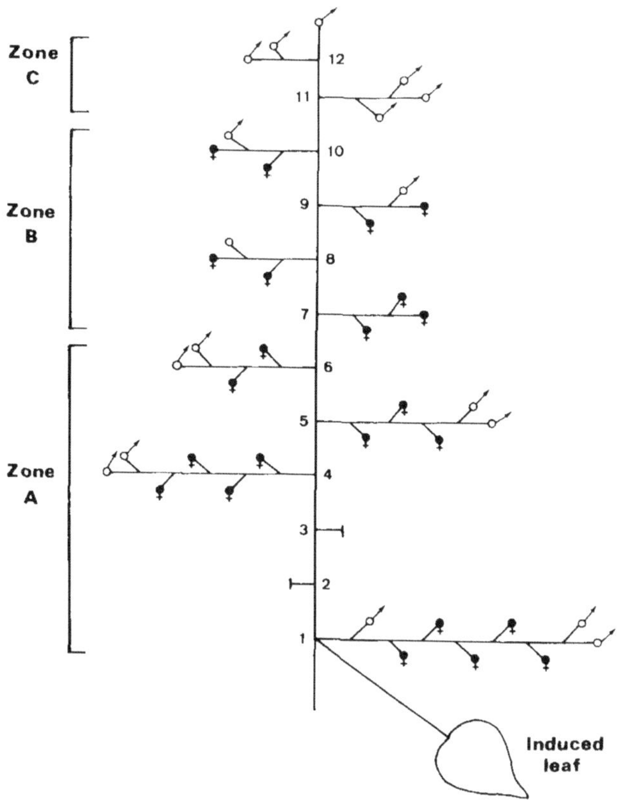

FIGURE 7. Distribution of male (♂) and female (♀) capitula (heads) on a plant of *Xanthium strumarium* induced by a single long night. Observations were restricted to the terminal bud of the main stem and the terminal and axillary buds of first-order branches. (O), undifferentiated meristem. Side axes are numbered acropetally from the induced leaf. Explanations of zones A, B, and C in text. (From Léonard, M., Kinet, J. M., Bodson, M., Havelange, A., Jacqmard, A., and Bernier, G., *Plant Physiol.*, 67, 1245, 1981. With permission.)

C. Cymose Type of Flower Groupings

In the formation of the first flower of the cyme the apical meristem is used up; growth of the primary axis ceases and is thus determinate. Continued growth of the inflorescence depends on precocious development immediately below the terminal flower of one or more axillary meristems which in turn form secondary branches each first terminated by a flower (Figure 2F, G, H). Tertiary- and higher-order axes are each produced from new subterminal axillary meristems. Growth of cymose inflorescences is thus sympodial, whereas that of racemose ones is monopodial. On an axis of any order in a cyme, flower formation follows a basipetal sequence, the reverse of that in racemose inflorescences.

A cymose inflorescence in which two axillary branches occur below the terminal flower is a "biparous" cyme or "dichasium" (Figure 2F); it is a "uniparous" cyme or "monochasium" when only one axillary branch develops. Uniparous cymes are "helicoid" when branches at successive nodes along the (false) axis are on alternate sides and "scorpioid" when branches are all on the same side (Figure 2G, H).

The case for *Nicotiana glutinosa,* studied by Diomaiuto-Bonnand, illustrates the morphogenesis of a typical cyme (Figure 8).[38] The evoked meristem in this species is entirely transformed, after a typical prefloral stage, into the terminal flower, tf, of the inflorescence. At the same time a precocious axillary meristem, a_1, is formed in the axil of the last leaf,

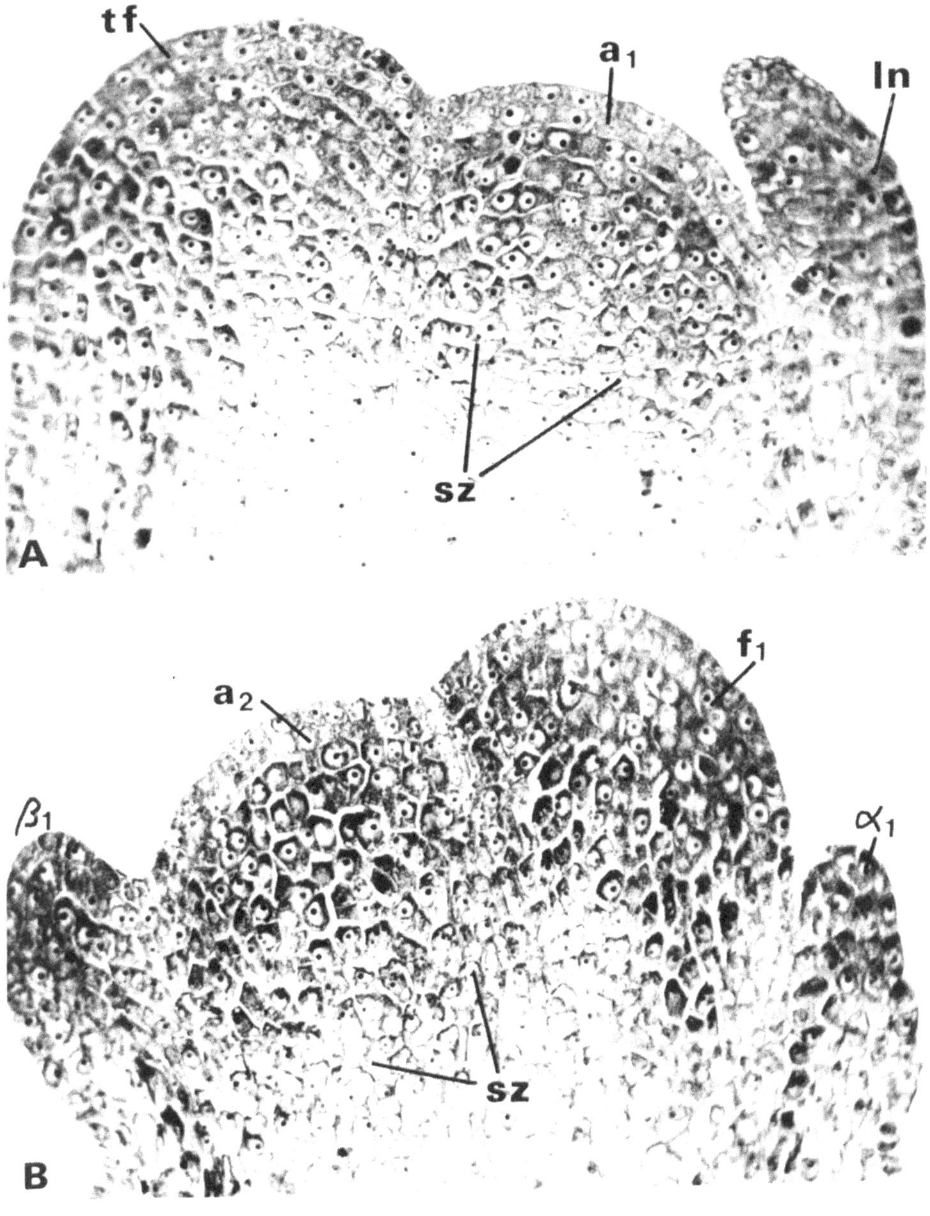

FIGURE 8. (A) Formation of the terminal flower (tf) of the cyme in *Nicotiana glutinosa*. The precocious axillary meristem a_1 is seen on the right in the axil of the last-formed leaf, l_n. (B) Production by meristem a_1 of the first axillary flower of the cyme (f_1) on the right. Note bracts α_1 and β_1 and the new precocious axillary meristem a_2 on the left. Note the shell zone (sz) at the base of a_1 and a_2. (Methyl green and pyronin stain.) (From Diomaiuto-Bonnand, J., *Rev. Cytol. Biol. Vég.*, 29, 1, 1966. With permission.)

l_n, produced before the floral transition (Figure 8A). Very rapidly a_1 initiates successively two bracts, α_1 and β_1, on the two opposite flanks. Then the axillary meristem a_1 is used up in the formation of the first axillary flower, f_1. A new axillary meristem, a_2, arises in the axil of β_1 (Figure 8B) and follows the same morphogenetic course as a_1 but in a perpendicular plane. The same process is repeated many times with successive axillary meristems being alternatively in two perpendicular planes. Thus, flowers tf, 2, 4, etc. are all above each

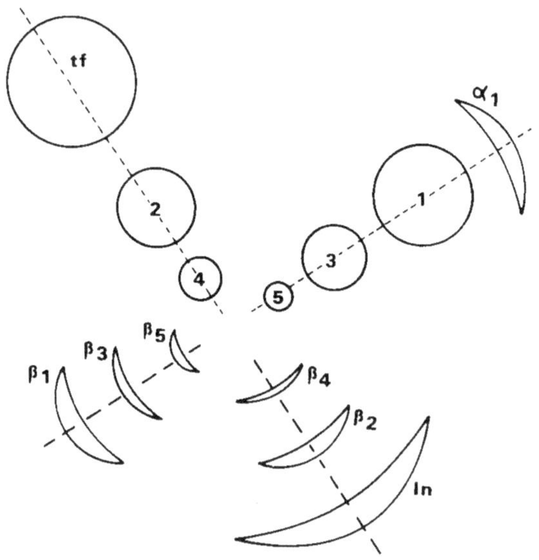

FIGURE 9. Diagram showing a top view of the uniparous scorpioid cyme of *Nicotiana glutinosa*. All flowers (circles), including the terminal one (tf), are on two orthostichies (dotted lines) located on the same side of the inflorescence. The last leaf, l_n, formed before the floral transition and all β bracts are on two other orthostichies (broken lines). (From Diomaiuto-Bonnand, J., *Rev. Cytol. Biol. Vég.*, 29, 1, 1966. With permission.)

other on the same orthostichy, whereas flowers 1, 3, 5, etc. are on another orthostichy. On the other hand, the group of leaf l_n and bracts β_2, β_4, etc. and the group of bracts β_1, β_3, β_5, etc. are both on two other orthostichies (Figure 9). The cyme of *N. glutinosa* is typically uniparous since there is normally no axillary meristem developing in the axils of bracts of the α series and it is scorpioid since the two floral orthostichies are on the same side of the inflorescence. Note that the size of bracts of the β series is decreasing rapidly from β_1 to β_n and that bracts of the α series after α_1 may not be formed at all (Figure 9). For each sympodial unit the terminal flower is displaced laterally from its terminal position by the active growth of the adjacent axillary meristem (Figure 8A, B).

Transitions to a biparous cyme commonly occur in *N. glutinosa* by development of a meristem in the axil of a bract of the α series. Return to the uniparous condition, which is also frequent, is by suppression of this axillary meristem.

Other plants, e.g., *Myosotis*, produce uniparous scorpioid cymes in which there are no bracts.[39] Tomato (*Lycopersicon esculentum*) is sometimes believed to belong to this category. However, another view expressed by Ecole is that the inflorescence meristem in tomato divides into two almost equal parts (dichotomy): one differentiating into a flower primordium and the other giving rise to the meristem responsible for further inflorescence development (Figure 10).[40] This last meristem divides again in a similar manner and the process is repeated five to nine times.

D. Complex Inflorescences

Some inflorescences, like other living structures, are difficult to classify.[41] Some flower clusters appear to be intermediate between the established types. For instance, a short raceme with only a single terminal and two lateral flowers may resemble a dichasium. However, a precise study of early stages of inflorescence organization reveals the acropetal initiation of

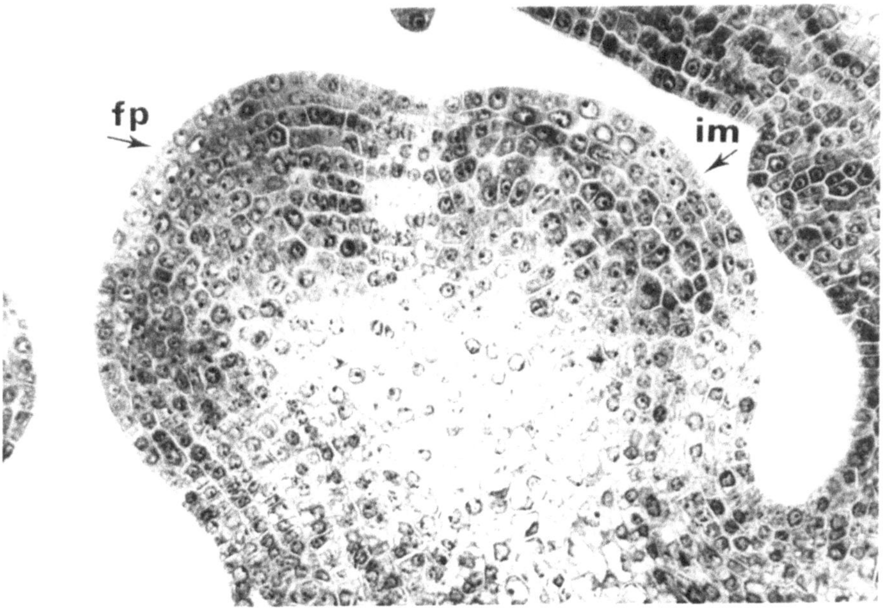

FIGURE 10. Inflorescence meristem of tomato (*Lycopersicon esculentum*). The meristem divides into two almost equal parts: one differentiating into a flower primordium (left part, fp), the other giving rise to the meristem responsible for further inflorescence development (right part, im). (Haematoxylin stain.)

flowers, typical of racemes, and thus eliminates the confusion.[42] Early abortion of the first or central flower in a dichasium, as occasionally found in some *Nicotiana* and *Begonia* species, results in only apparently dichotomous branching in the mature cymes.

Other complex flower clusters may result from the grouping of several inflorescences into a single unit. The spike of wheat and the panicle of oat are complex inflorescences which could be called a raceme of spikelets and a raceme of racemes, respectively. Umbels are usually compound, each primary peduncle (or ray) terminated by a secondary umbel, as in carrot. Species are also known which produce a capitulum of capitula. In several species the branches of compound inflorescences are cymes of various sorts. This is the case, e.g., in grapevine *(Vitis vinifera), Aquilegia vulgaris, Ruta graveolens, Syringa,* and tobacco (*Nicotiana tabacum*), where the inflorescence is basically a raceme of cymes.

Another complication may arise when vegetative and reproductive structures are closely associated, as in *Mercurialis annua,* where one of the lateral branches of the dichasium develops into a leafy shoot and the other into a cyme.[43]

III. GENERALITIES ON THE MORPHOGENESIS OF REPRODUCTIVE STRUCTURES

A. Sequence of Initiation

As explained above, the order of flower initiation in racemose and cymose inflorescences is, respectively, acropetal and basipetal, but cases are known in which the typical sequence is partly modified. One example is the spike of wheat in which, contrary to expectation, initiation of the upper bulges (spikelet primordia) of the double ridges does not proceed acropetally, but starts in the middle of the prospective inflorescence and then extends acro- and basipetally (Figure 3A). However, the lower bulges (leaf primordia) were produced acropetally in advance of the upper bulges when the plant was still vegetative. Another case is that of the capitulum in *Cosmos bipinnatus*. Work by Molder and Owens shows that after

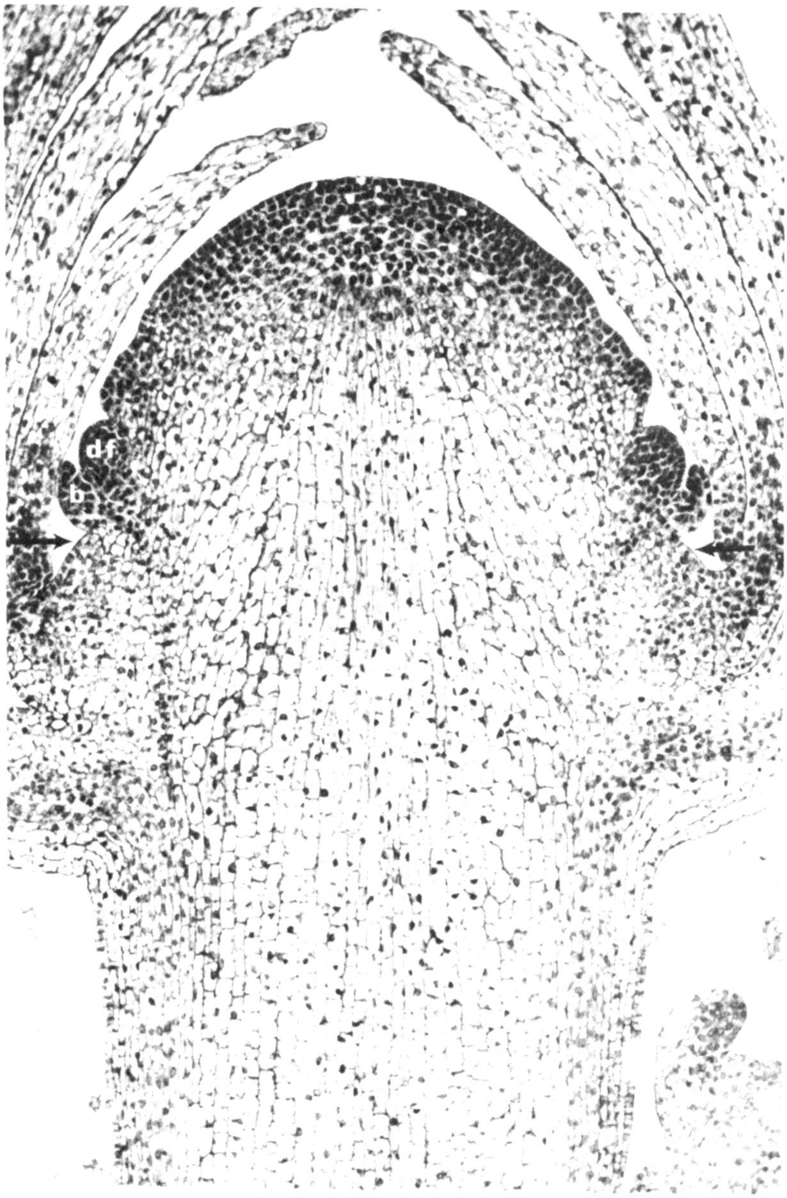

FIGURE 11. Inflorescence meristem of *Cosmos bipinnatus* showing initiation of disk floret (df) primordia and associated subjacent bracts (b). Arrows indicate sites of future ray floret initiations. (Methyl green and pyronin stain.) (Reproduced by permission of the National Research Council of Canada, from Molder, M. and Owens, J. N., *Can. J. Bot.*, 51, 535, 1973.)

two thirds of the inflorescence apex is covered with disk florets, the single marginal row of ray florets is initiated below the already-formed disk florets, in a region of the apex that is elongating rapidly just above the involucral bracts (Figure 11).[33]

Generally, the various floral appendages are also initiated acropetally (centripetally) in the order: sepals, petals, stamens, carpels. However, there are plants in which the sequence for some of the members is perturbed, i.e., it is basipetal (centrifugal). Centrifugally initiated stamens are known in species of 32 families.[44] In cacti, carpel primordia are initiated usually in advance of the first stamen primordia: carpel initiation delimits an abaxial flank meristem in the form of a ring from which the numerous stamens are initiated in basipetal sequence.[45,46] In *Anthriscus sylvestris*, floret morphogenesis starts with the initiation of five successive stamens. The first petal primordium becomes visible during initiation of the last stamens and is soon followed by the other petal protuberances and the gynoecium.[1] Perhaps the most impressive example of altered sequence is that of *Lithrum salicaria* as worked out by Cheung and Sattler:[47]

Inner sepals → Outer sepals → Outer stamens → Gynoecium → Inner stamens → Petals

Variations also occur during the formation of grass florets. Thus, while the palea arises in advance of stamens in wheat,[16] barley,[1] and corn,[25] it seems to appear after the stamens in oat.[22] On the other hand, the exact time of lodicule initiation is often difficult to determine.[1,16]

Some of these variations may not be significant, however, because flower morphogenesis is in some cases such as extremely rapid process that it may preclude a precise sequential description of changes under the dissecting microscope. The conflicting results found in the literature about the exact order of appendage formation in the Leguminosae flowers (discussed in Sattler[1]) illustrate this difficulty. As emphasized by Tucker, definitive conclusions can only be reached after examination of sections or of the developing flower bud using SEM.[48] Also, retarded development of small primordia of only a few cells, as is often the case for petals, may escape detection until their growth resumes at a later stage.

In cases where the centrifugal initiation of flowers or flower parts is well demonstrated, further investigations on the reason(s) of this curious situation are warranted. One may wonder, e.g., whether the centrifugal structures are, as suggested by Tucker, analogous to vegetative axillary buds whose initiation is known to be delayed by two to five plastochrons compared to their subtending leaves.[44] Or is it the result of sluggish adaxial (apical) growth combined with active abaxial growth in the zones of the reverted sequences or of another unknown reason?

B. Histogenetic Origin

As a rule during vegetative growth, leaves originate from periclinal cell divisions in the subepidermal layer of the meristem flanks (often also the epidermal layer in grasses), whereas buds arise by divisions in the third or fourth cell layer in the axil of young leaves. Thus, the depth at which the periclinal mitoses of initiation occur seems to typify the foliar or cauline nature of the nascent organs.

Histological studies have shown that flowers and inflorescence branches are invariably initiated by deep periclinal divisions and may thus be viewed as homologous to axillary buds. On the other hand, sepals, petals, and carpels are very often borne from more superficial divisions in the manner of leaves. For stamens the situation is less clear since they are initiated from the subepidermis in some plants, e.g., *Chrysanthemum*,[29,30] *Sinapis*,[10] *Portulaca*,[49] beet,[50] and from the subjacent layer in others, e.g., Gramineae,[16,20] *Catharanthus*,[51] and tobacco.[52] Thus, whether stamens should be considered as foliar or cauline structures on the basis of their histogenetic origin remains unresolved.

Use of periclinal cytochimeras, initiated by Satina and Blakeslee, allows a more unequivocal determination of the contributions of the different layers of the meristem to various organs.[53-55] With *Datura stramonium* this approach has shown that sepals and petals form primarily from activity of the subepidermis. Stamens, carpels, ovules, and placentae have a deep-seated origin, deriving essentially from the third layer. In peach (*Prunus persica*) the emerging picture is more complex.[56] Here the third layer contributes significantly to sepals and the pistil, but not to petals and stamens. Cells of the ovary wall derive from more than one layer and their arrangement within the tissue does not reflect their origin. In chimeras of cranberry (*Vaccinium oxycoccos*) the situation appears equally complex.[57] Whether this is normal for many plants, but undetectable in classical histological studies because of lack of markers of cell lineages, as suggested by Dermen and Stewart,[56] or is due to altered relationships between adjacent layers in cytochimeras is not known.

Thus, with the evidence at hand concerning histogenetic origin, it is not easy to label different floral organs "leaf-like" or "bud-like". However, concern for histogenetic layers may wane based on recent work by Green and coworkers (to be reviewed in Section V.B below) stressing other aspects of organ initiation.

C. Common Primordia

In most cases the primordia of distinct reproductive structures arise individually, but in some occasions it has been reported that the meristem produces "common" or "double" primordia which at a later stage divide into two different parts. Using both SEM and histological techniques on two species of Saururaceae, Tucker describes the production of common primordia for flowers and associated bracts.[48,58] The common protuberances bifurcate soon after initiation to form a flower meristem above and a bract primordium below. Interestingly enough in one of these species, *Houttuynia cordata,* common primordia are not produced at the beginning and end of inflorescence formation. Common primordia occur only between these stages when the inflorescence meristem is presumably most active; they seem thus to result from simultaneous initiation of the flower and its axillant bract, a case of ultraprecocious initiation of axillary buds.

Common primordia for floral appendages are also known, although such structures do not seem very widespread. A remarkable example has been described in Primulaceae, e.g., *Anagallis, Lysimachia, Samolus,* and *Cyclamen,* in which the flower meristems produce common primordia each differentiating into adaxial stamen primordia and abaxial petal primordia (Figure 12).[1,59,60] In this case, too, stamens may be viewed as ultraprecocious axillary buds of petals. Common primordia for nectaries (sometimes interpreted as being reduced petals) and stamens are also known in a species of Berberidaceae.[60a]

Claims for common primordia in older studies must be checked with modern techniques since, as in the case for tulip above, two adjacent primordia initiated in close succession may originally appear as a common primordium under the dissecting microscope.

D. Number of Floral Organs

The number for each type of appendages is fixed in the flowers of many species. As a rule, these flowers possess a small number of appendages of each sort, e.g., tulip, *Sinapis,* wheat, and tomato, but the flowers of other species normally have large and indefinite numbers of at least one class of appendages. For example, petal number varies from 57 to 184 in the glasshouse carnation and from 35 to 80 in 'Baccara' rose.[61,62] The number of stamens in *Begonia,*[63] *Papaver,*[64] *Michelia,*[65] *Portulaca,*[49] carnation,[66] *Ranunculus,*[67] etc. is also relatively high and is normally variable from flower to flower. These flowers are thus characterized by a relative lack of rigid control of flower morphogenesis compared to those having fixed numbers of parts.

Petal number in some horticultural crops, e.g., carnation and rose, is important in determining flower quality. Carnation flowers with too few petals are described as "hollow", whereas flowers with too many petals are called "bullheads".[61] Garrod and Harris showed that, besides the normal petals produced directly by the flower meristem in carnation, additional petals develop from secondary growth centers formed around the gynoecium.[61] Normally the number of additional petals is low, 3 to 5, but increases to values as high as 120 after some experimental treatments. The bullhead malformation of 'Baccara' rose is characterized by a similar proliferation of secondary and rudimentary flowers between sta-

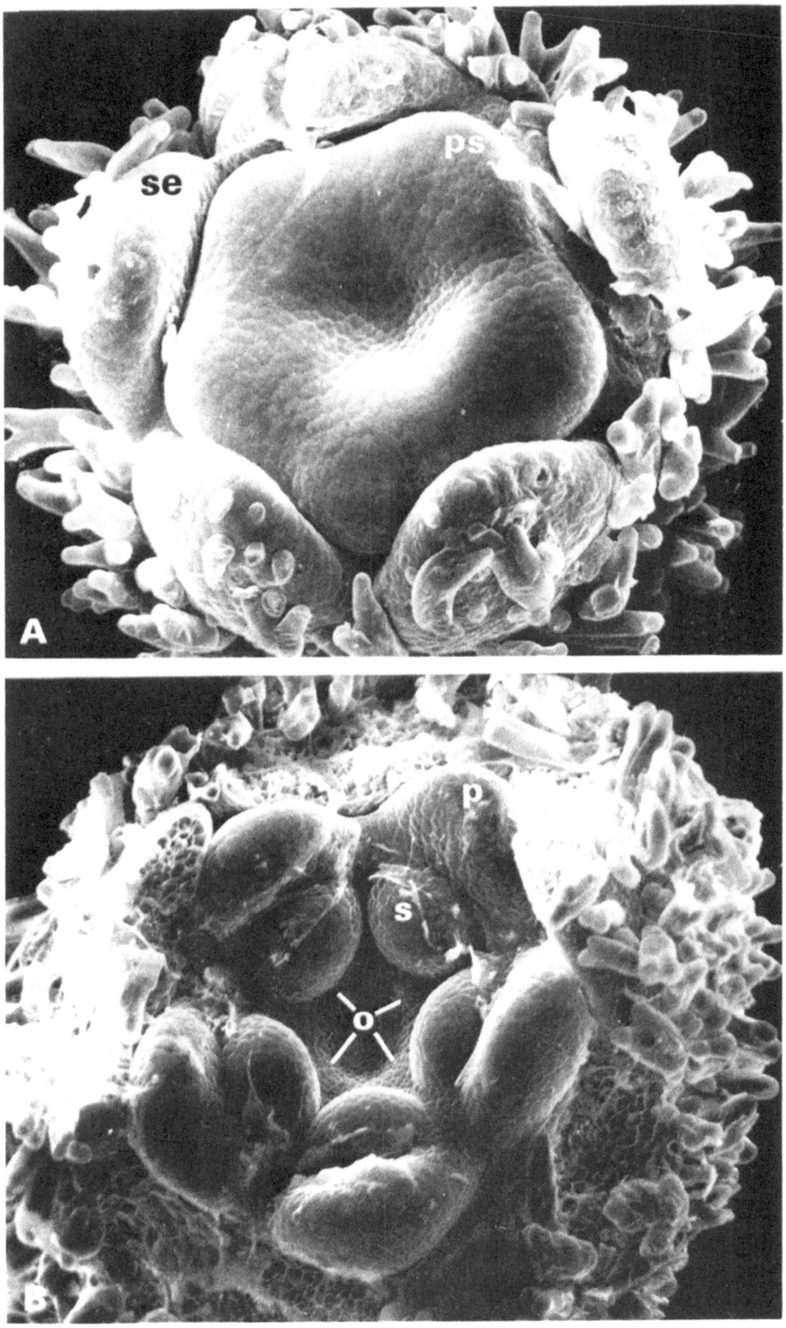

FIGURE 12. Production of petal-stamen common primordia by flower meristem of *Cyclamen orbiculatum*. (A) Initiation of common primordia (ps) alternating with sepal (se) primordia. (B) Common primordia have each differentiated into one abaxial petal (p) primordium and one adaxial stamen (s) primordium. Note initiation of ovary wall (o) in this older flower (sepals dissected). (From Sundberg, M. D., *Am. J. Bot.*, 69, 1707, 1982. With permission.)

Table 1
FREQUENCY OF FLORAL ABERRATIONS IN PLANTS OF *SILENE COELI-ROSA* GROWN AT HIGH TEMPERATURE (27°C)

	% of flowers with			
Floral whorl	**Wrong number of parts**	**Missing parts**	**Extra parts**	**Aberrations in one whorl only**
Sepals	28	23	5	1
Petals	39	32	7	3
Stamens	85	63	22	41
Carpels	32	25	7	4

Adapted from Lyndon, R. F., *Can. J. Bot.*, 57, 233, 1979.

mens and carpels (where a ring-shaped nonfunctional nectary is formed in normal flowers) and by high petal numbers.[62]

Modification of appendage number is common even in flowers usually considered to be uniform. Examples are tulip with 11 to 22 appendages instead of the usual 15,[4] tomato with variable numbers of stamens or carpels,[68,69] and female flowers of *Begonia* with bi- or tetracarpellate, instead of the normal tricarpellate, ovary.[63] Similarly in several Rosaceae, e.g., *Agrimonia* and *Fragaria*, tetramerous flowers are occasionally found among normally pentamerous ones.[1,42]

In *Silene coeli-rosa* flowers, normally with 5 sepals, 5 petals, 10 stamens, and 5 carpels, more than one fifth of all flowers observed by Lyndon has aberrant numbers of parts in at least one whorl.[70] The two stamen whorls are most often abnormal (Table 1) and there is a tendency for a greater proportion of abnormal flowers with parts missing than with extra parts. Very few flowers have an aberrant sepal number without having aberrant numbers in at least one other whorl (Table 1, last column). Although this suggests that abnormality in one whorl may lead to abnormalities in subsequent whorls, detailed examination of the abnormal flowers indicates that, except for sepal and petal numbers, the aberrations occur independently in the different whorls. Almost half of the plants with aberrant numbers of sepals show the same aberration in the petals. Thus Lyndon proposes that petal number in *Silene* is dependent, although not entirely, on sepal number, but numbers of parts in other pairs of floral whorls are not closely linked.

When the literature on aberrant numbers of flower parts is reviewed it appears that part of the variations is certainly due to meristem misfunctioning and another part, perhaps more important, to transformations of appendages from one type to another. All floral organs, in one species or another, can develop in the form of any of the other organs present in the normal flower.[71] The most frequent of these transformations are stamen petalody and pistillody. According to Meyer, petaloid stamens, sometimes referred to as "petalomania", are of such common occurrence (double flowers) that they can hardly be considered an abnormality.[71] In many flowers with extreme doubling the carpels as well as the stamens are affected. Sepalody and staminody are, on the contrary, relatively uncommon abnormalities.

What conclusions can we draw from these observations? First, the morphogenetic pattern is never absolutely fixed and the control of these processes is not as perfect as one might think given the importance of floral appendage numbers for classification of some species. Second, because extra appendages necessarily arise in unusual places in aberrant flowers, it is clear that the fate of a primordium cannot be entirely predicted from its site of initiation.

E. Changes in Phyllotaxis

Phyllotactic changes commonly occur just before or at the time of inflorescence or flower initiation in several species. In *Xanthium* there is a shift from a (2,3) contact parastichy pattern, typical of the basal vegetative part of the stem, to a (3,5) pattern, characteristic of the last leaves formed before flowering, and a (5,8) pattern during early formation of the terminal capitulum of male florets.[72] In the capitulum of *Helianthus*, another Compositae, the most obvious pattern of contact parastichies is (34,55) in its outer part, (21,34) in its median part, and possibly as low as (13,21) or (8,13) in its center.[73] Thus, after the normal increase in the phyllotactic complexity during the floral transition there is a decrease, at least in this case, during inflorescence ontogeny.

In some species producing a terminal raceme, such as members of the Cruciferae family, there is also an increased phyllotactic complexity. In *Sinapis,* e.g., the contact parastichy pattern changes from the usual (2,3) to one in which the primordia are much more numerous and crowded.[74]

In a majority of the Gramineae the distichous phyllotaxis of foliage leaves is continued unchanged by spikelet or bract arrangement within the inflorescences. In sugarcane, pearl (*Pennisetum typhoides*), and other millets, however, the distichous pattern of the vegetative shoot is changed into a high-order spiral arrangement of branches in the terminal panicles.[2,75]

Phyllotactic changes are also very spectacular in species bearing a solitary terminal flower. During flowering in *Ranunculus repens,* Meicenheimer noted a shift to a higher order of phyllotaxis.[67] The (2,3) contact parastichy pattern typical of leaves is changed into a (3,5) pattern at the sepal stage. Formation of a whorl of five petals establishes a (5,8) pattern with the sepals. Subsequent initiation of numerous stamens and carpels results in a (8,13) pattern.

With initiation of terminal flower, *Silene coeli-rosa* shifts from a decussate phyllotaxis (for leaf positioning) to spiral arrangement of floral appendages.[76,77] As seen in Figure 13A the first two sepals are already displaced from an ideal opposite arrangement: their divergence angle is 156° instead of 180°, typical of opposite primordia. The third sepal is displaced towards the first, whereas the fourth is formed on the opposite side of the meristem from the third, approximately midway between the first and second sepals. The fifth sepal is formed in the gap between the second and the third. All sepals appear in a spiral sequence which becomes obvious especially with initiation of the third one (Figure 13A). The divergence angles for the last three sepals average 141.7°, a value which is close to the ideal angle for spiral phyllotaxis, 137.5° (Figure 13B). Then, petals and stamens arise in such a quick succession that their exact order of appearance is sometimes difficult to determine. The five petals are initiated almost simultaneously and alternate with the sepals. The stamens are formed sequentially, first the five antesepalous (opposite to the sepals) stamens, and then the five antepetalous stamens (Figure 13C). Finally the carpels, seen as five flat areas on the meristem, appear to be initiated in a spiral sequence continuing the stamen sequence (Figure 13D). Divergence angles between petals, stamens, and carpels, while more variable than those between the last three sepals, are of the order of 141 to 148°. A very large angle of 175° separates the last stamen and the first carpel.

According to Lyndon, the positioning of all floral appendages in *Silene* is traceable to the sepals:[76,77] The petals appear midway between adjacent sepals; the stamens have a position compatible with the idea that they are axillary buds of sepals and petals, and the carpels lie on the same radii as the sepals. Lyndon also believes that the first two sepals are opposite at initiation, but are displaced secondarily towards one side of the meristem. Displacement is towards the leaf of the last pair (just below the flower) which possesses the most developed axillary bud (Figure 13A, B). The position and sequence of initiation of the last three sepals are determined by the older sepals, but the influence of an older primordium seems to last for only two plastochrons.[77] The nature of this influence is entirely unknown, but the value

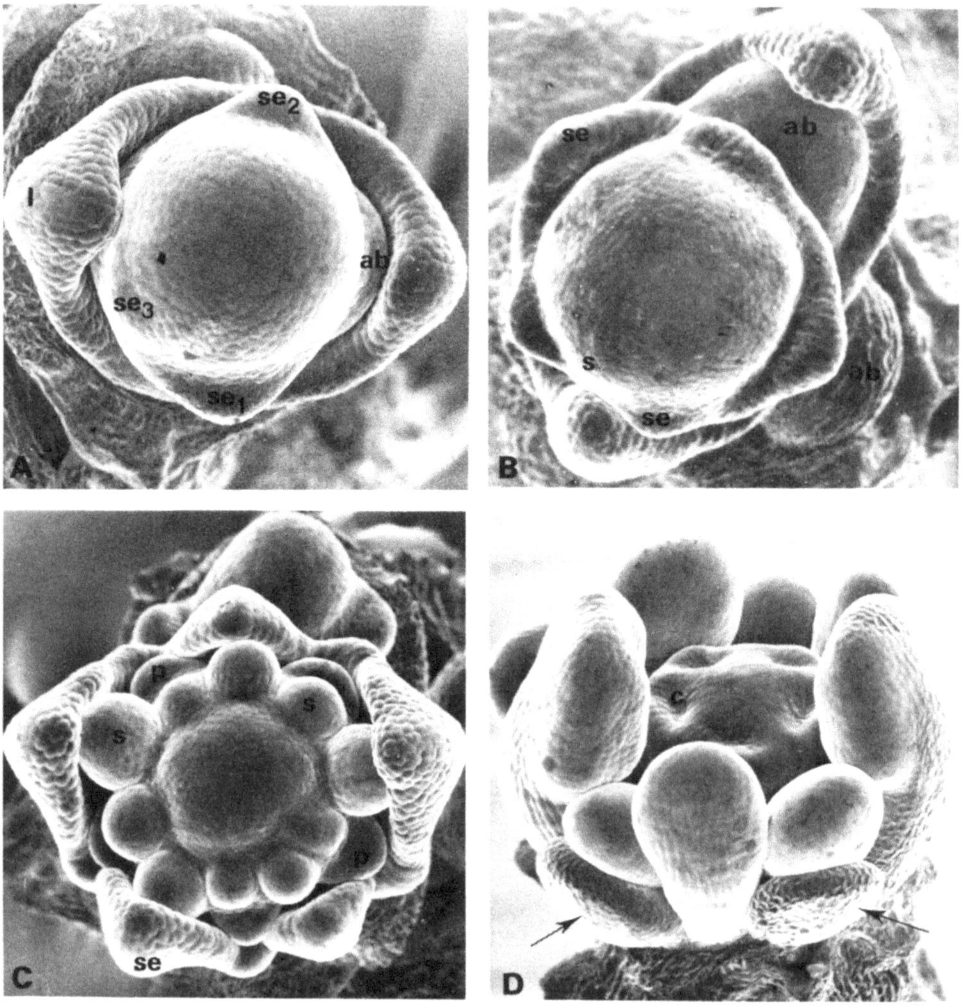

FIGURE 13. SEM micrographs of apical meristem of *Silene coeli-rosa* during formation of terminal flower. (A) At the stage of initiation of the 3rd sepal (se_3). Note sepals 1 and 2, the last pair of opposite leaves (l) and the axillary bud (ab) at the node below the flower. (B) Just at the start of the stamen (s) initiation. Five sepals are present. Note the axillary bud immediately below the flower (at 1:00 o'clock) and the next lower axillary bud (at 4:00 o'clock). (C) Just after petal (p) and stamen initiation. Stamens 1 to 5 are opposite the sepals and stamens 6 to 10 are opposite the petals. (D) Carpels (c) have been initiated and are seen as five depressions on the apical surface. At this stage, sepals have been removed. At the front of the flower the two petals (arrows) which are visible are now smaller than the stamens they subtend. Magnification is decreasing from A to D. (From Lyndon, R. F., *Ann. Bot. (London)*, 42, 1343, 1978. With permission.)

of such analyses for physiologists is that we may be encouraged to look to the literature on phyllotaxis for useful leads concerning flower morphogenesis.

Associated with the increase in phyllotactic complexity of the main shoot during the reproductive transition several growth parameters change in the apex. As seen in Table 2, some changes are common to all species investigated regardless of reproductive structure; the common changes include an increase in meristem volume and decreases in plastochron duration and plastochron ratio. Other changes vary with the species such as meristem shape which is modified in all cases except *Ranunculus* and internode growth (vertical growth) which increases in *Xanthium* and *Sinapis* and decreases in *Silene* and *Ranunculus* (as expected in cases of terminal flowers with very short internodes). The divergence angle, on the other hand, may vary depending on the phyllotaxis of the vegetative portion of the shoot.

Table 2
CHANGES[a] IN APICAL GROWTH PARAMETERS ACCOMPANYING THE INCREASE IN PHYLLOTACTIC COMPLEXITY OF THE MAIN SHOOT DURING THE REPRODUCTIVE TRANSITION

Species	Reproductive structure made by the apical meristem	Meristem		Plastochron		Young internode elongation	Divergence angle[c]	Ref.
		Volume	Shape	Duration	Ratio[b]			
Silene coeli-rosa	Single flower	+	Change	–	–	–	–	77
Ranunculus repens	Single flower	+	No change	–	–	–	0	67
Epilobium hirsutum	Raceme	+	Change	–	–	–	+	78
Sinapis alba	Raceme	+	Change	–	–	+	0	74
Xanthium strumarium	Capitulum	+	Change	–	–	+	0	72

[a] + = increase; – = decrease; 0 = no change.

[b] The plastochron ratio is the ratio of the distances of two successive meristem appendages from the center of the apex. This value or its logarithm represents the relative radial growth per plastochron.

[c] The angle between two successive meristem appendages in the tangential plane.

Another change, noted by Lyndon in *Silene* and believed to be essential, is a large decrease in size of primordia of floral organs at initiation: sepal primordia are only one third the size of leaf primordia and petal and stamen primordia about one half the size of the sepals.[77] A similar decrease which permits closer and closer packing of primordia during reproductive morphogenesis might well occur in many other species, as suggested by photographs of reproductive meristems,[1] but remains to be established by appropriate measurements.

The above modifications of the growth pattern in the apex that correlate with phyllotactic changes are often regarded as the determinants of these changes. In Labiatae like *Perilla*, however, several growth parameters such as internode growth, meristem size, and plastochron duration are altered during the floral transition without any accompanying modification of phyllotaxis of the main shoot,[13,79,80] so that the significance of these growth changes in other species remains doubtful.

In *Michelia fuscata* the dramatic decrease in size of the flower meristem during carpel initiation has no effect on carpel phyllotaxis. For Tucker, the predominant factor in the control of carpel position is the vascular pattern.[65,81] The vascular sympodia supplying the dorsal carpellary traces are radially aligned with the carpel parastichies and appear in the order of carpel initiation. Before a carpel is initiated its respective procambial strand is present immediately below the area of inception. In fact, a procambial cell is in contact with the periclinally dividing cell of the subepidermis which starts carpel initiation.

Factors controlling phyllotaxis are probably very complex and studies on this problem in inflorescences and flowers are still essentially concerned with the discovery of correlations, not causative factors. Further work of an experimental nature is required. Phyllotaxis must be studied at the stage of primordium initiation, not in mature flowers. Later ontogeny can obscure exact order and site of initiation so that spirals finally may seem to be whorls and vice versa.

Finally, it should be emphasized that the spatial relationships between appendages in some flowers do not seem very rigid. This is the case, for instance, in many species with centrifugally initiated stamens.[44] In these special flowers the first stamens are produced in the normal acropetal order and occupy positions consistent with the previous phyllotaxis of the flower. Subsequent stamens are initiated in the reverse direction, i.e., centrifugally, and usually arise in a sequence alien to the established phyllotaxis. The flower meristem seems to have lost strict control of appendage formation in such cases. A similar situation is reported frequently for flowers having numerous appendages. Under the binocular microscope, the number, arrangement, and sequence of initiation of stamens in *Juglans, Malva,* and *Papaver* appear different in different flowers,[1,64] but quantitative work of the kind reported above is necessary before concluding that there is no strict control of phyllotaxis in these flowers.

F. Fusion of Floral Organs

One of the most fascinating aspects of flower morphology is the fusion of parts. Morphologists usually contrast the "ontogenetic" with the "congenital" fusions.[82] Ontogenetic fusions occur between parts that are initially free and congenital fusions refer to structures that begin as a unit but are believed, for phylogenetical reasons, to be composed of multiple parts. The situation is not clear-cut since, as stressed by Baum and Cusick, the two types of fusion often occur simultaneously in the same organ and produce indistinguishable results.[82-84] Moreover, ontogenetic fusions occurring very early during flower development may easily be misinterpreted as congenital, although use of cytochimeras may help in demonstrating that ontogenetic fusions have really occurred.[56] In the rest of this volume, the term "fusion" will be restricted to the observable processes of ontogenetic fusions, whereas examples of so-called "congenital fusions" will be discussed under the heading of "zonal growth" which is what is actually occurring in these cases.

1. Corolla Tubes

Petal fusion can be transient or permanent. The case for *Downingia*, studied by Kaplan, illustrates the process of temporary margin-to-margin fusion of petals.[85] In this process, the epidermal layers of adjacent petal margins are "knit together" by interlocking of their cells so that early in development the corolla forms a hooded structure, enclosing the internal floral organs, but at anthesis the corolla breaks open along these temporary lines of union.

Permanent petal fusion is apparently a rare process, found only in species of Apocynaceae where it occurs when the corolla is about 400 to 500 μm in height.[86] As first shown by Boke, the margins of adjacent petal lobes become apposed and then fuse very intimately.[87] At later stages, the lines of fusion become undetectable under the light microscope. The fused parts develop into the upper portion of the corolla tube, i.e., that portion above the level of stamen insertion.

Corolla tube inception in *Pharbitis*, *Digitalis*, and species of the Solanaceae, as well as inception of the lower portion of this tube in Apocynaceae, occurs by a different mechanism.[88-90] As shown by Nishino, this results from very early growth between petal primordia giving continuity of marginal meristems of adjacent primordia.[88,89] This interprimordial growth is definitively a fusion process but, because it occurs so early, it unites meristematic instead of differentiated tissues and may be easily misinterpreted as a "congenital" fusion.[91]

2. Stamen Fusion

In Campanulaceae, Compositae, tomato, and other species, stamens arise as separate organs but, as they grow upward and differentiate into a filament and an anther, they become fused secondarily along their margins forming a tube encircling the style. In some species stamens are united for almost their entire length or for only part of it, and the stage at which fusion occurs also varies. Fusion may be more or less intimate involving only the merging of cuticles of contacting stamens, as in *Downingia* and *Rechsteineria*,[92,93] or interpenetration of elongated epidermal cells in other species (in a manner similar to petal lobe fusion in *Downingia*).[94]

3. Carpel Fusion

As shown by Baum, varied fusion processes are frequently encountered during gynoecium development differing in histological and cellular details as well as in the degree of intimacy attained by the united tissues.[83,84] As a rule the earlier the fusion during development, the less discernible is the suture. Examples of carpel fusion have been described in many species, including *Datura*,[54] *Aquilegia*,[95] *Pisum* (pea), *Solanum*,[1] etc.

In the case of *Catharanthus* (= *Vinca*), worked out by Boke and Walker, the two horseshoe-shaped carpels are initiated separately on opposite sides of the flower meristem.[51,96,97] They then grow towards each other until their distal tips touch and fuse forming a common tissue that will differentiate later into the stigma, style, and top part of the compound ovary (intercarpellary fusion). Somewhat later, the basal margins of each carpel primordium infold and fuse together to seal shut the locule (intracarpellary fusion). In the regions of contact, cells with distinctively epidermal features progressively lose their epidermal character after fusion. However, the rapidity and extent of this dedifferentiation process are variable. Even within the fused carpel tips the degree of intimacy attained by united tissues is different in the different regions: in the stigma and style former epidermal cells respond rapidly to contact, divide periclinally and redifferentiate into transmitting or secretory tissues. They appear to retain no evidence of their epidermal origin and it becomes extremely difficult to identify the line of fusion. On the contrary, cells contributing to the ovary react much less and the original fusion line remains detectable until anthesis.

Using the electron microscope Walker observed that, during fusion of carpel tips, there is an intense cytoplasmic activity in the regions where the walls of the epidermal cells become appressed.[97] The plasmalemma is wavy, apparently fusing in many places with vesicles originating from the numerous dictyosomes and rough endoplasmic reticulum present nearby. These organelles appear to be involved in a process of wall modification through the deposition of materials which effect adhesion of the contacting walls. After completion of fusion the fused cells have typically a wall of double thickness with "trapped" cuticles and no plasmodesmata. Also, secretory activity of the cytoplasmic organelles has ceased.

In the fusion of the false septum in the gynoecium of *Capsella* the cuticles trapped between the fused epidermal cells also persist, but in a fragmented state, until maturity.[98] As in *Catharanthus*, there are no plasmodesmata or other visible communication lines between the protoplasts of fused cells.

The processes of ontogenetic fusions deserve further study since they characterize essentially reproductive development as opposed to vegetative growth and because they raise significant questions concerning plant morphogenesis. Indeed, the nature of the stimulus for fusion, which is certainly not random contact of epidermal layers, is still entirely unknown. Also fusion, contrary to wound closures, does not involve callus formation and, as pointed out by Walker, does not impair the continuing differentiation of the fused floral parts and may even be a prerequisite for the further development of the united tissues.[96]

G. Zonal Growth and the Development of Perigynous and Epigynous Flowers

After their inception, corolla tubes in many species grow by a process called "zonal growth"; i.e., a complete ring of tissue below the level of attachment of the component parts is activated and by vertical growth produces a cylinder of tissue.[82] The activity of the annular intercalary meristem may extend to the bases of successive whorls of floral organs producing more complex tubular structures, e.g., a corolla tube on which stamens are inserted. Processes of this kind are responsible for development of the peri- and epigynous conditions of flowers.

Perigyny is characterized by the insertion of sepals, petals, and stamens on the margin of a cup-like structure surrounding the gynoecium (superior ovary); the latter is at the bottom of the cup. The position of the gynoecium on the concave receptacle thus seems lower than that of other floral appendages. Ontogenetic studies of perigynous flowers in *Agrimonia, Pereskia aculeata, Portulaca,* and *Eschscholtzia* indicate that successive floral members are initiated normally by the convex flower meristem, as if the flower was hypogynous.[42,45,49,64] At a relatively late stage of development, namely after carpel initiation, cup formation starts from the activity of a peripheral intercalary meristem located just beneath the bases of sepals, petals, and stamens. Elevation of the outer border of the receptacle results, whereas the central portion of the receptacle grows very slowly. Consequently, the entire structure takes the shape of a hollow cup.

In rose (*Rosa hybrida*) and almond (*Amygdalus communis*), other species with perigynous flowers, cup* formation starts much earlier during ontogeny, i.e., at the time of sepal initiation.[62,69] The bases of sepal primordia are carried upwards by growth at the periphery of the receptacle, resulting in a floral cup with relatively steep sides. Subsequently, petals and stamens are produced from the lower inner edges of the cup, not from the meristem itself. Finally, the carpel(s) arise(s) as protuberance(s) from the remainder of the flower meristem located at the floor of the cup.

Anatomical evidence is in line with this ontogenetic description of the perigynous condition. The vascular bundles supplying traces to the floral members run upwards, with xylem and phloem normally oriented, then turn at a point about half-way to the apex, proceed downwards with vascular tissues inverted, and finally give off branches to the carpels at the base of the cup.[45,62,100] Thus, the course taken by the bundles supports the view that the cup

* The cup in Rosaceae is sometimes referred to as the "hypanthium", an incorrect term.

in perigynous flowers is the result of an invagination of the floral receptacle due to growth at the periphery of the receptacle exceeding that in the center.

Whereas in the normal 'Baccara' rose flower carpels are produced only after receptacle invagination has begun, Lindenbaum and colleagues found that, in the bullhead, carpels are initiated when the receptacle is still flat.[62] The bullhead malformation is thus the result, at least partly, of the failure to achieve the perigynous condition.

Epigyny is a condition in which the gynoecium is completely concrescent with the wall of the cup and the ovary is apparently "sunk" in the receptacle (inferior ovary). The cavities of such ovaries arise by a process of zonal growth similar to that occurring during development of perigynous flowers, as exemplified by cucumber,[101] *Chrysanthemum*,[29,30] and *Downingia*.[102] Typically in epigynous flowers the activity of the peripheral intercalary meristem starts very early during flower ontogeny, at the time of initiation of the first floral appendages.

Contrary to the situation in perigynous flowers, the vascular bundles in many epigynous flowers diverge to the floral appendages in an acropetal sequence and do not descend to supply the carpels.[102] Interpretation of the significance of this important difference in vascular anatomy is a long-debated issue among morphologists (see Kaplan[102] for references).

Classically the structures formed as a result of zonal growth are said by comparative morphologists to be "congenitally fused". Cusick has hypothesized that each component of these structures, often recognized at early stages as superficial, separate mounds of tissue, has its own growth center in part submerged in the tissue of the meristem from which it arises.[82] Zonal growth might then be viewed as the merging of these initially independent growth centers. Following this line of reasoning, zonal growth is seen as a special kind of fusion requiring the early disappearance of fields of inhibition set up by the original growth centers. The occurrence of an initial phase, when the component growth centers are independent, is not at all obvious; in many cases, e.g., cucumber,[101] *Frasera*,[103] *Chrysanthemum*,[29,30] a circular primordium seems to occur first, and only later do the individual parts develop on top. We are still in difficulties when trying to delineate exactly the internal physiological boundaries of growth centers and this problem needs to be solved before Cusick's suggestion can be tested.

H. Zygomorphic Flowers

There are only a few studies of zygomorphy, but they agree in general that this condition is established very early during flower ontogenesis, i.e., even before petal initiation.[85] Indeed in pea, according to Sattler, the order of appearance of the five sepal primordia is from the abaxial to the adaxial side of the flower meristem instead of the spiral sequence usually found in actinomorphic flowers.[1] The five petals then appear in the following order: first the two keel abaxial petals and then almost simultaneously the last three petals comprising the standard and the two wings. The two keel petals grow independently for some time and then fuse along their common margin giving the false impression of a tetramerous corolla in the mature flower. A similar situation is found in *Digitalis*.[90]

In the orchid, *Habenaria clavellata*, zygomorphy is also detectable very early since the flower meristem expands laterally from the start of floral member initiation.[1] In the flowers of *Peperomia*, with no sepals and no petals, zygomorphy is organized during stamen initiation.[1]

Postinitiation events accentuate the initial zygomorphy. In Leguminosae, such as pea, different petals exhibit differences in the distribution and directions of growth, and finally may have very different shapes. In some cases, e.g., the corolla tube of ligulate ray florets in Compositae, postinitiation features alone seem responsible for zygomorphy since these florets are indistinguishable from the tubular actinomorphic disc florets during the early stages of ontogeny.

The fascinating difference between radial (actinomorphic) and dorsiventral (zygomorphic) patterns, probably related initially to subtle differences in the distribution of growth within the flower meristem, is worthy of more extensive investigation.

I. Incomplete and Unisexual Flowers

Floral appendages of one or another type may be normally missing in some flowers. Cases in which some or all perianth members are absent have been described above, e.g., in Gramineae and Compositae, although structures like the lodicules and the pappus in these plants are sometimes viewed by morphologists as homologous to the missing parts.

While hermaphroditic (or monoclinous) flowers possess both stamens and carpels, unisexual (or diclinous) flowers have only one kind of spore-producing appendages, either stamens or carpels. As stressed by Heslop-Harrison, a range of situations is found among unisexual flowers.[104] Many are morphologically bisexual in early stages of development, functional unisexuality resulting from a secondary unbalanced growth in androecium and gynoecium, one developing to the exclusion of the other. Such flowers have the rudiments of the missing sex, either staminodia or pistillodia, visible at maturity or at least at some stage of their ontogeny. Examples of this are *Xanthium*,[34] maize,[25] *Cucumis sativus* (cucumber),[101,105] *Rhamnus*,[1] *Asparagus*,[106] and *Phoenix*.[107]

In monoecious cucurbits, like cucumber, the mature male (= staminate) flowers exhibit a three-lobed undeveloped pistillodium, whereas the female (= pistillate) flowers have three rudimentary staminodia.[101,105] There are no anatomical differences between these two types of flowers during their early development. For instance, Galun and Porath show that the rudimentary organs incorporate [^{3}H]thymidine into DNA during their early growth.[108] Potentially male flower buds of cucumber, cultured in isolation, develop a normal pistil, whereas potentially female buds of squash show a greater development of stamens in vitro than in vivo.[109,110] This indicates that the rudimentary primordia of the missing sex are potentially capable of full development.

In other species that have initially bisexual flowers the stage at which the male and female flowers diverge morphogenetically may be very different. In maize and *Phoenix dactylifera* (date palm) development of the nonfunctional sexual organs ceases at a relatively early stage,[25,107] as in cucumber, whereas in *Asparagus officinalis* this occurs at a fairly late stage.[106]

In other unisexual flowers only one class of spore-producing appendages, either stamens or carpels, is formed during ontogeny. Rudiments of the missing class are not detectable, suggesting that in such cases, e.g., *Cannabis sativa* (hemp),[111] *Begonia*,[112] *Carex*,[113,114] *Juglans, Populus*, and *Quercus*,[1] the flowers do not pass through a stage in which they are morphologically bisexual.

In *Begonia* male flowers have a four-part perianth, whereas female flowers possess a five-part one.[63,112] The character of the perianth also differs between male and female flowers in *Quercus*,[1] hemp,[104] etc. Sex determination thus seems to take place before initiation of perianth primordia, i.e., very early during flower morphogenesis. However, *Begonia, Quercus*, and other monoecious species occasionally may produce hermaphroditic flowers.[63,71] Also, dioecious plants like hemp may form either hermaphroditic flowers or flowers of the opposite sex,[71,104] leading to the idea that all unisexual flowers have the potential for development of both sexes and that sex determination may be altered during the course of flower ontogeny. Plasticity, i.e., absence of early irreversible determinative events, is known to characterize many other morphogenetic processes in plants.

In monoecious species, such as maize and *Xanthium* described above, the male and female flowers are grouped in separate inflorescences differing in their position on the plant (Figure 7). A similar situation is found in monoecious cucurbits, as shown by Nitsch and coworkers.[115] In *Cucurbita pepo*, e.g., the axillary male flowers are restricted to lower nodes (Figure 14A). Then the plant passes through a stage in which axillary male and female flowers are formed and, finally, only female flowers develop. The female phase continues until death of the vine.

Besides the classical monoecious type of cucumber plant, several genotypes have been

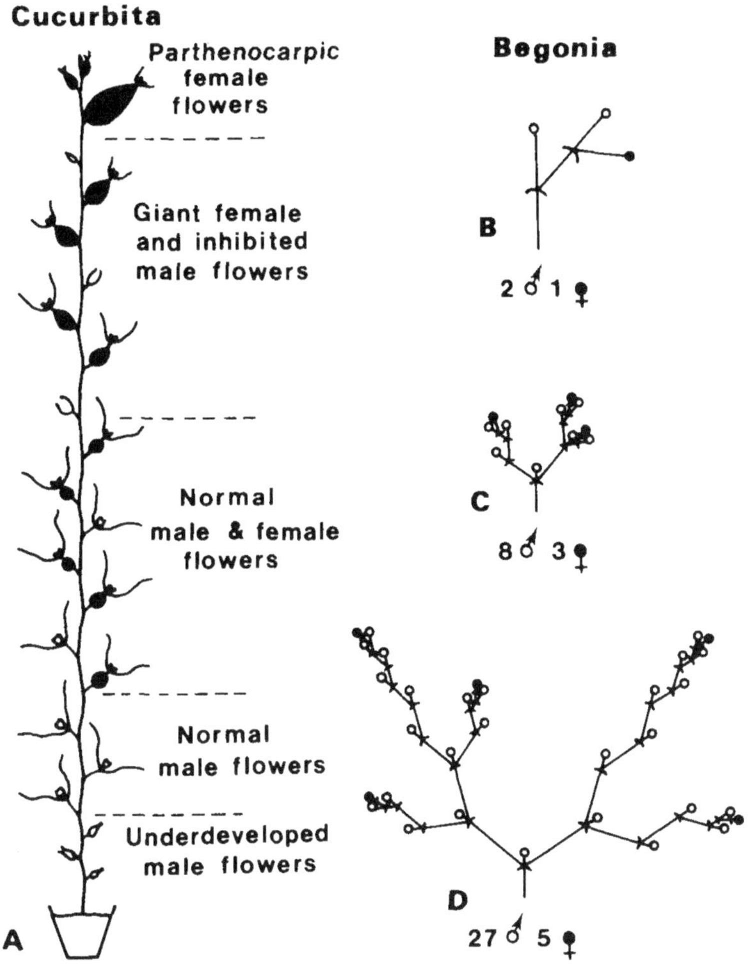

FIGURE 14. Schematic representation of the sequence of flower types in monoecious species. (A) *Cucurbita pepo* cv. 'Acorn'. (B) *Begonia franconis*. (C, D) *Begonia semperflorens*. In all cases there is a shift from maleness to femaleness. In *Cucurbita* (A), the male phase is restricted to the initial stage of plant development; the female phase extends over the greatest part of the life of the vine which continues to produce female flowers until death. In both *Begonia* species (B to D), the male zone extends over the greatest part of the cyme and the production of female flowers is exclusively associated with the final stage of inflorescence development. (From Nitsch, J. P., Kurtz, E. B., Jr., Liverman, J. L., and Went, F. W., *Am. J. Bot.*, 39, 32, 1952 for *Cucurbita;* from Berghoeff, J. and Bruinsma, J., *Z. Pflanzenphysiol.*, 93, 303, 1979 for *Begonia franconis;* from Matzke, E. B., *Am. J. Bot.*, 25, 465, 1938 for *B. semperflorens*. With permission.)

developed which form: (1) primarily female flowers even at lower nodes (gynoecious lines), (2) primarily male flowers at all nodes (androecious lines), (3) male and hermaphroditic flowers (andromonoecious lines), (4) hermaphroditic and female flowers (gynomonoecious lines), (5) male, hermaphroditic and female flowers (trimonoecious lines), and (6) hermaphroditic flowers only (hermaphroditic lines).[116]

In most Compositae the capitulum is gynomonoecious, i.e., the ray florets are female and the disk florets are hermaphroditic.

In monoecious *Begonia* male and female flowers also occupy different and specific locations in the same inflorescence. In *B. franconis* the basic inflorescence is a uniparous scorpioid cyme of three flowers in which the first two initiated flowers are male and the third female (Figure 14B).[112] Transitions to the biparous condition as well as return to the uniparous one may occur in this species, as described above in *Nicotiana* (Section II.C), giving rise to cymes bearing four or five flowers, including two female and two or three male, but in all cases female flowers are always the last formed. A similar, although more complex, situation is described in *B. semperflorens* by Matzke.[63] With the number of flowers per inflorescence varying from 6 to 37 and the ratio of male to female flowers from 1.75 to 6.50, 31 different patterns of arrangement of male and female flowers can be found, but female flowers are invariably formed at the final developmental stage (Figure 14C, D).

The different numbers of perianth parts in male and female flowers of *Begonia* or the different morphologies of male and female inflorescences in *Xanthium* and maize are called "secondary sex characteristics". The size of male and female flowers may also differ in many species, whereas in others the numbers of flowers produced by male and female individuals are different. The reader interested in morphological differences related to sex should consult the paper by Lloyd and Webb.[117]

J. Polymorphic Flowers

Diclinous plants are not the only species having more than one type of flower. Di- or trimorphism is also characteristic of heterostylous species in which the length of the style in relation to other parts of the flower, often the stamens, differs in the flowers of different plants. In buckwheat one type of flower has styles shorter than stamens, whereas in the other type the styles are longer than the stamens.[1] In *Lithrum salicaria* three types of flowers occur on different individuals as a result of differential elongation of outer stamens, inner stamens, and style.[1]

Other cases of dimorphic flowers are found in cleistogamous species which produce both open (or "chasmogamous") and closed (or "cleistogamous") flowers. The closed flowers never achieve anthesis but are self pollinated. Closed flowers are characterized by reduced stamen size and/or number and corolla size.[118] In *Lamium amplexicaule* there is a progressive change from cleistogamous to chasmogamous flower production along the main shoot, with flowers of intermediate morphology in between.[119] Lord views these variations in floral morphology as a function of reproductive shoot development as the counterpart of "heteroblastic" leaf development in maturing vegetative shoot systems.[119] Another case of heteroblastic reproductive development, described in the previous section, is the obligate succession of male and female flowers along a vine in cucurbits (Figure 14A) and in the inflorescence in *Begonia* (Figure 14B, C, D).

In species like *Solanum dulcamara*, the sequence of sepal initiations is clockwise in some flowers and counterclockwise in others.[1] The imbrication of sepals in mature flowers may also show handedness, as in *Hibiscus furcartus* which possesses clockwise and counter-clockwise flowers.[120]

A unique floral dimorphism characterizes plants of the genus *Wackendorfia*.[121] The flowers of some individuals have the style and one stamen sharply deflected to the right; in other individuals the style and stamen are deflected to the left. In species of *Cyanella* and *Cassia*, both types of these so-called "enantiomorphic" flowers are produced by the same plant and even in the same inflorescence.

These selected examples of flower polymorphism, added to the great variety of random variations of flower structures in many species, demonstrate the extraordinary plasticity of flower development. Despite all that flower morphology is basic to species identification and classification. But it is erroneous to believe that flower morphogenesis in a given species is invariable. Rather it is a process giving statistically predictable results; the dispersion around the mean can be quite instructive to morphologists and to physiologists as well.

K. Termination of Activity in Reproductive Meristems

In many racemose inflorescences the apical meristem is indeterminate. Generally, in these cases it aborts more or less rapidly thereby terminating growth without having formed a terminal flower. Less frequently, it may form spines or hairs.[122] This is not a general character, however, since in other species producing simple or complex racemes, e.g., lupin or tobacco, the reproductive meristem is finally used up in the formation of a terminal flower. Racemes with or without a terminal flower are even found in the same species, sometimes in the same individual, e.g., in *Primula malacoides.*[123] In cereals some species produce a terminal spikelet (wheat, Figure 3C), whereas others (barley) do not.

The capitulum is a racemose inflorescence commonly believed to possess a terminal floret. Yet a small axial area at the tip of the reproductive meristem in some Compositae, e.g., *Cosmos,* may remain free of florets.[33] In other Compositae the most apical floret may sometimes exhibit anomalous features, e.g., it may be incomplete or devoid of a subtending bract (Figure 6). Or the axial area may be occupied by a bract without an axillary floret.[124] According to Lawalrée, these abnormalities are probably related to insufficient residual meristematic material at the tip of the meristem for production of a complete floret and accompanying bract.[124]

In plants producing a cymose inflorescence the apical meristem is entirely transformed into the first flower of the inflorescence and, in this sense, it reacts exactly as in plants producing a solitary terminal flower.

Within flowers the fate of the meristem also varies with the species. In *Catharanthus, Michelia* and *Phoenix* a residual meristem persists after carpel initiation and gradually vacuolates and senesces.[51,65,107] In other species the flower meristem is entirely used up in the formation of parts of the gynoecium, either the carpel margins or placentae in *Anagallis,*[59] *Impatiens,*[125] *Downingia,*[102] *Portulaca,*[49] *Cyclamen,*[60] etc., or the ovules in wheat,[16] *Chrysanthemum,*[29,30] *Piper,*[126] etc. In a few species a single carpel, occupying an apparently terminal position, is derived from the remaining meristem cells in the center of the flower, e.g., almond and *Drymis lanceolata.*[99,127]

In the leafy cacti of the genus *Pereskia,* Boke reports that the residual flower meristem begins to expand late in ontogeny when anthesis approaches and mainly during fruit development; it forms the columella and fills whatever space is not occupied by the seeds.[45]

IV. GROWTH AND DEVELOPMENT PATTERNS OF REPRODUCTIVE STRUCTURES

A. Sequence of Development

In many racemose inflorescences the sequence of flower development follows the acropetal order of flower initiation with flowers opening first at the base. Similarly, in many capitula and umbels florets reach anthesis earlier at the periphery than in the center. In wheat the first-formed spikelet primordia, i.e., those located about the middle of the spike, are always morphologically the most advanced (Figure 3).[15,19] They initiate florets first, produce more florets and more grains, and give the heaviest grains.

Many species having a raceme with a terminal flower behave differently. For example, in *Agrimonia eupatoria* and *Campanula persicifolia,* flower development proceeds normally in an acropetal direction except that the terminal flower takes precedence over the few flowers located immediately below it.[42,128] Preeminence of the terminal flower, which is initiated last, is also found in the complex raceme of tobacco and *Caulophyllum.*[38,60a]

In cymes, the first-formed (or terminal) flower is generally the first to open, but lateral flowers may open in acropetal, basipetal, or other sequences. Thus, depending on the species and environmental conditions, the order of flower anthesis within the inflorescences may follow variable patterns.

Similar observations have been made for floral appendages. Petals, e.g., although initiated before stamens and carpels, may have a very retarded development compared to other flower parts. They may remain very small until shortly before anthesis at which time they then expand rapidly. This is the case in *Ranunculus* and *Solanum,*[1] *Nigella,*[129] *Papaver,*[64] *Sinapis,*[10] *Silene* (Figure 13D),[130] etc. In *Chenopodium album* retarded development is a property of the stamens.[1] Thus the order of development of floral members does not necessarily follow the order of initiation. These observations, generally made with the naked eye or under the dissecting microscope, should be complemented with measurements of the growth rates of different flower components, a topic that is discussed in Section C below.

B. Trend towards Synchronous Development

Cross-pollination can be achieved only by chronological adjustment of flowering of the pollen donor and pollen receptor.[116] Synchronized maturation of reproductive structures is also critical for successful self pollination. Thus, reproductive development should be characterized by a trend towards synchronous initiation, differentiation, and growth of flowers or flower parts. Indeed, it is a common observation that parts of inflorescences and flowers are initiated in close sequence at rates higher than those recorded during production of vegetative parts. Whorled or nearly whorled phyllotaxis frequently encountered in flowers is also presumably a way to achieve better coordinated development since whorled appendages as a rule differentiate and grow synchronously. Recall also that most cases of synchrony at the cell level have been found in relation to sexual reproduction: first, in the subapical tissues during bolting in rosette plants, second in the shoot apical meristem at evocation (Volume II, Chapter 4), and then during pollen and embryo sac formation in the anther and nucellus, respectively.

In wheat Kirby observed that there is a trend towards synchronization of development in the different spikelets of the same spike.[19] Although spikelet 8 is initiated 8 days prior to the terminal spikelet, the latter starts floret production only 5 days after the former. Also, the different florets of a given spikelet tend to synchronize in their development despite being initiated in succession.[131]

In tobacco the first (terminal) flowers of all lateral cymes reach anthesis almost synchronously despite successive initiation of these cymes in an acropetal order.[38] This clear tendency towards synchrony is lost, however, in less vigorous individuals in which the first flowers of lateral cymes develop in a basipetal direction.

In South Australia inflorescences of *Acacia pycnantha* are initiated in every month of the year on new shoot growth but, as shown by Buttrose and coworkers, differences in the rate of development cause inflorescence buds produced several months apart to reach anthesis in the same month (Figure 15).[132] Other species in which late flowers develop more rapidly than — and thus tend to synchronize with — early flowers are tomato,[74] *Phoenix,*[107] *Impatiens,*[125] etc.

Near synchronous development not only occurs within an individual plant, but is also often achieved between the different individuals of the same species through identical response to environmental parameters and subsequent endogenous regulation, as will be explained in subsequent chapters.

Cases are known in which the anthers develop and ripen in advance of the pistil(s) of the same flower (protandry), e.g., *Malva, Epilobium,* carrot, onion, *Gerbera, Cocos nucifera* (coconut), etc. Others, such as aroids, *Atropa*, hemp, and date palm, are characterized by the reverse situation (protogyny). In *Annona hybrida,* e.g., the anthers ripen about 26 hr after the pistils become receptive.[133] Such mechanisms, commonly assumed to prevent self pollination, require the closely controlled synchronous — but sequential — development of stamens and pistils.

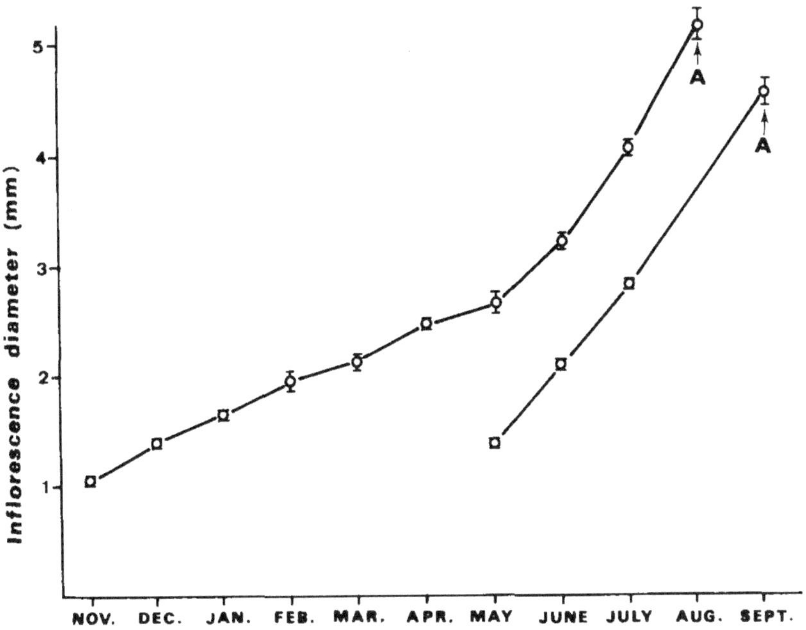

FIGURE 15. Increase in diameter of early- and late-initiated inflorescences in *Acacia pycnantha* grown in South Australia. Recall that August in Australia is late winter. Vertical bars represent standard errors; A, anthesis. (From Buttrose, M. S., Grant, W. J. R., and Sedgley, M., *Aust. J. Bot.*, 29, 385, 1981. With permission.)

C. Growth Rates

Precise descriptions of growth patterns within inflorescences and flowers are very few because growth measurements in such complex and changing structures are difficult and tedious. As shown by the pioneer work of Erickson, the developing flower bud of *Lilium longiflorum* constitutes a marvellously integrated system.[134] The whole bud and perianth members have a typical exponential growth from early ontogeny almost to anthesis. Growth of the anthers is also exponential, but can be divided into an early period of rapid increase in length and weight and a late period of slower growth. The ovary, on the contrary, does not grow exponentially, but linearly with respect to the logarithm of bud length.

In wheat Williams showed that the early growth of each leaf during the vegetative phase is also exponential, but the relative growth rate, R, then rises to approximately twice that during the exponential phase before falling to zero.[73,135] The only component of the wheat spike that exhibits such a growth pattern is the palea (Figure 16). There is no evidence of an early phase of exponential growth in spikelet, glume, lemma, and ovule primordia. From the earliest growth stages the R values for these parts fall continuously with time. As in *Lilium,* stamens and carpel behave differently. While R for the stamens decreases continuously and rather rapidly to zero, the carpel exhibits biphasic growth with two periods of high R value, one just after carpel initiation and then during later growth of the two styles. In both *Lilium* and wheat stamens have a much reduced R well before anthesis, at a stage when the gynoecium is still growing.

Other plants in which R in the apex declines during reproductive development are *Chrysanthemum*[136]and *Silene*.[130] In the last species R (per day) in the meristem increases from 0.67 to 1.06 just before sepal formation, then decreases progressively during floral member production and is about 0.37 at the time carpels are initiated. Parallel to this decrease there is a dramatic increase in the rate of appendage initiation, indicating that production of

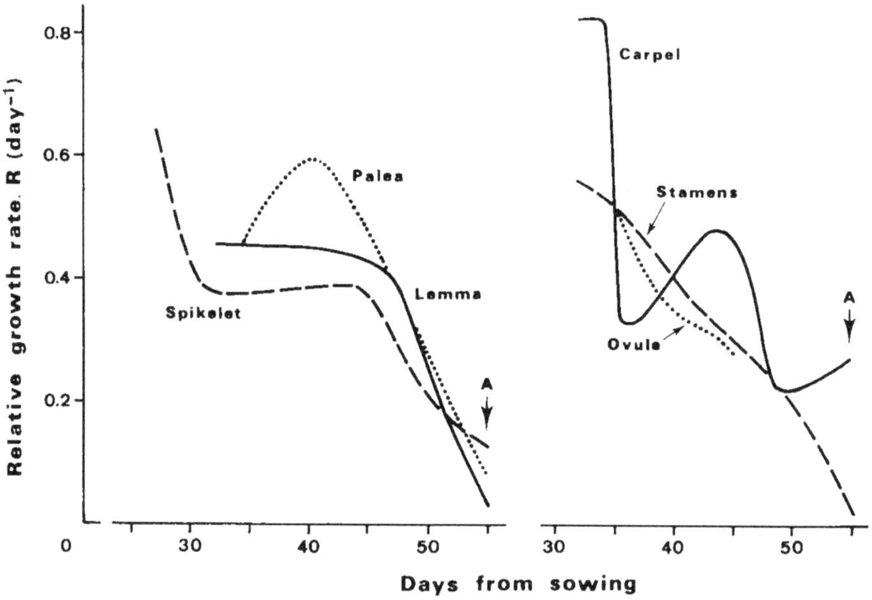

FIGURE 16. Relative growth rates for a spikelet and parts of its basal floret in wheat (*Triticum aestivum*). The time scale is different for the different structures. A, time of anthesis. (From Williams, R. F., *Aust. J. Biol. Sci.*, 19, 949, 1966. With permission.)

primordia is not a simple function of the whole meristem growth rate. In further work on this question one should also focus on specific zones within the meristem, namely the zones that directly initiate the floral members.

The primordia of all appendages in *Silene* have a higher relative growth rate than the reproductive (flower) meristem itself, 0.65 for sepals and stamens, and 0.56 for petals against 0.37 for the meristem.[130] The primordia of the same whorl of appendages, say all sepals or all petals, have similar R values. These differences are traceable back to the moment of organ initiation. Thus, the growth rate of these primordia appears to Lyndon to be programmed at their inception.[130] A similar conclusion will be reached in Chapter 6 for other species on the basis of allometric relationship between various floral members. Note that there is no parallelism in *Silene* between R values and primordium size at initiation since petal primordia are larger but have a lower R than stamens (see also Figure 13D).

By manipulating the temperature conditions prevailing during flower initiation and development in *Silene*, Lyndon was able to alter the rates of primordium initiation and growth rates and sizes of primordia on initiation.[130] Yet the form of the flower, as exemplified by the relative sizes of the primordia at carpel initiation, remained constant in the different temperature regimes. This puzzling observation implies that form is not primarily determined by these parameters, but is essentially controlled by subtle interactions between the various floral members compensating in some way for perturbations in growth parameters.

Sex expression might also be related to growth rate. In monoecious *Picea sitchensis* (Sitka spruce), before appearance of distinct morphological differences between bud types, Tompsett found that the apical meristem of future female buds has a higher growth rate than that of future male buds.[137] This possibly relates to the observation of Wareing that female strobili in *Pinus sylvestris* are formed first and on leading shoots, whereas male strobili are produced several years later on apparently less vigorous branches.[138] Also in several annuals, e.g., *Xanthium*, hemp and spinach, development of female flowers requires stronger photoinduction than that of male flowers.[35,139-141]

The growth of the flower axes in dicotyledonous species is also exponential.[142,143] Elongation of this structure in these species, as well as in grasses, is due to the activity of an intercalary meristem located just below the reproductive apex and extending in many plants over several centimeters.[11] In discussing the characteristics of this meristem, Sachs has proposed calling it a "primary elongating meristem".[11]

In many species, e.g., tulip and Compositae, the important elongation of the axis below the inflorescence or flower contrasts with the near cessation of internode growth within the inflorescence or flower itself. This puzzling situation implies strict spatial and/or temporal control of axis growth, but the mechanisms involved are still poorly understood.

During stamen development the filament arises relatively late due to the activity of an intercalary meristem located below the already differentiated anther region. Greyson and Tepfer found that this meristem in *Nigella* is diffuse along the filament, contrary to other intercalary meristems of determinate organs that are as a rule localized.[144]

Workers have also looked for correlations between changes in the growth rate of different parts of the reproductive structures. Correlations among phenomena do not distinguish between cause and effect, but suggest systems that should be the subject of further investigations. Thus, in wheat, the relative elongation rate for the spike falls at the time spikelet growth becomes well established and the same happens for spikelets when the growth of the two lower florets is established.[135] According to Nicholls, initiation of the terminal spikelet is correlated with appearance of first floret primordia in wheat and rye and with that of second-order panicle branches in oat.[145] In barley, wheat, and perhaps other cereals the onset of rachis internode elongation coincides with first floret initiation.[145]

In *Silene* the growth rate of all flower parts, including the meristem, sepals, petals, stamens, and pedicel, decreases markedly when the carpels have all been initiated.[130]

The mechanisms underlying these interesting correlations will be further investigated in Chapter 6. Precise characterization of normal and abnormal development of reproductive structures requires that quantitative analyses of the sort described here be extended to other species and to a variety of situations, such as unisexuality, zygomorphy, peri- and epigyny, etc.

D. Secondary Displacement of Parts

The spatial relationships between plant parts that are achieved at the time these parts are initiated may become profoundly disturbed by later processes of intercalary growth. Such growth renders interpretation of the mature structures very difficult without a detailed ontogenetic analysis.

In many Solanaceae, like tomato, the last leaf and terminal reproductive axis of each sympodial unit of the shoot are produced at about the same level. These two structures are secondarily displaced from each other by intercalary growth so that the inflorescence is finally inserted well below its corresponding leaf.[40] This explains why in such cases the mature inflorescences appear in "abnormal" extraaxillary positions. Mature inflorescences in grapevine are also in extraaxillary positions and this again results from a process of secondary displacement due to intercalary growth.[145a]

Complications may also arise from secondary displacements of parts within the inflorescence. In *Nicotiana glutinosa*, e.g., the bract β_1 and axillary flower 1 which are closely related in space at the time they are initiated by meristem a_1 (Figure 8B) can finally be found at very different levels in the mature cyme. This is due to one of these parts being displaced by intercalary growth into the next sympodial unit of the inflorescence, i.e., into the branch derived from the activity of meristem a_2.[38]

In *Potamogeton richardsonii* the tepals are normally initiated at the flanks of the flower meristem and are followed by stamen primordia. Soon afterwards, upgrowth occurs between and at the base of each developing tepal and stamen so that, as shown by Posluszny and Sattler, the tepals are finally inserted at the stamen connective in the mature flower.[146]

E. Incomplete, Aborted, and Abnormal Development

Since several examples of arrested or diverted development were discussed in Volume II of this series, only a summary of these cases is given here.

1. Incompleteness and Abortion

As explained in Volume II, Chapter 3, Section III.C.4, the "curd" of cauliflower is an amplified prefloral inflorescence axis arrested in its development between the typical vegetative and reproductive stages. On the other hand, the thorns in *Bougainvillea* and *Carissa* and tendrils in grapevine are aborted or incomplete inflorescence axes caused by exposure to unfavorable conditions to flowering. Hackett and Sachs have observed that thorn formation in *Bougainvillea* may be induced at many intermediate stages of inflorescence ontogeny, e.g., at bract or bracteole (showy bracts) formation or even after florets have been initiated.[147]

Blindness in the rose results from the abortion of the apical meristem which occurs, depending on the cv. and/or growing conditions, at the vegetative stage or after sepal and petal initiation.[148-150] As a consequence of abortion of the main meristem, an axillary meristem starts growing.

Reproductive meristems of many cultivated chrysanthemums grow slowly in unfavorable daylength conditions and generally fail to develop to anthesis. As shown by Popham and Chan and Cockshull, the stage at which development is arrested varies with cv. and environmental conditions.[151,152] In some cases an enlarged prefloral meristem bearing only involucral bracts is formed but no florets are initiated. Development then ceases at a stage similar to that of Figure 4 in Volume II, Chapter 3. In other cases florets (usually only ray florets) are produced, but cover only part of the meristem surface. Such incomplete capitula are known to horticulturists as "crown buds". Apical dominance is lost with the formation of these buds so that they are finally surrounded by shoots that function for a while vegetatively. These arrested reproductive meristems have the potential to form normal florets as shown by the fact that they can be rescued from abortion by removal of the surrounding leafy shoots or by transfer to favorable daylength conditions.[153]

In some conditions a large unemployed area may persist at the summit of the meristem of male capitula in *Xanthium*.[154,155] Development seems arrested at stage 5 or 6 on Salisbury's scale (Volume I, Chapter 1, Figure 2A).

Limited availability of floral stimuli is apparently responsible for the arrested morphogenesis in all the above examples.

A spontaneous variant (mutation) of *Aloë,* observed by Resende, also exhibits incompleteness: the inflorescence axis and bracts are normally initiated and show advanced development without floret initiation.[156] In this case incompleteness is of genetic origin.

In other cases abortion is apparently due to overproduction of reproductive structures in favorable conditions. Thus, in wheat, each spikelet produces six to eight florets but only two to four of these develop to anthesis. The other floret primordia normally abort.[15,19] In barley the apical meristem of the spike and the last-initiated lateral spikelets abort usually before ear emergence, so that only about 20 spikelets remain for grain formation out of the 40 which are initiated.[157] Similarly in many corn varieties only one of the two florets of each spikelet is functional and many spikelets at the tip of the cob remain rudimentary.[25] Generally, yield in corn is dependent upon the development of one or two cobs when the plant has primordia of potential cobs in all leaf axils.[25] As another example, soybean yield is only half of its real potential, partly because of a high rate of flower abscission.[158]

In many other species grown in natural or experimental conditions, it is also common to observe that parts of the reproductive structures that are normally initiated do not reach complete development and abort. Whether this is caused by "sluggish" organogenesis, lack of synchronized development or of vascularization in these parts, or to other reasons is not clear in most cases and should be investigated.

In some cases, abortion of a specific part is a normal taxonomic character of the plant in question, e.g., the fifth stamen in the 5-merous flowers of *Digitalis*.[90]

2. Abnormalities

Deviations from the normal morphology of inflorescences and flowers even in a single species are so varied that it is impossible to describe them in detail. Generally speaking three types of reproductive anomalies can be distinguished:

1. Those relating to inflorescence structure, i.e., increase or decrease in the number of bracts and/or branches, changes in receptacle shape and in inflorescence internode length, etc.
2. Those relating to flower structure, i.e., changed numbers of parts, fusion of normally free parts or independent growth (dialysis) of normally concrescent parts, abnormal symmetry, virescence (development of one or several floral organs into leaf-like structures), etc.
3. Those in which a new axis, reproductive or vegetative, develops inside a flower, i.e., "proliferations": this new axis arises in place of a floral appendage or from its axil or it is a continuation of the original flower axis; of special interest are the proliferations in which a vegetative bud continues the growth of the flower axis and results in the return to vegetative growth of the flower meristem, i.e., a "reversion".

Variations in the number of floral organs and sex reversal phenomena have already been mentioned in the previous sections. Cases of inflorescence reversion in dicots and spikelet proliferation in grasses were considered in Volume II, Chapter 5, Section VII.C. Various other malformations will be described in several chapters below.

Although many of these anomalous structures, e.g., virescent flowers or proliferations, are consistently observed in nature and can be produced at will in appropriate experimental conditions, they represent often an unpredictable proportion of the plant population and are very difficult to standardize. Consequently, most of the work in this area is a pure description of mature structures. Without better control over their production, a detailed study of how they arise is extremely difficult and their exact significance will remain unclear.

A system which seems better suited for environmental control is the cv. 'Buisson fleuri' of *Impatiens balsamina,* worked out by Simon, in which, contrary to the wild type, the shoot apical meristem is transformed under the influence of inductive SD into a terminal double flower.[125,159] With a progressive lengthening of photoinduction in this plant there is a progressive formation of successive whorls of floral appendages of the terminal flower before return to vegetative growth. With as few as three SD there is a change from the alternate phyllotaxis of the vegetative shoot to the whorled phyllotaxis of the flower, but the lateral appendages formed by the apical meristem remain green leaves (Plate 1, A*). With four to six SD the shape of some appendages is intermediate between those of leaves and petals (virescent petals) (Plate 1, B, C, and D). Normal red petals, stamens, and gynoecium are produced after a minimum of 8, 12, and 20 SD, respectively (Plate 1, E, F, G). In all cases because the SD treatment was transient, the floral appendages are followed by virescent petals and/or normal foliage leaves. In these virescent petals green patches committed to leaf development may occur side by side with red patches committed to petal development, suggesting that determination is expressed at the cell level rather than at the organ level (Plate 1, B, C, D). A similar conclusion has been reached by Steeves concerning leaf determination in ferns,[160] and could be reached from numerous observations of organs with a double nature, e.g., stamens bearing ovules, in teratological flowers.

* Plate 1 will follow page 38.

F. Flower Bud Dormancy

Many bulbous plants have an annual rest period in their life history. In some species, including *Allium, Lilium,* and *Gladiolus,* this dormancy period occurs shortly after the formation of the storage organ and before the onset of stem, leaf, and flower formation. In contrast, in other species as in *Tulipa, Hyacinthus,* and *Narcissus,* the dormancy period occurs after floral organ formation and is associated with flower maturation and elongation of the scape.[4,161-163]

Frequently, buds of many woody species also enter into a dormant state. In numerous spring-flowering woody plants, flowers are initiated within the differentiating resting bud before autumn and, often at an advanced stage of development, undergo dormancy. The apparent winter rest period generally includes a sequence of different physiological states: first, a correlative dormancy when the bud itself does not appear to be intrinsically dormant but is prevented from bursting by some stimulus arising elsewhere in the plant; second, an innate dormancy when bud burst is delayed as a result of conditions arising within the bud itself; and third, an imposed dormancy when bud burst is prevented by unfavorable environmental conditions.[164-166] Flower bud dormancy also occurs in several alpine and tropical species.

The developmental stage of the flower at the onset of dormancy is not the same for all species. Primordia of the inflorescence are present in the dormant bud of *Pyracantha coccinea lalandii*, but flower differentiation has not yet started.[167] In *Ilex cornuta,* the female flower is only partially differentiated in the winter dormant bud and carpels are not observed until spring growth occurs,[168] while in *Tulipa, Hyacinthus,* and *Narcissus*, the apparent dormancy period occurs after differentiation of all the floral organs. In *Soldanella minima* and *Saxifraga exarata,* the most advanced buds have well-developed anthers that never contain pollen grains; the sporogenous tissue is at the pollen-mother-cell stage and microsporogenesis coincides with termination of rest as also reported for peach (*Prunus persica*) and coffee (*Coffea arabica*).[169-172] In contrast, anthers of dormant flower buds of *Saxifraga crassifolia* and *Draba aizoides* contain mature pollen grains.[169,173]

Generally, growth is not completely stopped during the dormant state, but slowed. In peach buds indeed the floral cup enlarges continuously during the rest period when temperature does not fall below minimum growth requirements.[174] Bulbs, too, are never physiologically at rest. This is an important fact both from a practical and a theoretical point of view as stressed by de Hertogh.[163]

V. CELLULAR AND MOLECULAR CHANGES

Studies on cellular and molecular changes during inflorescence and flower development, except for the processes of micro- and megasporogenesis (which are outside the scope of this book), are few and fragmentary.

A. Cell Proliferation

In *Silene coeli-rosa* the cell-doubling time, which markedly decreases during the floral transition (see Volume II, Chapter 4, Section II.B), lengthens during flower ontogeny from 17 hr in the evoked meristem to 27 to 28 hr in sepal and stamen primordia and 48 hr in the flower meristem distal to stamens.[130] After the carpels are initiated the cell-doubling time increases to a week or so. Similar results were obtained in lupin and there is preliminary evidence that such a rise also occurs in other species, e.g., *Sinapis* or *Anagallis,* where the mitotic index (which is a reliable indicator of the rate of cell division in shoot meristems) falls considerably after start of flower initiation.[175-177]

B. Polarity of Cell Growth and Division and the Genesis of Forms

Based on their investigations on the regeneration of a vegetative shoot by mature detached

Plate 1. Apical part of the shoot of plants of the cv. 'Buisson fleuri' of the SDP *Impatiens balsamina* exposed to 3 (A), 4 (B), 6 (C, D), 8 (E), 10 (F), and 14 (G) inductive cycles and then returned to LD. Progressive formation of the terminal double flower is seen by production of whorled leaves (A); partial petalization of leaves as indicated by changes in blade shape and patches of red color on the margins (B to D, arrows); formation of normal petals followed by virescent petals and normal foliage leaves (E, F); and formation of normal petals and stamens followed by a progressive return to leaf production (G). (Courtesy of L. Simon.)

Plate 2. A proliferous flower induced in *Anagallis arvensis* by shifting the plant to 9-hr SD after a single 22-hr LD. A vegetative shoot develops in the center of the flower. se = sepal; p = petal. (Courtesy of J. Brulfert.)

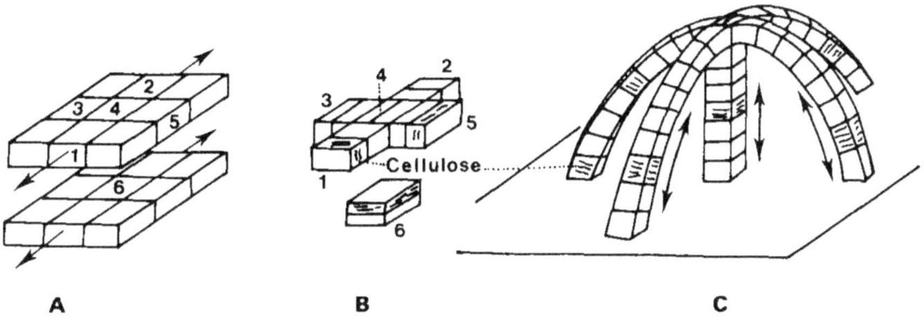

FIGURE 17. Diagram showing polarity changes sufficient to convert important biophysical properties of the starting structure (A) to a skeleton form of a new lateral axis (C). (A) Two sheets of cells, all with parallel structural polarity as shown by arrows. Arrows pass through rings of cellulose in each cell. Only 6 of the 18 cells are involved in the new organ. At first, three kinds of decisions are made involving the 6 cells. Cells 1 and 2 divide transversely anticlinally and continue their original cellulose polarity. Cells 3, 4, 5, and 6 change their polarity. Three cells (3, 4, and 5) in the top layer divide in the longitudinal-anticlinal plane, and the cellulose polarity is also altered to be parallel to the new wall. Cell 6 in the lower layer also changes polarity by 90°, but in a difference sense (periclinal division). Cellulose synthesis is also altered to conform. These changes lead to the configuration in B. A fourth decision is that all subsequent growth and division will be conservative, i.e., will maintain the existing polarities set up in B. Such activity will convert B to C. (Reproduced, with permission from Green, P. B., *Annu. Rev. Plant Physiol.*, 31, 51.© 1980 by Annual Reviews Inc.)

leaves of *Graptopetalum*, Green and co-workers have proposed a model of organogenesis in higher plants that involves several critical polarity changes in the cells of the parent leaf epidermis as well as in those of the subsurface layers.[178-180] This work is summarized here since the formation of floral organs presumably embodies the same basic principles as that of their vegetative counterparts.

Classical accounts of organ initiation put emphasis, as seen in Section III.B above, on the right-angle shift, from anticlinal to periclinal, of the place of cell division in some cells of one subsurface layer (exemplified by cell 6 in Figure 17). This shift in itself is insufficient to bring on a new direction of growth, and a second shift related to the growth direction of these cells, i.e., the orientation of cellulose microfibrils in their walls, is also necessary. These two 90° shifts are probably coupled and together they set up a new polarity in the core of the nascent primordium (Figure 17C). In classical histological studies the epidermis is believed merely to comply by continued anticlinal divisions following the growth of the interior rather than being the initiator of polarity changes. Green and colleagues demonstrate that this view is erroneous.[178-180] Although some epidermal cells continue to divide transversely anticlinally as before and also maintain their original cellulose orientation (cells 1 and 2 in Figure 17), there are others at the site of organ initiation which exhibit a 90° shift in their division direction. These cells, e.g., cells 3, 4, and 5 in Figure 17, now divide longitudinally anticlinally and the cellulose orientation in their walls is altered. The epidermal cells with these changed properties divide and elongate to form the left-right arch of the primordium of Figure 17C, whereas the cells with a conserved polarity account for the production of the front-back arch in the primordium. Cells with polarities differing by up to 90° are in fact essential to give radial symmetry to a new surface.

Once the polarity shifts in both the epidermis and subepidermis have occurred, harmonious growth of the nascent primordium requires that all subsequent growth and division be conservative, i.e., maintain the established polarities. Such activity will convert B to C in Figure 17.

Green attaches great importance to the polarity changes in the epidermis and, because this reorganization may precede the changes in the underlying cell layers, he views the morphogenetic role of the surface layer at least as significant as, if not more important than, that of the interior layers.[179] Equally important is the discovery that the coupling between division direction and cell growth polarity (cellulose "hoop" reinforcement in the wall) is facultative, not obligatory. Some epidermal cells involved in organogenesis may indeed exhibit a shift in their division direction not accompanied by a change in cellulose pattern. Interestingly enough, this "anomalous" behavior is related to cell shape.[180] Finally, there is evidence, at least in some cases, that cellulose reorientation is the earliest detectable polarity change in the epidermis and that cellulose alignments are altered on a cell-by-cell basis.[178,180]

There are reasons to believe that cytoplasmic microtubules participate in the control of oriented cell growth and division.[179] Indeed, there is often a close coupling, in time and space, of the orientation of cortical microtubules and of cellulose microfibrils in the wall (but see Reference 181 for a contrary observation). Also, the orientation of the division plane is set, or at least anticipated, by a cortical preprophase band of microtubules. For Green the cortical array of microtubules most likely determines the structural polarity of the growing cell.[179] Polarity shifts might thus be effected by changing the forces acting on microtubule orientation, but the nature of these forces is still poorly known. Dermen noted that the orientation of cell divisions in meristems is influenced by cell size and shape.[182] Similarly, the periclinal divisions in one subsurface layer, typical of organ initiation, occur just after a change in cell size and/or shape in the reproductive meristems of *Lolium*,[20] *Chrysanthemum*,[29] peach,[56] *Saururus*,[48] *Piper*,[122] etc.

One factor capable of altering the orientation of cell division and growth in a tissue is an externally applied pressure acting through the stresses it creates.[179,183-186] In a growing meristem internal stresses arise normally from the division and enlargement of component cells, and the stress pattern is a function of the overall shape of the meristem. In such a hydrostatic structure divided into many compartments by a continuous system of elastic cell walls, one may expect that the orientation of new walls (i.e., cell divisions) and cell growth is controlled by local stress pattern.[179,185,186] Lintilhac proposes that in a cell exposed to anisotropic stress, either tension or compression, a shear-free plane exists perpendicular to the direction of the largest principal stress exerted on that cell.[186] Some shear-sensitive cytoplasmic component, presumably the microtubules, is believed to align in a predictable manner with respect to the shear-free plane and thus causes the cell plate and the new wall to occupy a position perpendicular to the principal stress passing through the cell. Similarly, cell shape may be altered by stress and, as mentioned above, a polarity shift in cellulose reinforcement in the wall may occur when cell proportions are changed.[178-180] Changed division and cellulose orientation in turn will influence the effectiveness of stress and this can lead to a change in shape of the whole structure, in this case the meristem, which will alter the overall stress pattern. Mechanical stresses thus are seen as a special kind of developmental information in that a particular cellular configuration can directly affect the next, and generate a continuity of forms such as those frequently observed during inflorescence and flower ontogeny.[179,187]

These observations and conjectures, interesting as they are, are not without limitations and difficulties. First, organ initiation occurs very close to the tip of many vegetative and reproductive meristems. Tissue polarity of the parent axis is perhaps not yet definitively established in these sites, and certain aspects of symmetry conversion during organ formation by these meristems might be different from what was found during shoot regeneration by adult leaves.[178] Second, the model of Green (Figure 17) is intended mainly for inception of a new radially symmetrical axis when initiation of dorsiventral organs might involve different principles. In pea, the initiation of a leaf results not so much from the establishment of a new axis but, as shown by Lyndon,[188] from an outgrowth forming on one flank of the meristem without any shift in the polarity of cell division in the epidermal cells. In subepidermal cells, however, there is a loss of oriented divisions just before emergence of the primordium; later on, in the recently emerged primordium, the divisions in these cells adopt the same orientation as in the epidermis, suggesting that the latter imposes its polarity on underlying cells and is thus responsible for maintaining the overall polarity of the new organ.

Third, it is known that organ initiation proceeds as well on conical- and dome-shaped meristems as on flat ones; it may even occur on concave meristems, as in epigynous flowers.

In all cases histologists have recognized that initiation involved periclinal divisions in one subsurface layer. These divisions are attributed to mechanical stresses by Lintilhac,[186] but this is difficult to accept since the stress pattern in meristems of such different shapes is likely to differ considerably. In general, one wonders whether mechanical stresses are sufficiently specific to govern alone the very precise and localized changes in cell growth and division required for the genesis of the complex phyllotactic pattern recognized in the inflorescences and flowers. In an attempt to link their view on organogenesis and current phyllotactic theories, Green and Lang postulated that the long-recognized influence of nearby primordia on an initiation site involves a physical (stretch) or chemical (production of a substance affecting the polarity of cell division) formative effect on adjacent tissues.[180]

As pointed out by Lintilhac and Vesecky, concepts which are almost exclusively based on biophysical influences and totally avoid reference to gene instructions account mostly for meristem behavior during periods of essentially repetitive organogenetic activity.[187] Changes in gene expression are known to occur when the type of appendages produced by a meristem is changed, e.g., during transition from leaf to flower initiation (Volume II, Chapter 4) or during flower ontogeny (Section C below). During these transitions changes in gene expression (and the ensuing changes in metabolism) are precisely integrated, in space and time, with changes in biophysical parameters. The formidable and exciting challenge we face is to describe how this happens and how it is controlled.

C. RNA Content, Protein Complement, and Gene Expression

Simple examination of stained sections of developing reproductive structures in several species has revealed that RNA content increases in cells involved in organogenesis (Figures 8 and 11).[8,10,33,38,50,125] Considine and colleagues have substantiated this observation using cytophotometry.[189] They show that there is a linear increase in RNA content during fertile floret development in wheat, but no increase at sites for sterile floret initiation. This temporal trend in RNA accumulation starts apparently at the earliest possible stage of ontogenesis, i.e., even before the periclinal cell divisions accompanying floret initiation.

Several studies have established that the protein complement of flowers and individual flower parts is specific and differs qualitatively and quantitatively from that found in the vegetative organs of the same species.[190-194] This implies that different sets of genes are active in different mature plant parts, and hence during reproductive development there is a precise sequence of changes in gene expression in localized cell populations. The nature and function of the proteins specific to each class of flower members are generally unknown as is the proportion of proteins specific to each organ relative to the proteins common to all organs. Furthermore, despite its extreme importance for plant biology, the mechanisms controlling changes in flow of genetic information during flower ontogeny have not received much attention. The most comprehensive, but still very fragmentary, piece of work is that of Durand and coworkers in the dioecious *Mercurialis annua*.[195] Male flowers of this species are characterized by three anodic peroxydase isozymes not present in female flowers nor in leaves, whereas female flowers possess isoesterases not found in male flowers. Further work on the isoperoxydases typical of male flowers indicates that they are localized only in stamens. Using an histoimmunological technique Kahlem showed that these 'marker' isozymes appear very early during male flower ontogeny, i.e., possibly even before initiation of stamens.[196] It would be of great interest to know whether or not this change precedes the altered cell division pattern that gives rise to stamen primordia. In young stamens the anodic peroxydases are rather uniformly distributed, but their presence becomes restricted at later stages essentially to the sporogenous tissue and tapetum. These peroxydases specific for stamens were also found in 26 out of 34 species of angiosperms and gymnosperms, including monoecious, dioecious, and hermaphroditic species.[197,198] Kahlem proposed for further consideration that these isozymes are involved in the regulation of the level of hormonal regulators connected with the sporogenous functions of stamens.[196]

Male and female flowers in *Mercurialis* also differ in their pattern of isoaccepting tRNAs and corresponding aminoacyl-tRNA synthetases. As shown by Bazin and associates the female tRNAs for several amino acids are more numerous and/or more abundant and the corresponding synthetases are generally more active.[199] The significance of these observations is unclear, but the fact that all subspecies of female tRNAs are more aminoacylated by synthetase preparations than their male counterparts points to a common posttranscriptional modification of all female tRNAs, e.g., methylation.

Investigations along these lines should be extended in *Mercurialis* and other species to provide a description of flower development in molecular terms and a basis for better comprehension of macroscopic phenomena.

CONCLUSIONS

Morphological aspects of reproductive development are often neglected by physiologists. For example, they frequently equate initiation and development of structures or mix altogether all stages of flower development. Through these unfortunate practices confusion has arisen since it is clear that the various ontogenetic stages that can be distinguished in reproductive development react very differently to environmental and internal factors. Many data obtained while disregarding these facts are difficult, if not impossible, to interpret and therefore of little value. The descriptions of a few cases of flower or inflorescence ontogeny given in this chapter exemplify the sort of information that is required when a physiological study is envisaged.

The morphological literature concerning reproductive development is rich in observations of great interest to the physiologist as our survey has shown: sequences for initiation of parts in inflorescences or flowers sometimes differ from those for growth and further development of these parts. There are many variations in the numbers of flowers or of their parts and in their size or shape. There is the intriguing problem of the complex and changing arrangement (phyllotaxis) of flowers or of their parts and the fascinating question of the fusion between flower parts that have been initiated independently. There is also a clear trend towards synchronous development of reproductive structures in several species, etc. All of these problems are awaiting a physiological and biochemical analysis.

Unfortunately the morphological literature is also full of contradictory and confusing theories concerning the "nature" of the flower and its parts. Many morphologists have long accepted the concept that the flower is the homolog of a compressed determinate shoot and the floral parts are modified leaves and axillary buds. Evidence supporting this view comes from comparative morphology, vascular anatomy, ontogeny, and teratology. Pieces of it have been mentioned above but, as already stressed, interpretation of the data is often less than straightforward and homologies between reproductive and vegetative members rest on insecure grounds. Considering functional aspects, the different classes of floral members have a specific chemical composition and functioning and are thus characterized by the activity of different sets of genes. Following this view floral parts are distinct entities which, despite partial similarities, differ from each other and from their vegetative counterparts. Instead of trying to draw unconvincing homologies between them it is, as Cutter puts it, "more stimulating to consider them developmentally as expressions of different physiological states of the apex."[200]

Chapter 2

EXPERIMENTAL SYSTEMS

TABLE OF CONTENTS

I. PLANT MATERIALS

For obvious economical reasons the physiology of flower development is largely explored in ornamental and fruit crops. Although several environmental factors can greatly affect their flowering response, flower initiation and development in these cultivated plants are generally not under the absolute control of the environment. Wild types of many cultivated plants are in fact more dependent on the environment, especially photoperiod, for flowering control. The suppression of photoperiodic dependency for reproductive development in cultivated varieties may have been an unconscious objective in breeding programs since photoperiodic controls would be impediments to year round production and limit cultivation to certain latitudes.

For fundamental work, however, an absolute developmental control is valuable because it generally allows synchronization of plants, a prerequisite in studies concerned with the detection of short-term events and the establishment of their sequence. In some species flower development is fortunately under relatively strict control of the environment — photoperiod, light flux, or temperature. One must discover the stage(s) of floral development that is (are) controlled and how that (they) varies(y) according to the growth conditions and the position of the flower on the plant. Under unfavorable conditions, the reproductive structure will generally senesce and, after a critical time, it may be irreversibly targeted for abortion. Hence, maintaining the plants under these unfavorable conditions with the hope that they all will accumulate flower buds at the same critical stage and that synchronization will be achieved may be unrealistic. In a given set of conditions, a precise knowledge of the stage at which floral development is halted, and the lag period before abortion is inevitable, is thus required before planning studies with these species.

In some work, synchronization of plants has been reset before experimentation by a careful selection of individuals all at the same stage of floral development.[201] For example, in Kinet and co-workers' studies of inflorescence development in tomato, zero time for each experiment is macroscopic appearance of the inflorescence.[202,203] In order to obtain sufficient numbers of plants to this stage, in a short period, very large populations must be grown. Plants are examined at 2-day intervals, and at each examination new batches of "synchronized" individuals are selected for experimentation. This procedure has made it possible to determine with some precision the stage of development most sensitive to environmental and chemical parameters.

Depending on the kind of experiment to be performed, plants must have certain attributes. Studies of growth and development of floral organs in vivo are generally with many different species, selected according to the type of organ considered. Cucurbits, along with some other monoecious and dioecious species, have contributed most to knowledge of factors which control sex expression and development of stamens and pistils. These plants are particularly useful because the pathway of development of their flowers can be affected, within limits, by environmental parameters and chemical treatments.[204] *Cucumis sativus,* in particular, offers an invaluable advantage in that several genetic lines, differing more or less in female or male tendency, are available (Chapter 1, Section III.I).

Species with genetically based, variable flower morphology are indeed promising tools. Besides the case of *Cucumis,* several studies have been with single gene mutants affecting one or more organs in the flower. Examples of this are *Nigella damascena* where doubleness behaves as a monogenic character recessive to the wild single form,[205] and tomato where the single gene recessive mutant, stamenless-2, differs phenotypically from the normal only in stamen structure.[206] Abnormal flowers, too, could be of particular value because several abnormalities are inherited and their expression subject to modification by environmental conditions.[71]

Much less attention has been paid to flower parts other than stamens and pistils, although petal size, number, and pigmentation can be important factors in determining flower quality of ornamentals. Experimental control of perianth characters has been investigated in rose, carnation, *Impatiens,* etc. In chrysanthemum, bulbous plants such as tulip and narcissus, and other cut flowers, the regulation of reproductive axis elongation is a focal point of investigations.

In vitro culture techniques afford a potentially useful tool for investigating the control of flower morphogenesis. Ease of flower (inflorescence) excision, and possession of primordia that are relatively large and separate from other organs, are valuable plant attributes for such studies. Unfortunately, in vitro studies have been successful so far in only a few species including cucumber,[109,207] *Aquilegia formosa,*[208,209] *Viscaria (= Silene) candida,*[210,211] *Anagallis arvensis,*[212,213] muskmelon,[214] tobacco,[215] and *Nigella damascena.*[216]

Bulbous plants are useful for studying the onset and breaking of rest in flower bud development. A further advantage, exploited in commerce, is that dormant bulbs are convenient for handling and treatment.[4] Flower dormancy is also a regular event in the developmental cycle of several woody perennials, e.g., rhododendron, coffee, and various fruit trees. However, studies with many woody species are unwieldy owing to the large size of adult plants; hence, experiments under controlled environmental conditions are rare. Another difficulty with these species is that flowering is an annual event in nature; development may proceed for many months, often in two calendar years. Repetition of experiments in the same year is thus impossible. Furthermore, seed-grown woody species usually exhibit a long juvenile period (Volume I, Chapter 7) and the investigator may have reached retirement before their first flower is formed! Fortunately accelerated maturation in some species may be obtained by favorable cultural practices or a longer growing season. By growing plants of *Malus hupehensis* (tea crabapple) continuously under favorable greenhouse conditions first flowering occurs 9.5 months, instead of 4 to 7 years, after germination.[217] Another way of overcoming some of the difficulties inherent with woody plants is the use of cuttings which keep essentially the same adult characteristics as field vines or fruit trees, but are more amenable to experimentation. In orange, grapevine, and peach, progress has been made in producing such test plants.[218-221]

The choice of one species rather than another can also be dictated by some additional attributes. Plants with inflorescences may be advantageous because a range of different-aged flowers, all of the same genotype, can be readily harvested from the same individual.[215] The number and arrangement of the flower parts may also have some importance. There are many possibilities: a reduced instead of a large number of primordia; cyclic instead of helical arrangement of parts; separation instead of cohesion of parts within one whorl; etc. Large size of the floral bud is desirable in order to facilitate handling when microsurgical operations, such as bisection, are to be performed. Finally, plants easily grafted on partners may be valuable. For instance, further elucidation of the control of sex differentiation could be possible by intergraftings among different sex types.[222]

CONCLUSIONS

Physiological investigations of flower development at the most fundamental level are facilitated if a species has the following attributes:

1. **Developmental control of one or several stages: e.g., starting and stopping development with some easily adjusted environmental parameter, such as daylength, temperature, etc., then, synchronization of development becomes possible, which is a great aid to morphological and biochemical studies and combinations of the two**
2. **Genetic diversity, giving endogenous controls of development, such as in monoecious and dioecious species**

3. **Ease of flower and/or inflorescence excision, and relatively large primordia that are separate from other organs; quite useful for microsurgical and in vitro studies**
4. **Small size, essential in experiments involving many treatments, and a relatively short period of development which allows repetition of experiments**

Clearly, there is no ideal warhorse for the study of all aspects of flower and inflorescence development. Particular materials are far more appropriate and others ill adapted to investigate some specific problem. The choice by an investigator of the plant(s) most suitable for his work can only be made after a careful analysis of all facets of the research project. The importance of this step, preliminary to the experimental work itself, should not be underestimated. On the other hand, economic significance of development to anthesis is such that the species often chooses the scientist rather than the other way round. Therefore, it is more a question of chance when the applied physiologist has any of the above attributes available to him in his studies.

II. MEASUREMENT

Flower and/or inflorescence development is a complex epigenetic process; hence it requires a protracted period for study through many successive morphological stages each of which may have its particular criteria for measurement. According to the species under investigation and the special interest of the investigator, different characters of the flower and/or the inflorescence are taken into account. This section is divided into two broad parts dealing with the general systems of measurement that can be used with all species and the specific ones that are used for particular developmental aspects or for certain economically valuable plants; it is in the latter category that one gives primacy to qualitative features of flower development.

A. General Methodology

The general systems of measurement usually provide estimations of the rate of development of the reproductive structures and their ability to achieve normal development to anthesis. There seems to be, in many species, a clear-cut relationship between the rate of flower development (in relation to vegetative growth) and flower failure. That is, when flower or inflorescence growth rate, relative to that of other tissues, falls below some unknown value, flowers abort. It does not seem to be an ''all-or-none'' relationship, nor is flower failure very rapid; rather we seem to be dealing with (something like) flower bud senescence proceeding more rapidly than growth and differentiation, or perhaps it is merely a gradual decline in meristematic activity, until no growth or differentiation can be observed, followed by abscission zone formation and death of the distal reproductive tissue.

The simplest measurement of flower development is the determination of the proportion of plants reaching a predetermined stage, e.g., anthesis. This index has reduced value if flower initiation does not occur in all plants. A more fitting measure is the percentage of plants within a population that advance from the initial to the selected stage of development. A limitation of this index is its reduced sensitivity with species exhibiting great homogeneity in their flowering behavior. The score(s) will then be close to either 100 or 0%.

Frequently, the time to reach a predetermined stage of flower development is recorded. Obviously this measure is only valid when time is computed subsequent to flower initiation, and thus depends upon the production of batches of plants all of which have flower primordia present. The time of initiation itself may be accurately determined by examination of meristems under the dissecting microscope, but this involves destruction of the samples. An easier, nondestructive and often sufficient index is the time elapsed between two precise and easily identified stages of flower development, e.g., the number of days from macroscopic appearance of the flower bud to anthesis. This measurement is very useful for differentiating among treatments in which anthesis is reached in all plants. However, if the early stages of development (preceding the initial stage selected) are quite responsive to environmental parameters, this index will be in error and/or incomplete.

Floral stage systems, such as those presented for assessment of flower initiation (Volume I, Chapter 1), are in fact best adapted to measure rate of development of flowers or inflorescences; the earliest examples are found in the papers by Blaauw and co-workers, reviewed by Hartsema and Beijer.[4,223] In their studies with several bulbous crops, strict attention to developmental stage of inflorescence and flower (as well as other structural features) as a function of bulb storage treatment served as an indispensable indicator of the effectiveness and timing of such treatments. Initially, the phases of development were designated by use of Roman numerals (I, II, . . .), but later Beijer reclassified these stages using alphabetical letters and subscripts selected to correspond to the morphological stages of development.[223] The two designations have been used interchangeably in the literature on bulbous plants; they are compared in Table 1. Some of these stages for tulip are shown in Figure 1 of Chapter 1.

Subsequently, most investigations with other species assigned numerical values reflecting increased development. These numerical systems can be used with confidence provided all plants produce flowers and the time of flower initiation is known. At some arbitrary time after the start of the experiment, the shoot apex or axillary flowering position of each plant within an experimental batch is examined and, based upon its developmental stage, is assigned a number. The resulting values are then averaged. This method is very sensitive, but since it is destructive, requiring dissection of plants during the early stages of flower/inflorescence development, large numbers of plants must be used if development is studied as a function of time as well as treatment.

Flower stage systems have been devised for many Gramineae, including *Agropyron repens*,[224] rye,[27,225] barley and wheat (see references cited in the paper by Waddington et al.[226]), maize,[227] etc. Usually they are based essentially on the development of the most advanced spikelet primordium. In wheat the appearance of the terminal spikelet is a key stage because it indicates complete differentiation of the inflorescence axis.[228] For quantitative apical development in 'Petkus' winter rye, Gott and co-workers proposed a scale including measures of the length of the apex; their system produced nearly straight lines on a graph when the stages were plotted as a function of time.[225] This makes possible simple arithmetic averaging of data to determine rates of development as a function of treatment. This is not true if the measurements of development as a function of time are in some complex exponential relationships. The scales of Aspinall and Paleg and Waddington and co-workers for barley and wheat differ from that for rye in that only reasonably distinct morphological stages are used; they are not influenced by apical dimensions (Table 2).[226,229] In wheat the scoring system of Friend and associates is the number of days to reach a predetermined stage of development by plants grown under optimal conditions.[230]

In stage systems concerned with the development of a single flower, a whorl of floral parts usually represents a single developmental stage. Sometimes, however, each individual floral organ represents one or even several stages. The best illustration of these different possibilities is given by the scoring system for *Aquilegia*.[208,209] In this species, since the five sepals are initiated separately, each one is considered to represent a developmental stage. A whorl of petals, eight whorls of stamens, and two whorls of staminodia represent individually a developmental stage. Finally, carpel development alone represents six stages. Floral bud development in this species is thus characterized by 22 developmental stages.

Numerous other scoring systems have been designed and some will be referred to in later chapters of this volume; e.g., for *Bougainvillea* (Figure 1),[231] rose,[149,232] chrysanthemum,[233,234] *Phaseolus* (Figure 2),[235] *Gladiolus*,[236,237] *Vitis*,[238] and *Azalea*.[239] The scoring

Table 1
COMPARISON OF THE FLORAL STAGE SYSTEMS DESIGNED FOR BULBOUS SPECIES BY BEIJER AND BLAAUW AND CO-WORKERS

Stage of development of the flower/inflorescence	Stage symbol according to	Stage symbol according to Blaauw and co-workers for				
		Tulipa	*Hyacinthus*	*Narcissus*	*Iris*	*Gladiolus*
Vegetative meristem	I	I	I	I	I	I
Onset of flower formation, meristem doming	II	II	II	II	II	II
First flower primordium of inflorescence initiated	Pr		III			III
Spathe leaf of flower or inflorescence initiated	Sp			III		
Bract initiation	Br		IV			IV
Bracteole initiation	Bo					V
Petals of the first whorl initiated	P1	III	V	IV	IV	VII
Petals of the second whorl initiated	P2	IV	VI	V	V	VIII
Stamens of the first whorl initiated	A1	V	VII	VI	III	VI
Stamens of the second whorl initiated	A2	VI	VIII	VII		
Initiation of the carpels	G	VII	IX + X	VIII	VI	IX
Initiation of the trumpet (paracorolla) in narcissus	Pc			IX		

From Beijer, J. J., *Meded. Landbouwhogesch. Wageningen*, 46(5), 3, 1942. With permission.

Table 2
SCALE OF APICAL DEVELOPMENT IN BARLEY (*HORDEUM VULGARE* L.)

Stage of development of the apex	No.
Single ridge on side of apex	
Apex not elongating	1
Apex elongating	2
Double ridge on apex	3
Upper ridge enlarging	4
Lateral spikelets visible as simple mounds	5
Glume initials visible	6
Lemma initials visible	7
Stamen initials visible	8
Awn initials visible	9
Awns longer than spikelets	10
Anthesis	11

Reprinted from Aspinall, D. and Paleg, G., *Bot. Gaz. (Chicago)*, 124, 429, 1963. With permission of The University of Chicago Press. Copyright 1963 by the University of Chicago.

system established for *Phaseolus* is applied for all stages of differentiation of floral parts; subsequently bud length is used as the criterion of developmental progress.[235] In *Vitis vinifera*, Srinivasan and Mullins constructed a code for flowering in which the various stages are related to changes in the shape of organs or to the addition of new structures.[238] They consider successively the formation of the anlagen, the inflorescence primordia, and the flowers. The scale is derived from SEM.

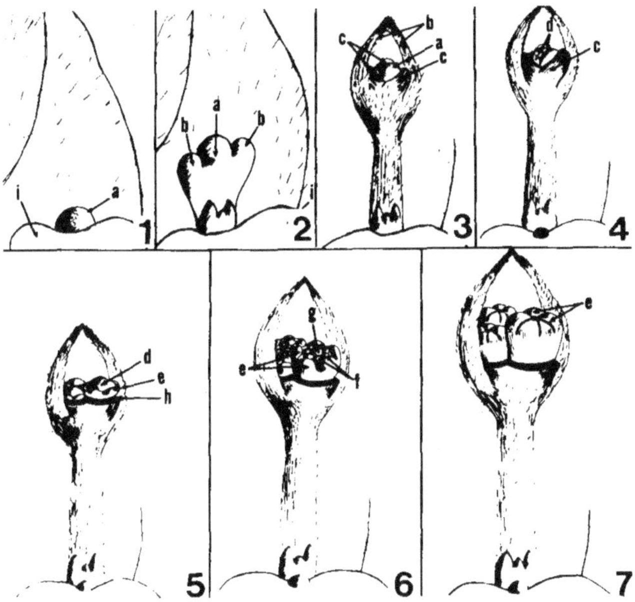

FIGURE 1. Scale of developmental stages for the inflorescence of *Bougainvillea* 'San Diego Red'. Stage 1: origin of the inflorescence as a dome-shaped meristem. Stage 2: the meristem is subtended by a pair of bracts; lateral determinate axes are present in the axils of bracts but not shown. Note indeterminate vegetative bud subtending inflorescence. Stage 3: initiation of three bracteoles (showy bracts). Stage 4: initiation of three florets. Note second vegetative bud. Stage 5: initiation of five perianth lobes on each of three florets (bracteoles pulled away). Stage 6: initiation of stamens and pistils. Stage 7: perianth lobes completely cap each floret, and bracteoles (pulled away) are about 2 mm high. Stages 8, 9, and 10 (not shown) are based on bracteole height: 5 mm for stage 8; 8 mm for stage 9; and 10 mm for stage 10. (a) Meristem; (b) bract; (c) bracteole; (d) floret primordium; (e) perianth; (f) stamen; (g) pistil; (h) scar from removal of bracteole; and (i) leaf petiole scar. (From Hackett, W. P. and Sachs, R. M., *Proc. Am. Soc. Hortic. Sci.*, 92, 615, 1968. With permission.)

In *Chenopodium, Amaranthus,* and various Gramineae, the morphological development of the inflorescence bears a reasonably close relation with its length, over a wide range of environmental conditions.[240-242] Apical size, which can be accurately measured, may thus provide a sensitive index of inflorescence development in these plants (see Volume I, Chapter 1, Figure 4). Flower/inflorescence axis diameter or length can also be readily determined in many species and used as a continuous variable in mathematical functions of development vs. time (or treatment). However if growth is in three dimensions, as in most inflorescences, great changes in volume are encompassed by relatively small changes in any one dimension. In such cases weight becomes the measurement of choice. The introduction of microbalances of high sensitivity has greatly facilitated use of this technique. If tissues are freeze dried before dissection, the inflorescence can be isolated and its dry weight determined with relative ease. Fresh weight of small primordia is subject to large error owing to rapid and variable drying during the dissection and weighing processes.

Node of first flower or inflorescence reaching a selected stage has been used in some species. In *Bougainvillea* the node at which the terminal bracteoles of the inflorescence have developed to at least 1 cm in length is recorded. At this stage of development the inflorescence proceeds to anthesis.[147] In *Pisum* node of first open flower is counted.[243] Since this value may reflect variations in the node of first initiated flower or inflorescence, it may be a

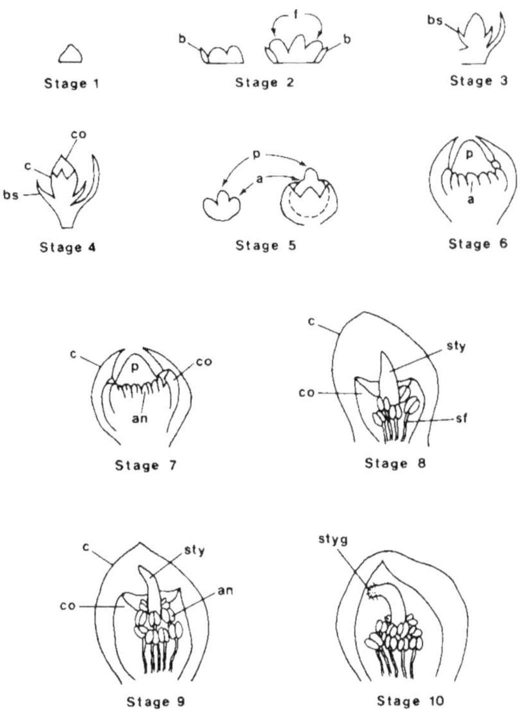

FIGURE 2. Scale of developmental stages for the flower of *Phaseolus vulgaris* P 47. Stage 1: initiation of first flower primordium (triad primordium). Stage 2: first flower primordium gives rise to two lateral flower primordia each of which is subtended by a bract. Stage 3: first lateral flower primordium forms a pair of bracteoles. Stage 4: initiation of calyx and corolla. Stage 5: initiation of androecium and pistil. Stage 6: androecium breaks into concretions. Stage 7: connectives appear in concretions, recognizable as ten sessile anthers. Stage 8: anthers have short free filaments; pistil is differentiated into ovary and a long straight style. Stage 9: style is curved at its distal end. Stage 10: stigma is differentiated. (b) Bract; (f) flower primordium; (bs) bracteoles; (c) calyx; (co) corolla; (a) androecium; (p) pistil, (an) anthers; (sf) stamen filaments; (sty) style; and (styg) stigma. (From Ojehomon, O. O., Zehni, M. S., and Morgan, D. G., *Ann. Bot. (London),* 37, 871, 1973. With permission.)

misleading index of development, except when initiation occurs at each node as in *Bougainvillea.* A more fitting measure is the difference between the node of first flower developed up to the selected stage and the node of first initiated flower. The number of nodes with flowers at the selected stage is recorded in other studies.

An additional index is the number of flowers per inflorescence. There are, however, several limitations to this method, also used to measure flower initiation, that are cited in Volume I, Chapter 1, Section II.B. The ratio of flowers reaching a precise stage to the total number of flowers would certainly be an improvement insofar as developmental index is concerned.

B. Specific Methodology

1. Sex Expression

In studies devoted to sex expression, the ratio of male to female flowers is frequently used. This ratio serves as a measure of sex tendency. Frankel and Galun stressed that this measurement may lead to erroneous deductions when male and female flowers are not randomly distributed on the plant.[116] This is usually the case in monoecious species which bear male and female flowers on different zones of the plant (Chapter 1, Figure 14). Preventing these plants from reaching a particular zone or collecting results before this zone is attained changes the ratio of male to female flowers though there is actually no modification of sex. This is, e.g., the case of *Begonia franconis* as shown in studies by Berghoef and Bruinsma, who found that the effects of environmental factors and growth regulators on sex expression are mediated mainly through their action on inflorescence structure.[112,244]

As a measure of the trend of sexuality during development in a monoecious *Cucurbita pepo,* a species where the type of flower bud produced shifts from maleness to femaleness as node number increases (Chapter 1, Figure 14A), Nitsch and colleagues use different criteria: (1) the node number at which the first female flower appears; (2) the node number at which the zone of 100% female flowers starts; and (3) the average percentage of female flowers per ten consecutive nodes.[115] When these percentages are plotted against node number, approximately straight lines are obtained. The angle which the lines make with the abscissa is used as an "index of feminization". The higher the index number, the faster the production of female flowers. In cucumber, Shifriss and Galun showed that the number of nodes to the first female flower is a fair measure of sex tendency.[245] It is no less effective than the ratio of male to female flowers that requires counting numerous flowers, a tedious and time-consuming procedure.

In *Zea mays*, degree of sex reversal of the apical male inflorescence has recently been recorded using a scoring scale.[246]

2. *Breaking of Rest; Assessing Frost Resistance*

In a number of species, particularly deciduous fruit trees, stages of phenological development generally defined by an event, such as form or color of the bud, have been used to design rating scales for floral bud break or to separate the frost-susceptible period into subperiods that can be identified easily.[247,248]

3. *Floral Injury*

de Munk devised a scale to assess floral injury in tulip.[249] This scale has been modified by Moe and collaborators who distinguish several stages from 0 when there is no injury symptom, all floral organs being normal, to stage 9 characterized by the abortion of all floral parts.[250]

4. *Quality*

In ornamental crops, numerous traits have been used to assess the extent of development of a floral bud or of an inflorescence, e.g., the size, fresh or dry weight of the flower or inflorescence or of their individual parts, the number of nodes per inflorescence, the number of floral parts, their pigment content, the length and diameter of the floral axis, etc. Several of them, e.g., size, weight, form, and pigmentation, underly the grading systems used by commercial growers. We shall not deal in detail with these features which are of primary importance to horticulturists concerned with flower quality. Readers interested in this field are referred to manuals which are devoted to a particular species (e.g., chrysanthemum, carnation, etc.). One recent book edited by Larson may also be useful as well as the reviews by Halevy and Mayak on postharvest physiology of cut flowers where the importance of stage of development at harvest time is discussed.[251-253]

Uniform anthesis for all flowers in an individual plant may be an essential attribute for some floricultural crops, e.g., chrysanthemum, azalea, cyclamen, etc., where the entire plant is sold. In azalea, a scale from 1 to 10 has been devised to assess this quality. The value of 10 is given where uniformity is judged excellent, and 1 to plants with very poor uniformity.[254] On the other hand, crop uniformity, i.e., uniformity of the population of plants, is an essential attribute in many horticultural and agricultural species and requires excellent growing conditions as well as some specific genetic qualities, e.g., a determinate flowering tendency in mechanically harvested tomatoes.

CONCLUSIONS

One can profitably categorize the problems of measurement according to the methodologies used. In doing so, we find that regardless of the system chosen, two aspects of development are emphasized: quantitative ones, in which investigators are interested primarily in assessing rate of development or in comparing the efficacy of a range of treatments, and qualitative ones, in which some critical event (e.g., anthesis, sex expression, etc.) is the key feature.

In many respects methods to evaluate flower development are comparable to those used for initiation. Since the measurements in both cases are usually based on morphological changes which occur at the apical meristem after flower initiation, they are in fact more suitable to estimate development. However, since the characters used as an index are frequently influenced by environmental conditions prevailing before initiation, especially by general plant vigor as well as specific inductive treatments (see Volume I, Chapter 1, Section II), changes in flower development result not only from conditions following initiation, but also from prior treatments.

Methods for measuring flower and inflorescence development must be selected according to the species under investigation, the aim and precision of the work, number of plants available, etc. The best measurements can be made if batches of plants, all of which have initiated flowers, are available. The time of floral initiation — or the attainment of an easily identified stage of flower development — should be accurately determined and used as the starting point for subsequent measurements.

Scales of floral stages provide the most sensitive and reliable measurement of development especially when plants are grown in controlled environments.

In studies of sex expression, nonrandom distribution of male and female flowers may strongly influence results and can lead to erroneous deductions. The internally determined (genetic) patterns of development can override experimental treatments. Simply put, we must know the species or cultivar characteristics.

Chapter 3

CONTROL BY NUTRITION

TABLE OF CONTENTS

I. CARBOHYDRATE/NITROGEN RELATIONSHIP THEORY

As stated in Chapter 2 of the first volume of this series, the carbohydrate/nitrogen hypothesis of Klebs and Kraus and Kraybill was applied to the problem of control of flower initiation, but their work was mainly with plants in advanced stages of reproductive development and thus better applied to flower or inflorescence development.

A. Rate of Development and Abortion

By modifying N fertilization of tomato plants grown under a given light intensity and at a given temperature, Kraus and Kraybill altered the endogenous C/N ratio and found that flowering and fruiting are promoted neither by highest N nor highest carbohydrate levels, but by some balance between them.[255] Later studies on tomato corroborated these results.[256,257] Also in soybean and wheat a significant increase in the C/N ratio was reported after flower initiation at a time when floral development is well advanced.[258-260]

Such a simple hypothesis hardly accounts, however, for all the contradictions concerning N. In various species, for instance, the development of the flowers and inflorescences is promoted by a high N supply (see Section II in this chapter), and even in a single species opposite data have been reported. In tomato, again, many workers observed that flowering and fruiting are accelerated by a high N level and have challenged the conclusions of Kraus and Kraybill.[261-264]

A serious problem in evaluating this and all other controversies concerning N nutrition results from the variable amounts of nitrate and ammonium fertilizers used. Quite often investigators seem to be unaware that the absolute amounts and ratio of the two forms have great influences on physiological systems.[265,266] Also, light conditions have a great impact on the C/N relationship. Kraus and Kraybill stressed that available carbohydrate through photosynthesis and storage constitutes as much of a limiting factor as the available N.[255] An analysis of the literature on tomato, with this view in mind, is quite revealing. Howlett's experiments, showing that manure treatments reduce the number of flowers reaching anthesis in the early clusters, were with plants grown during fall and winter and most likely suffering from severe carbohydrate deficiency.[257] In contrast, the results showing that a high N supply promotes flowering were with plants grown in favorable light conditions and apparently without carbohydrate deficiency. Thus when light is not limiting, the higher the N supply, the better the development of the inflorescence. Modern agricultural production practices in field- and greenhouse-grown crops have confirmed these findings many times over. For example, Bunt and Ryan and co-workers clearly showed that high N supply may be detrimental to flower/inflorescence development of winter-grown tomatoes, whereas high N during spring or summer promotes development.[267,268]

B. Sex Expression

Workers of the 1920s and 1930s found that the C/N ratio is also correlated with sex expression. It was frequently observed that environmental conditions favoring low carbohydrate levels result in repression of male organs in hermaphroditic as well as in monoecious and dioecious species, but have relatively little effect upon female organ development.[257,269-272] On the other hand, N deficiency has little effect upon anther development and pollen fertility, but markedly represses the growth of pistils.[63,257,273-276]

Schaffner suggested that the reversal of sex is probably brought about by a physiological condition dependent in some way on the C/N ratio in tissues.[270] This idea received support from the work of Talley who found that in hemp the ratios of the various classes of carbohydrates to N are greater for male than for female plants.[277] Since in this study analyses were performed after opening of the most mature flowers, the question arises as to whether the differences between sexes are causal physiological factors or only incidental accompanying factors.

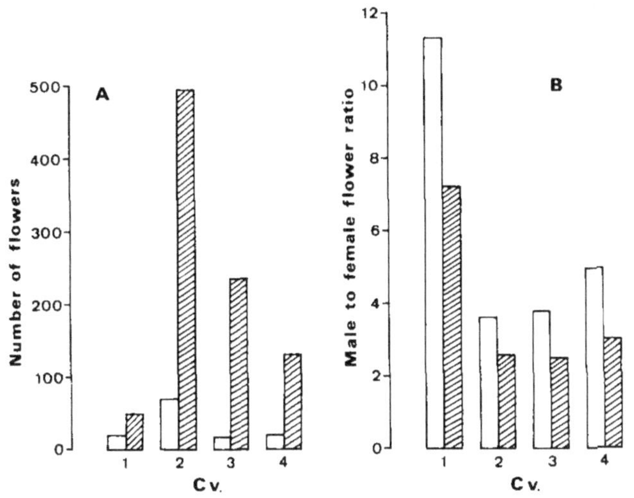

FIGURE 1. Effect of N level on (A) flower number and (B) ratio of male to female flowers in four cvs. of *Begonia semperflorens*. cvs.: 1 = 'Dwarf'; 2 = 'Erfordia'; 3 = 'Prima'; 4 = 'Luminosa'. Open columns: plants are grown in sand. Hatched columns: plants are grown in a heavy-manured clay loam to which N is added. (Adapted from Matzke, E. B., *Am. J. Bot.*, 25, 465, 1938.)

Correlations of carbohydrate content and final sex expression in *Mercurialis perennis* and *Lychnis dioica* revealed that vegetative plants with a higher sugar content eventually differentiate into male plants.[278-280] In strawberry high carbohydrate content is associated with the production of perfect flowers, while low carbohydrate content correlates with the development of female flowers.[269] However, a general problem with all of these studies is that the status of the meristematic regions of the male and female plants at the time of flower formation is not likely to be properly reflected by analyses of the whole shoot.

Since in the 1930s it was firmly established that photoperiod is a potent factor determining sex expression, and hormonal theories of control of development were on the rise, Schaffner's hypothesis and the role of nutrition in general have been largely ignored. This has been particularly true with the accumulation of contradictions concerning N effects. Recent studies demonstrate, however, that mineral nutrition, especially N availability, affects the metabolism and the distribution of growth regulators within the plants, which should influence flower development and sex expression. Hence, if the C/N relationship theory has been considered erroneous by some people, it could be because it accounts for only a part of a more complex reality. The importance of both mineral nutrition and photosynthesis and of their interactions for flower and inflorescence development is hardly deniable. In the following pages the influence of mineral nutrition will be reviewed; the role of photosynthesis will be treated in Chapter 9.

II. MINERAL NUTRITION

A. Nitrogen

1. Rate of Development and Abortion

As already discussed for tomato, N has a great influence on flower development. In *Xanthium* high N levels greatly promote development of both male and female inflorescences,[281] and the number of days from visible flower buds to open flowers is increased in N-deficient plants of *Perilla* and *Kalanchoe*.[282] In addition, flower number is a function of N level in several species including *Begonia semperflorens* (Figure 1),[63] cucumber,[274] gher-

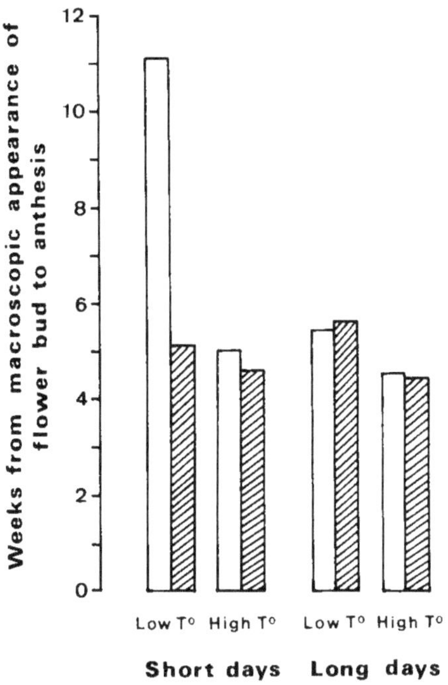

FIGURE 2. Effect of N level, temperature, and daylength on flower development in *Dianthus caryophyllus* cv. 'Puritan'. Open columns: low N treatment (10 mg/ℓ). Hatched columns: high N treatment (250 mg/ℓ). Plants are supplied with nutrient solution by subirrigation at 36- to 48-hr intervals. In the high T° treatment the temperature is always maintained above 15°C. For the low T° treatment night temperature is allowed to drop to 10 to 12°C. There is little difference in the maximum T° attained during the daytime in these two treatments. SD is 8 hr of natural daylength. For the LD treatment supplementary illumination is given at either end of the period of natural daylength, extending the length of day to 17 hr. (Adapted from Blake, J. and Harris, G. P., *Ann. Bot. (London)*, 24, 247, 1960.)

kin,[276] *Perilla, Sinapis,* and *Kalanchoe* (Volume I, Chapter 2, Table 1),[282] soybean,[283] *Hydrangea,*[284] azalea,[285] etc.

In the wheat cv. 'Arawa', spikelet numbers are increased by higher N levels between the double-ridge stage and ear emergence.[286] Since the influence of high N on spikelet numbers is less than the effect on the number of grains per ear, it follows that the proportion of fertile florets and then the degree of grain setting within spikelets are affected by treatment.[286] In the cv. 'Hilgendorf', comparable florets reach a higher stage of development 20 days after the double-ridge stage in plants maintained at high than at low N.[287] High N should thus result in more florets in each spikelet reaching a stage at which they are capable of becoming fertilized. The rate of floral development in barley is also a function of mineral nutrition. As for wheat, the developmental differences arise after double-ridge formation.[229]

Interactions between N and other environmental parameters have been observed. In carnation, low N produces a marked delay in anthesis only under SD and low temperature conditions (Figure 2).[288] Usually, the effects of N are more pronounced under conditions otherwise unfavorable to flowering. In the SDP *Xanthium*, the rate of development of terminal male inflorescences of N-deficient plants is reduced at photoperiods near the critical;[289] with each half-hour increase in length of the dark period from 8.5 to 10 hr, the rate of development of the male inflorescences increases and approaches that of inflorescences on plants grown in a complete nutrient solution. Thus, in some unexplained way a longer dark period may compensate for low N levels in the development of inflorescence primordia. Exposure of tulip bulbs to ethylene during storage causes a predisposition to flower bud blasting. Applying calcium nitrate as a N source during forcing markedly reduces ethylene-induced flower abortion.[290] Time of application is an important parameter in tulip since supplying N during the growing season in the field, i.e., before flower initiation, increases the proportion of blasted flowers. A possible role of Ca cannot be dismissed, however, in these experiments.

Inhibition of development by high N supply has been found in some species. A high concentration of soil nitrate decreases the yield and the quality of flowers in Creole lily;[291] in *Sinapis* a high N supply reduces the rate of development of the flowers as reflected by the number of days from visible flower buds to open flowers (Volume I, Chapter 2, Table 1).[282]

2. Sex Expression

The effect of nutritional conditions on the development of the stamens and pistils has received much attention. In the 19th century it was generally acknowledged that a rich soil, high in available N and moisture, tends to promote femaleness, whereas a poor soil promotes maleness.[111] The best evidence for these effects is from work with monoecious and andromonoecious species. Growing *Cucumis sativus* plants in a soil of high N content increases the total number of flowers of both sexes, but the increase of female flowers exceeds that of male ones.[274] In LD male flowers increase by 45.5% on the plants grown in high N and female flowers by 55.1%. In SD the increases are 3.1 and 20.7%, respectively. A similar effect of soil N on sex ratio has been found in *Cucumis anguria* and *C. melo*.[276,292] In *Begonia semperflorens* Matzke found a shift towards femaleness with high N supply (Figure 1).[63] In four different cvs. flower number is much greater on plants grown in a heavy manured clay loam (to which N is added) than on plants grown in sand, but the increase in female flowers is disproportionately large. Remember, however, that in each inflorescence of *B. semperflorens* the first flowers are male and the terminal flowers female (Chapter 1, Figure 14C, D); thus, preventing the inflorescence from reaching the terminal female phase would cause an apparent shift to maleness and could account for the results with plants grown in low N.

A peculiar situation has been described in *Cleome spinosa* where stamen or pistil abortion may occur in the hermaphroditic flowers.[293] When the stamens abort, the pistil is usually well developed and functional when the pistil aborts, the stamens are often highly developed and functional. Flowers of the unifunctional type alternate from the base of the raceme to the apex, passage from one zone to the next being marked by intergrading flower types. Murneek found that varying the N supply does not suppress the cyclic occurrence of the different sex types in *Cleome*, but modifies the ratio of flowers of the two sexes; the proportion of female flowers formed increases as a function of N supply.[273]

Nitrogen nutrition also influences sex expression in some dioecious species. Early studies with hemp gave conflicting results possibly because N availability and other environmental parameters were not accurately controlled (see discussion in Heslop-Harrison[111]). In an attempt to control N availability more precisely, Tibeau grew hemp plants in solution culture.[275] Plants were supplied with different levels of N by adding or reducing nitrates to the complete nutrient solution. Despite the small number of plants used, impressive results were obtained. In the control plants with normal N there was a 65:35 female to male flower ratio. Plants supplied with high N (eight times normal) differentiated female flowers only and all those without N differentiated male flowers only. Tibeau also showed that the time at which N starvation begins strongly influences the plant response.[275] All plants transferred

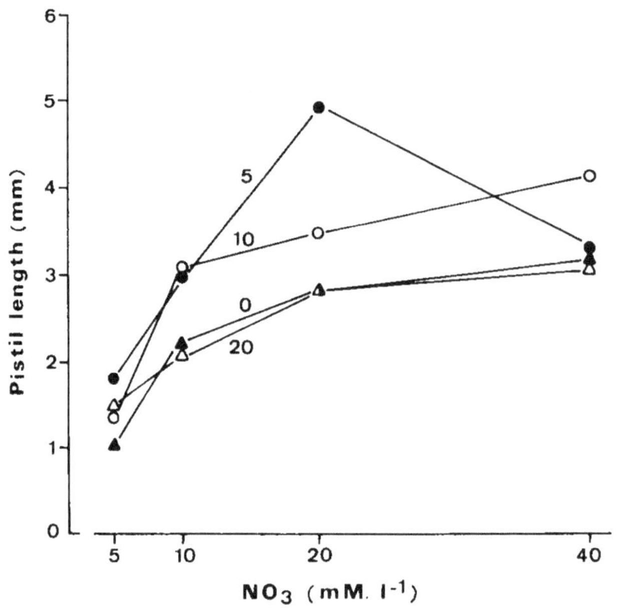

FIGURE 3. Interaction between nitrate and ammonium, added to a N-free basal medium containing 10^{-6} *M* zeatin and 10^{-6} *M* NAA in the development of *Cleome iberidella* pistils during 14 days. Nitrate and ammonium added as $NaNO_3$ and NH_4Cl, respectively. The figures above the curves are the NH_4 concentrations (m*M*/ℓ). (From de Jong, A. W., Smit, A. L., and Bruinsma, J., *Z. Pflanzenphysiol.*, 72, 227, 1974. With permission.)

to a high-N solution after 27 days on N-deficient media differentiated female flowers only, whereas plants transferred after 44 or more days of starvation differentiated male flowers only.

Increasing N supply also enhances femaleness in *Humulus japonicus* (hop).[270] In tomato, N deficiency has little effect upon anther development but represses the female organs.[294]

Further evidence of the importance of N for the development of the gynoecium is provided from in vitro experiments with excised floral buds or excised pistils. Provided that kinetin is added to basal medium, *Aquilegia* buds initiate carpels on high N media, but not on low N.[295] In *Nigella*, low N concentrations prevent the full growth potential of excised pistils. The optimal nitrate level for ovule formation and pistil growth is 30 to 40 m*M*.[296] However, when ammonium at concentrations above 10 m*M* is added to nitrate media, ovule formation is inhibited. Stimulation of the growth of the floral bud and the pistil by nitrate and inhibition by high ammonium levels has also been found in *Cleome iberidella*.[297] In this species, the best combination turns out to be 20 m*M* NO_3^-/l + 5 m*M* NH_4^+/l (Figure 3). In *Begonia franconis* growth of the ovary and the perianth of the female bud is poor with either nitrate or ammonium as the sole N source. Optimum growth is found at 5 to 10 m*M* NO_3^- and 5 m*M* NH_4^+.[298]

B. Other Elements

Clear effects of K and P on the development of flowers and inflorescences have been demonstrated, but generalizations are impossible. Depending on the species, promotive, inhibitory, and no effect were recorded.

Potassium deficiency reduces the number of inflorescences and flowers in dill and *Perilla* and prevents growth of *Cleome iberidella* and *Begonia franconis* flower buds grown in vitro.[282,297,298] Insufficient K produces shorter flowering rose shoots and smaller flowers than

Table 1
FOLIAR K CONTENT AND FLOWERING IN *SOLANUM SISYMBRIFOLIUM*[a]

	Foliar K content[b]	
	Low	**High**
Inflorescences per plant	11.0	9.4
Flowers per inflorescence	10.8	9.8
Total flowers per plant	118.4	91.6
Female sterile flowers	41.6	4.4
Fully developed flowers (% total)	64.9	95.2

[a] Results collected over a period of 10 weeks.
[b] Foliar K contents reflect soil K levels.

From Wakhloo, J. L., *J. Exp. Bot.*, 26, 425, 1975. With permission.

normal; with severe K deficiency flower buds often fail to open.[299] In tomato Besford and Maw found that low K reduces the number of flowers reaching anthesis,[300] an observation made earlier under somewhat different circumstances.[268]

In contrast, high K levels are unfavorable to flower production in azalea and narcissus.[285,301] In *Solanum sisymbryfolium,* K level does not affect flowering time, but high K in the leaves is correlated with a depressed number of inflorescences and a reduced number of flowers per inflorescence (Table 1).[302] Increased flower production with low K is mainly due to initiation of female sterile flowers. A correlation between foliar K level and the production of female sterile flowers has also been found in *S. xanthocarpum.*[303]

Potassium nutrition apparently does not affect flower and/or inflorescence development in *Xanthium* and 'Croft' lily.[289,304]

Phosphorus deficiency alters floral development in various species including *Sinapis*, dill, *Perilla, Xanthium,* and *Cleome iberidella.*[282,289,297] In tomato, 10 days of P deficiency reduces the number of flowers that develop to anthesis on the first truss and delays anthesis of the first flower by 7 days.[305] In *Begonia franconis* female flower buds turn black and die within a few days when grown in vitro on a medium without P.[298]

In azalea and chrysanthemum high P is unfavorable to flower development,[285,306] while in soybean the number of days to flowering is unaffected by P.[307]

Several other elements may affect floral development. In 'Croft' lilies, many flower buds blast in Ca-deficient cultures.[304] In chrysanthemum and carnation, however, liming decreases flower production and quality.[308,309] In *Begonia franconis* the best growth of the female floral bud on an agar medium occurs without Ca.[298]

Omission of Mg, Fe, or S was found to delay or reduce the floral response in some plants.[282,310] In *Sinapis*, however, S deficiency increases the percent of plants with open flowers and reduces the number of days from visible buds to open flowers.[282]

Copper, boron, and molybdenum deficiencies also delay flowering and impair flower and/or inflorescence development in various species.[234,308,309,311-315] A common symptom is male sterility, best described in wheat,[311—313,315] but also observed in some other plants.[311,314,315] Graham showed that Cu-deficient wheat plants develop small anthers commonly less than half as long as those of normal plants and of about one tenth the volume. Cross-pollination experiments demonstrated the nonviability of Cu-deficient pollen and the viability of the ovule.[312] Time of Cu application to wheat plants grown in Cu-deficient soil is critical for restoring male fertility (Figure 4).[313] Cu supplied during the 12 weeks after sowing produces

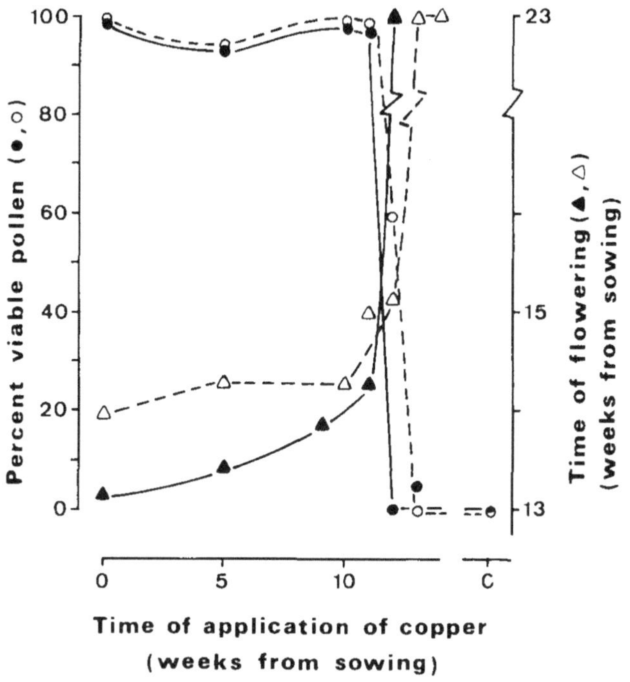

FIGURE 4. Time of flowering and viability of pollen in main culm (solid lines) and first tiller (broken lines) of wheat plants (*Triticum aestivum* cv. 'Halberd') grown on a Cu-deficient soil and supplied with $CuSO_4$ solution at different times after sowing. Flowering points at 23 weeks indicate failure to flower. C = control plants which were not supplied with $CuSO_4$ solution at time of sampling. (Adapted from Graham, R. D., *J. Exp. Bot.*, 27, 717, 1976.)

viable pollen, but later applications are largely ineffective. Plants treated at 12 weeks produce viable pollen in tillers only, the main culm being male sterile. These data parallel the grain yield data and are a mirror image of the time to flowering data (Figure 4). The vegetative yield, measured by the dry weight of the shoot, is relatively unaffected by the time of Cu treatment, suggesting that the effect on reproductive development is stage dependent rather than an overall plant vigor. This idea is also supported by the fact that the loss of response to Cu is sudden and occurs at the same physiological age in both the main culm and first tiller (which reaches anthesis 1 week after the main culm). The critical stage beyond which Cu addition is ineffective is the early stage of stem elongation, when meiotic figures are usually found in pollen mother cells of normal inflorescences. This suggests that Cu deficiency interrupts microsporogenesis at or near meiosis.[313]

CONCLUSIONS

Mineral nutrition clearly affects flower development.

In many species high N increases the rate of floral development and the number of flowers. Responses to N depend upon experimental conditions: they are usually more pronounced under conditions otherwise unfavorable to flowering. In various monoecious species, but also in some dioecious and hermaphroditic species, femaleness is increased with high N supply. In vitro culture of excised floral buds or pistils also reveals the importance of N for gynoecium development.

The most important comment we can add here is that, in general, insufficient attention has been given to the physiological significance of the NO_3^-/NH_4^+ ratio in

fertilizer solutions and soil mixes. There are a number of studies demonstrating that the ratio of oxidized to reduced N affects plant development. A more thorough investigation of this subject seems warranted with special emphasis on flower development.

As expected, other elements, such as K, P, Ca, Cu, and Bo, influence flower development. The available evidence shows that responses depend upon species, experimental conditions, the criteria used to evaluate flower development, etc. Specific target sites, if they exist for the elements in question, are unknown. In various species however, male sterility is associated with Cu, Bo, and Mo deficiency which would impair microsporogenesis at or near meiosis.

Several studies show that metabolism and distribution of growth regulators are affected by mineral nutrition suggesting that some symptoms reported could result from modifications in the hormonal status of the plant.

Chapter 4

CONTROL BY LIGHT

TABLE OF CONTENTS

I. INTRODUCTION

The importance of light for flower development has long been recognized. In numerous early studies, where there is no clear distinction between the effects of daylength and light flux, interpretation of the data is difficult. Now, it is clearly established that both factors can influence, independently, flower development. In nature they are linked in that both contribute to the daily light integral, an important parameter for various species.

II. DAYLENGTH

Most studies on the effects of daylength on flowering have concentrated largely on those stages prior to and including the initiation of flower primordia at the stem apex; these have been reviewed in the first volume of this series. In early work on photoperiodic responses flower development was also found to be under the control of daylength in many species. For example, Garner and Allard observed that flower initiation occurs after a minimal number of inductive cycles, but when plants are returned to noninductive photoperiodic conditions flower development is inhibited.[316,317]

A frequent problem in determining the photoperiodic requirement for flower development is that an appropriate daylength is usually not required during the whole process. In many species photoperiod influences development only for a limited time, just for the completion of some precise stage. If the flowering response of a plant is evaluated from a stage subsequent to its phase of sensitivity to daylength, it is obvious that its photoperiodic requirement will be overlooked.

A. Rate of Development and Abortion: Photoperiodic Response Types

Depending on the species or on the cv., absolute (qualitative) or facultative (quantitative) photoperiodic responses may exist. Plants with an absolute photoperiodic requirement for flower development never develop any flower to anthesis when grown under an inappropriate daylength; as is the case for initiation, they have a clear cut critical daylength which separates promotive and inhibitory photoperiods. High arctic plants of *Saxifraga rivularis* provide a good illustration of such a situation (Figure 1).[318] In facultative photoperiodic species, flower opening is delayed in unfavorable photoperiods.

In a number of species the optimum daylength may change with the stage of development of the reproductive organs and, in fact, may differ for initiation and further development of the flower. As a consequence, classifications based upon photoperiodic requirements for the complete flowering process have been proposed.[319,320] Vince-Prue recognized four major groups;[321] we have added a fifth for completeness:

Group 1. Plants with the same photoperiodic requirement for initiation and flower development
Group 2. Plants with a photoperiodic requirement for initiation but day neutral for flower development
Group 3. Plants which are day neutral for initiation but with a photoperiodic requirement for development
Group 4. Plants with different photoperiodic requirements for initiation and flower development
Group 5. Plants which are day neutral for both initiation and development

The critical photoperiods for initiation and anthesis may thus differ. This is particularly evident in groups 2, 3, and 4.

Initiation and flower development have the same absolute photoperiodic requirements in a number of species (group 1), including the SDP Biloxi soybean, *Cosmos sulphureus*,

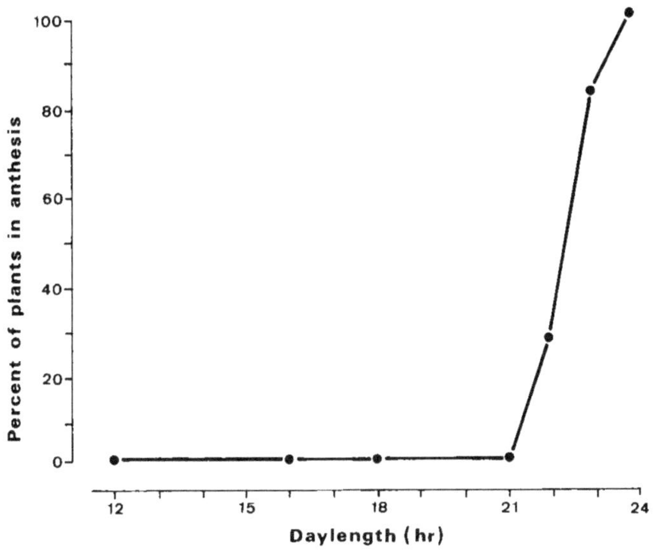

FIGURE 1. The response to daylength for flower development in *Saxifraga rivularis*. Under 20-hr LD, 5 plants out of 20 developed flowers to anthesis but 4 of these 5 plants had only partially developed flowers which had less than one half the dry weight of a normal flower. The flowering response was assayed 28 days after the start of the various photoperiodic treatments. (Reprinted with permission from Teeri, J. A., *Ecology*, 57, 374, 1976. Copyright© 1976, The Ecological Society of America.)

Impatiens balsamina, Chenopodium rubrum, Lemna paucicostata 6746, *Vigna unguiculata*,[322-327] and the LDP *Anagallis arvensis, Lotus corniculatus, Salvia splendens* 'Feu de la Saint Jean', *Lemna gibba* G_3, *Abelmoschus esculentus, Sinapis alba*, etc.[59,328-331] Development of flower buds of *Lemna paucicostata* is a direct function of SD induction. In cultures exposed to one to five photoinductive cycles followed by LD, development proceeds up to some maximum by the third LD, but the floral buds do not develop further. They disappear by the fifth LD regardless of the floral stage previously reached. Abortion of floral buds does not occur if cultures are maintained under SD conditions. In *Sinapis* all individuals subjected to a single LD when 2 months old produce numerous flowers (Volume I, Chapter 3, Table 1) which abort before opening. A minimum of 4 LD is required for at least one flower to develop to anthesis on some plants; with 16 LD all plants produce flowers that go to anthesis with the number increasing as a function of LD induction (Figure 2). In the SDP *Vigna unguiculata* (cowpea), Lush and associates demonstrated that initiation and development have the same critical daylength.[327]

Other plants belonging to Group 1, such as the SDP *Xanthium strumarium* and *Chrysanthemum morifolium*,[139,153] the LDP *Silene coeli-rosa* and *Lolium temulentum*,[76,242] have a qualitative requirement for only one step and exhibit a quantitative response for the other. In *Xanthium* inflorescence initiation absolutely requires SD. When the plants are exposed continuously to photoinductive cycles development of both female and male inflorescences is very rapid and, at the end of a month, seeds are almost mature.[139] Plants subjected to one single photoinductive cycle also develop inflorescences but at a very slow rate. They can produce ripe seeds after 7 or 8 months.[332] Thus, in *Xanthium*, unfavorable daylength may not arrest inflorescence development (at least of the female head). *Lolium* and *Silene coeli-rosa* behave in the same way whereas late-flowering (autumn and winter) cvs. of chrysanthemums have a quantitative SD requirement for flower initiation but absolute requirements for SD for the further development of the capitulum.

Early-flowering (summer) chrysanthemums differ from late-flowering cvs. in that lateral

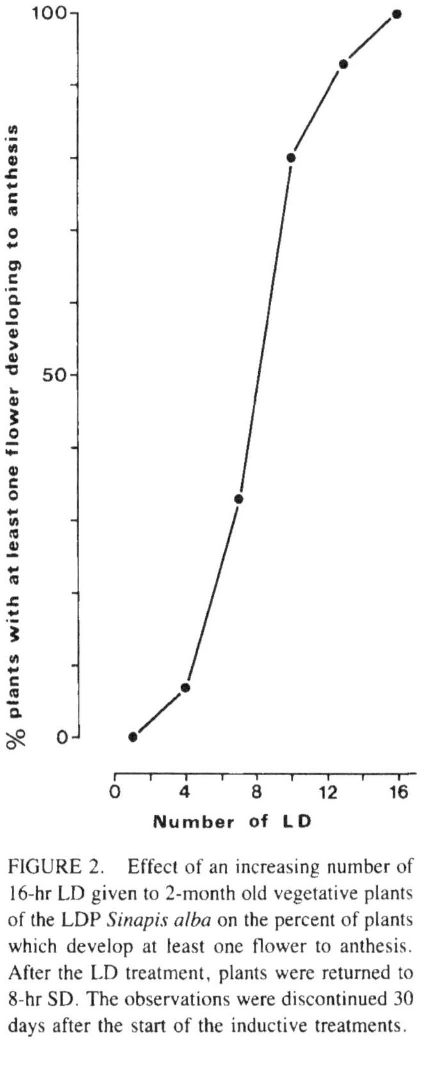

FIGURE 2. Effect of an increasing number of 16-hr LD given to 2-month old vegetative plants of the LDP *Sinapis alba* on the percent of plants which develop at least one flower to anthesis. After the LD treatment, plants were returned to 8-hr SD. The observations were discontinued 30 days after the start of the inductive treatments.

buds, and often the terminal buds, develop to anthesis in LD. However, both flower initiation and development occur more rapidly in SD indicating that both processes have the same quantitative photoperiodic requirement.[333] Numerous species exhibit a similar behavior. In most barley and wheat cvs. ear initiation is advanced by LD, but there is no minimum daylength below which it does not occur. Inflorescence development is usually faster in LD with no critical photoperiod below which development is prevented (Figure 3).[28,334-337] However, the number of spikelets formed is higher under SD than LD; that is, there is an inverse relation between daylength and spikelet number.[18] Conditions accelerating flower initiation and inflorescence development in wheat reduce spikelet number by hastening the formation of the terminal spikelet in the ear.[337] In rye spikelet number is also reduced under conditions which promote inflorescence development.[225] In carnation a long photoperiod decreases the number of leaf pairs below the flower and the time to flower initiation.[338] Macroscopic appearance of the flower bud is also hastened, but the period between bud appearance and anthesis is almost unaffected.[338,339] Flower development thus has a LD requirement during early stages but is day neutral by the time the bud is visible. In *Azalea* there is a great diversity of response to daylength. Bodson showed that in the semiearly cv. 'Reinhold Ambrosius' floral initiation and early development of the inflorescence are hastened in 12-

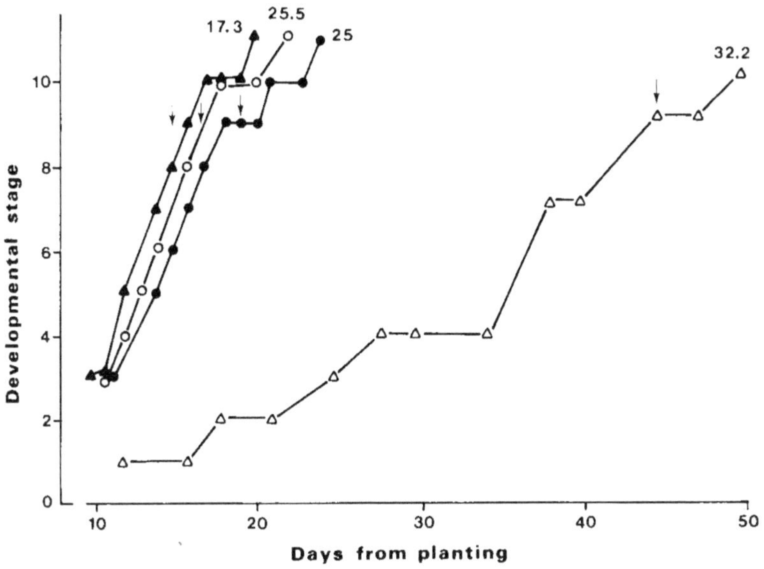

FIGURE 3. Effect of various light regimes on the time taken for the most advanced spikelet on the ear of 'Clipper' barley (*Hordeum vulgare* L.) to reach and complete each stage of development. The developmental stages are assessed using a slight modification of the scale of Aspinall and Paleg (Chapter 2, Table 2). Stage 8 = stamen initials just visible, Stage 9 = stamen initials clearly visible as three well defined mounds. The arrows indicate the time when the production of primordia ceases. The figures above the curves are final numbers of spikelets. — ○ — = 16 hr per day in high intensity light (200 $\mu E/m^2s$); — △ — = 8 hr per day in high intensity light; — ● — and — ▲ — = 8 hr per day in high intensity light with an additional 8 hr of low intensity incandescent light (10 $\mu E{\cdot}m^{-2}/s$) given, respectively, immediately before and immediately after the high intensity light. (From Cottrell, J. E., Dale, J. E. and Jeffcoat, B., *Ann. Bot. (London)*, 50, 57, 1982. With permission.)

or 16-hr days in comparison to 8-hr days, whereas in the late cv. 'Knut Erwén' initiation and development proceed more rapidly in 12-hr days than in longer or shorter photoperiods.[239] In 'Alaska' and 'Redwing' azaleas 8-hr SD, applied to plants with visible flower buds and before the cold storage period required for dormancy breaking, induce faster rate of flower opening than 14-hr LD, suggesting that the flower buds are in a more advanced developmental stage at the end of the SD treatment.[340]

The situation where flower development is day neutral though initiation is under photoperiodic control (Group 2) has not been unambiguously described. Only in *Fuchsia hybrida* does flower development seem to be daylength indifferent though initiation requires LD.[341]

The Peruvian P47 variety of *Phaseolus vulgaris* is a plant in which flower development but not initiation is under the control of daylength (Group 3). Ojehomon and co-workers showed that in this variety the first flower primordium is initiated simultaneously in 11- and 15-hr photoperiods. Subsequent development in both daylengths proceeds at the same rate up to the time of anther and pistil differentiation, after which differences are readily detected (Figure 4).[235] In SD the development of the bud continues up to anthesis 40 days after sowing whereas in LD flower development slows down before differentiation of the stigma and the bud never attains more than 4 mm in size. Abscission occurs about 43 days after sowing.[235] Also, in two early oilseed cvs. of sunflower, time to flower initiation is not significantly influenced by photoperiod although development is faster in 11-hr SD than in 18-hr LD.[342] *Bougainvillea* and *Caryopteris* × *clandonensis* were also reported to display a similar behavior.[321] However, no proof was provided that initiation is totally indifferent

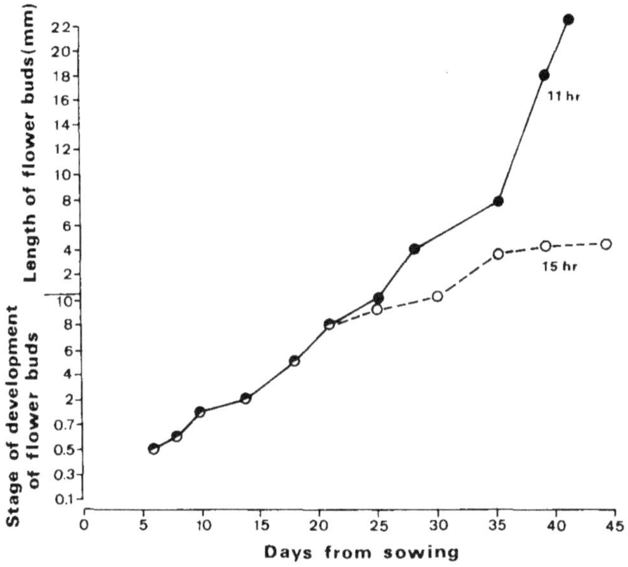

FIGURE 4. The rate of development of the first initiated flower bud in *Phaseolus vulgaris* P47 grown in 11- and 15-hr photoperiods. The flowering stages are described in Chapter 2, Figure 2. (From Ojehomon, O. O., Zehni, M. S., and Morgan, D. G., *Ann. Bot. (London)*, 37, 871, 1973. With permission.)

to daylength in *Caryopteris*.[343] In two lines of *Pisum* it was found that flower initiation is insensitive to photoperiod while flower development is promoted by LD treatments.[243]

In a limited number of species floral initiation and further development were shown to have opposite photoperiodic requirements (Group 4). *Delphinium ajacis* and *Callistephus chinensis* are plants in which LD promote initiation and SD favor development.[344-346] In the perennial grass *Bouteloua eriopoda* initiation is favored by SD and development by LD.[347] Almost all plants flower in constant 13-hr daylength (Figure 5). In shorter photoperiods flower initiation occurs but the reproductive apex degenerates. Degeneration begins in some plants when the inflorescence is at a fairly advanced stage; in others it occurs so early that a seemingly empty flag leaf sheath is formed. In continuous LD (>13 hr) flowering is partial, nonuniform and greatly delayed, such that the shoot of most plants dies while the apex is still in the vegetative state. In a northern ecotype of *Poa pratensis*, SD and low temperatures are independently capable of causing flower initiation, while normal flower development requires LD and is stimulated by high temperature.[348]

Plants which are day neutral for both initiation and development (Group 5) are rather common though precise investigations sometimes reveal slight quantitative photoperiodic requirements for either one or both steps. Flower development in these species is usually most influenced by temperature or light intensity; details will be given in later sections.

B. Abnormalities

In the preceding section the classification according to photoperiodic response type was essentially based upon two different developmental criteria: the occurrence or absence of flowers reaching anthesis (qualitative response) and the rate of flower development as measured by the time from initiation to anthesis (quantitative response).

Noninductive photoperiodic conditions, besides preventing or delaying development, also induce reproductive abnormalities which generally indicate that a given primordium in a reproductive structure (bract, flower, or floral organ primordium) is not committed at ini-

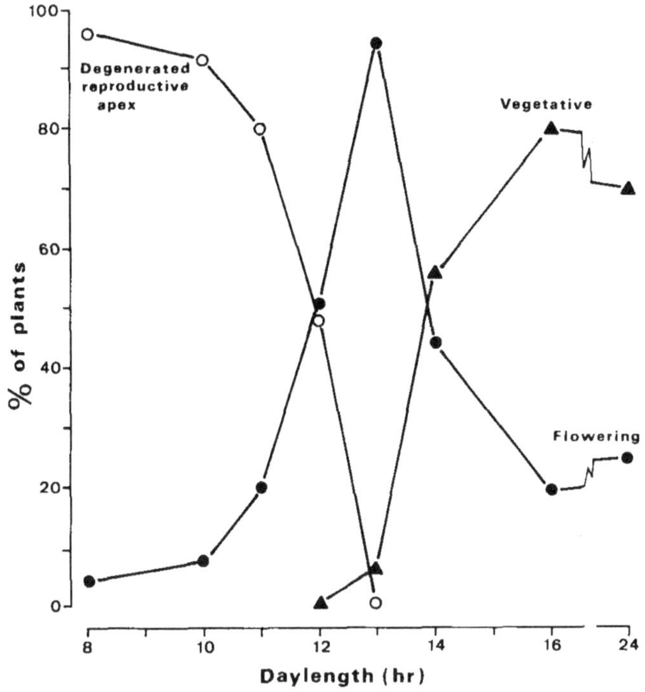

FIGURE 5. Flowering response in the perennial grass *Bouteloua eriopoda* as a function of daylength. Plants are classified in one of the three following groups: (1) plants with a reproductive apex which degenerates early, (2) plants which flower normally and (3) plants in which the apex degenerates while vegetative. (Reprinted from *Bot. Gaz. (Chicago)*, 136, 41, 1975, by Schwartz, A. and Koller, D., by permission of the University of Chicago Press. Copyright 1975 by the University of Chicago.)

tiation and that developmental paths remain optional for some time (see also Volume II, Chapter 5, Section VII).

1. Inflorescence Branching

There have been several reports of branched inflorescences in plants grown under unfavorable photoperiod. Branch number in the terminal male panicle of maize a quantitative SDP for initiation, is highest in plants kept under LD conditions (SD with a 2-hr night interruption) and is reduced by SD (Figure 6).[349] In another quantitative SDP, *Amaranthus caudatus*, apical inflorescences are branched in LD and not in SD. When plants are transferred from LD to SD, the earlier the transfer, the lower the number of branches.[350]

Inflorescence branching also occurs in LD cereals and grasses grown in SD and in the LDP *Sinapis* when subjected to SD at a low irradiance after an initial treatment in LD.[242,351-355] In these species branching usually results from the transition of flower primordia into axillary axes or spikelets instead of normal flowers or florets.

In *Chrysanthemum*, subsidiary inflorescences develop in the axil of involucral bracts of the capitula in LD conditions; this may be viewed as an example of branched inflorescences.[153] Branching of the terminal male inflorescence of the SDP *Xanthium* also occurs in LD. After a single SD one or two axes appear in the axil of involucral bracts. These axes are composed of one small terminal and two small lateral male inflorescences rather than of a single male inflorescence as in continuous SD.[139]

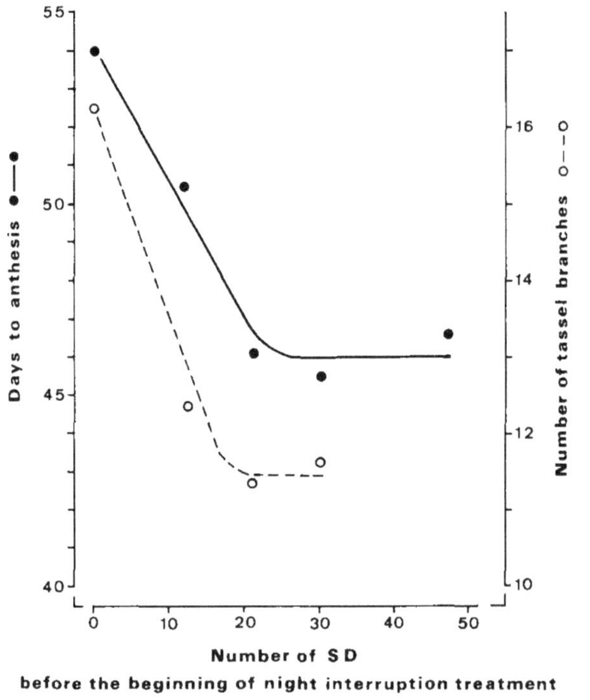

FIGURE 6. Effect of different numbers of 8-hr SD experienced before the beginning of a night-interruption (LD) treatment on time to anthesis (first release of pollen from the male inflorescence) and on number of tassel branches in the SDP *Zea mays*. Night interruption treatment consisted of 8-hr SD accompanied by 16-hr dark periods interrupted in the middle by 2 hr of 400 lx incandescent light. (From Moss, G. I. and Heslop-Harrison, J., *Ann. Bot. (London)*, 32, 833, 1968. With permission.)

2. *Foliaceous Bracts*

Foliaceous bracts frequently appear in capitula of Compositae when, after minimal induction, plants are returned to unfavorable daylengths.[153,356-360] Floral bracts may also become leafy organs on male inflorescences of *Xanthium* plants which have received a single inductive SD and then grown under LD.[139] In grasses under unfavorable daylength glumes and lemmas may assume a leafy form.[361-363] In the LDP *Sinapis* foliaceous bracts frequently develop when a LD treatment is interrupted by SD (with low irradiance) (see Volume II, Chapter 5, Figures 3 and 4); bracts are absent in normal inflorescences of this species.[354,355]

3. *Inflorescence Proliferation and Proliferous Flowers*

Proliferation of the inflorescence has been observed in both LD and SD grasses exposed to unfavorable photoperiodic conditions during inflorescence development.[242,362,364,365] This anomaly results from the production of a leafy shoot by the apical meristem of a developing spikelet. In *Poa bulbosa*, bulbils (completely vegetative buds that serve as propagules) develop in place of normal florets under SD and at temperatures below 20°C.[366]

Proliferous flowers have been found in several species grown in noninductive or just subcritical daylengths.[59,355,360,363,367-369] In some plants the potential for proliferation is maintained until relatively advanced stages.[59,363,367] The work of Brulfert shows that the flower buds in *Anagallis* have reached different stages of development when reversion to vegetative growth starts: in some buds this occurs after formation of sepals, in others after formation of sepals, petals and stamens or even after the gynoecium is formed (Plate 2*). In all cases reversion is due to an axial proliferation, that is the flower meristem is apparently transformed at any stage of flower ontogeny into a vegetative meristem. When reversion occurs after the gynoecium is formed, then the vegetative meristem arises at the top of the central placenta.[59] In the SDP *Impatiens balsamina* Krishnamoorthy and Nanda show that the tip of the placenta is able to resume vegetative growth, in continuous light, even after the formation of fertile anthers and abortive ovules in the ovary.[367] In the LDP 'Opal' spring wheat, following transfer from LD to SD, an inflorescence axis bearing glume-enclosed spikelets of several florets, may emerge from the upper portion of the ovary between the stigmas. In some cases the inflorescence occupies the normal ovary site and the carpel is reduced to a subtending leaf-like appendage.[363]

4. *Metamorphosis of Floral Organs*

All organs of the flower may develop leaf-like characters (phyllody or virescence) or take on forms of other floral organs (sepalody, petalody, staminody, or carpellody). These abnormalities have long been recognized in numerous genera.[71] They have been produced experimentally in a few species by manipulating daylength. Virescence of all the floral organs has been induced in *Impatiens balsamina* cv. 'Buisson fleuri' by a transfer to continuous light after induction in SD (Chapter 1, Section IV.E.2).[125] In the LDP *Sinapis*, virescent sepals are found when LD-induced plants are given a SD treatment under low light intensities.[370] Petaloid stamens occur in chrysanthemum inflorescences that develop in LD,[153] and in the 'Opal' spring wheat various degrees of transformation of stamens into ovaries occur when plants are shifted from LD to SD at the time of differentiation of stamen primordia on the most mature florets of the main shoot.[363]

C. Sex Expression and Fertility

1. *Sex Expression*

The influence of daylength on sex expression was first recognized by Tournois.[371,372] He observed that about 50% of early planted male plants of hop have bisexual flowers at early stages of plant development whereas at later stages, at a time corresponding to the normal flowering season when days lengthen almost all flowers are male (except for some plants which produce normal female inflorescences). Tournois also found that summer-grown hops maintained in 6-hr SD by enclosing them daily in dark boxes for 18 hr show bisexuality. He further reported that hemp responds to daylength in the same way as *Humulus* and concluded that, in these two dioecious species, SD enhance female tendency. These results were matched by subsequent workers using the same species and other studies also demonstrated that daylength is a potent factor determining sex expression in various monoecious species.[111,116] SD were found to decrease the proportion of male to female flowers in various Cucurbitaceae. For example, Nitsch and associates showed that lengthening the day from 8 to 16 hr results in delayed appearance of the first female flower in *Cucurbita pepo* cv. 'Acorn'. *Cucumis sativus* and *C. anguria* behave similarly.[115] These were the first experiments separating the effects of daylength from those of temperature, which also markedly affects sex expression (see next Chapter).

In maize Heslop-Harrison found that SD induce the development of the gynoecial primordia in tassels which would otherwise be wholly male (Table 1).[373] SD also enhance the female tendency in *Xanthium*,[35,139] *Ambrosia trifida* and *A. elatior*,[374,375] *Heteropogon contortus*,[376] etc. Jones experimented with two-branched plants of *Ambrosia trifida:* one branch was held in 6-hr SD while the other was exposed to 15-hr photoperiods.[375] The SD branch showed accelerated development of the lateral female flowers and suppression of the terminal male raceme whereas the LD branch produced a terminal male bud, but no axillary female flowers.

Promotion of female organ development in SD is not, however, of general occurrence.

* Plate 2 follows page 38.

Table 1
EFFECT OF DAYLENGTH AND NIGHT TEMPERATURE ON SEX EXPRESSION IN THE MONOECIOUS SDP *ZEA MAYS*

Night temperature[a] (°C)	Daylength (hr)	Mean number of female flowers in the tassel
22	8	5.67
	21—22	0.00
10	8	12.50
	21—22	1.17

[a] Day temperature is regulated above 22°C.

From Heslop-Harrison, J., *Proc. Linn. Soc. Lond.*, 172 Session, 108, 1961. By permission of the Council of the Linnean Society of London.

Several studies with cucumber stressed that the situation can be very complex. Besides interactions with temperature and irradiance, there is a strong daylength × cv. interaction. Sex expression in several androecious and gynoecious lines is unaffected by daylength,[377,378] whereas in 25 Japanese monoecious cvs. some were found to be daylength insensitive, some revealed increased femaleness in SD and others increased femaleness in LD.[379] Several other species are known in which LD favor female sex expression. Examples are muskmelon and spinach.[292,380,381] In *Dahlia variabilis* cv. 'Newby' a night break for 2 hr in the autumn or spring increases the number of ray florets (female) compared to that in natural days. The number of disc florets (hermaphroditic) is unaffected by the interruption of the night (Figure 7).[382]

Among the species in which female flowers are favored by SD most are SDP for initiation; the others are DNP with some SD tendency. The species in which female organ development is promoted by LD are generally LDP for initiation. In an attempt to reconcile the apparent discrepancies between the observations on the effects of daylength on sex expression, Heslop-Harrison suggested that the degree of femaleness is increased when plants are maintained under inductive photoperiodic conditions.[111] In *Begonia × cheimantha,* however, where flower initiation is promoted by SD and low temperatures, the number of female flowers which develop is higher under marginally inductive conditions.[383] Frankel and Galun stressed that it is not clear whether or not an 8-hr photoperiod is optimal for flowering in *B. × cheimantha*. If it is too short, female flower formation could be suppressed as in oriental cucumber varieties.[116] A close look at Heide's paper suggests that the results are influenced by the pattern of male and female flower distribution in the inflorescence of *B. × cheimantha*. This pattern is almost identical with that of *B. semperflorens* (Chapter 1, Figure 14C, D). Heide reported that at 15°C, under 8-hr days, only 1 plant in 7 formed two female flowers during the experimental period. The rest of the plants developed large, repeatedly branched inflorescences with male flowers only. On the other hand, under near-critical conditions for flower initiation (21°C and LD), only 2 to 3 male flowers were produced before female flower formation occurred.[383] It is thus possible that in the experiments with *B. × cheimantha,* the terminal female phase was delayed in SD and low temperatures and could not be recorded before the experiments were terminated. Thus the results might reflect the pattern of sex distribution and time of appearance of flowers rather than environmental effects on sex differentiation.

Another species in which photoinductive conditions seem to be unfavorable to female sex

A B

Average number of	A	B
Ray florets (R)	56	166
Disc florets (D)	45	41
Ratio R/D	1.24	4.05

FIGURE 7. Effect of daylength on the production of ray florets (female) and disc florets (hermaphroditic) in the gynomonoecious species *Dahlia variabilis* cv. 'Newby'. It is unclear whether or not initiation is influenced by photoperiod in this cv. (A) Inflorescence from a plant grown under natural daylength during autumn. (B) Inflorescence from a plant being given the same conditions, but with a 2-hr night break of fluorescent light. (From Canham, A. E., *Acta Hortic.*, 14, 109, 1969. With permission.)

expression is the gynomonoecious SDP *Chrysanthemum morifolium*. In most early flowering cvs. that develop a capitulum not only in SD but also in LD, there is an increased number of female ray florets relative to hermaphroditic disc florets in LD given by a 4-hr light interruption at low irradiance in the middle of each dark period.[333] The interruption of SD induction by 10 LD during inflorescence initiation also increases the total number of florets and the number of female ray florets in relation to the hermaphroditic disc florets.[384] This last study, however, does not show unequivocally whether the daylength contribution is based upon a true photoperiodic reaction or whether it is, in part at least, due to the somewhat higher total light energy received by plants under LD conditions. An increased daily light energy supply enhances the number of ray florets in *Chrysanthemum* principally when given during the initial period of floret formation.[385]

Hence, the rule that female sex expression is favored by inductive conditions seems largely valid despite the possible exceptions of *Begonia* and *Chrysanthemum*. Further support for this view has come from the demonstration that cucumber scions grafted on fully induced *Sicyos angulatus* stocks (a qualitative SD cucurbit) develop larger numbers of female flowers than scions grafted on noninduced stocks (Table 2). Scions on induced stocks also develop their first female flower at a lower node and the time to the first anthesis is shorter.[386]

2. Fertility

In some species fertility is modified by photoperiod; male fertility is usually most affected.

Normal tapetal and pollen development fails to occur in the SDP *Cosmos sulphureus* and Biloxi soybean when they are transferred to LD after a marginal inductive treatment.[359,387]

In 'Proctor' barley, a LDP, a 2 week SD treatment at the time of awn initiation on the main shoot induces sterility.[388] Batch and Morgan showed that the SD effect is truly photoperiodic, since LD are obtained by extending 8 hr of daylight in the greenhouse with 8 hr of incandescent light at low intensity. SD consist of 8 hr of daylight followed by 2 hr of incandescent light. Plants in SD have small and shriveled anthers. Male sterility is the main cause of floret sterility as demonstrated in an experiment in which florets on one side of an ear are cross pollinated with pollen from plants held in LD whereas florets on the other side are left to be self pollinated. Grain numbers on the self pollinated side of the ear are very low, whereas those on the cross-pollinated side are considerably higher (Figure 8). This result indicates clearly that SD induce male sterility and not some kind of failure of the female gametophyte.[388]

In the SDP *Xanthium* and maize SD shift the balance of sexuality towards femaleness and also diminish male fertility.[281,373] These responses are truly photoperiodic. The effects of SD are eliminated in maize when the long dark period is interrupted with a 2-hr interval of light of low intensity.[349] In *Xanthium* the LD treatment is natural SD with supplementary illumination at low irradiance.

Male fertility is increased in a few species when plants are grown under unfavorable daylength conditions before the beginning of inductive photoperiodic treatment. In the LDP *Silene pendula* a high proportion of the flowers formed on plants held in LD from germination is male sterile. Full pollen fertility is obtained only when the plants are exposed to 40 to 50 SD in early life.[389] In the SDP *Rottboellia exaltata*, 12 weeks of LD, before SD induction, are insufficient to induce fertility of the male flower of the sessile spikelets occurring at each node of the inflorescence. However after 17 weeks of LD (before exposure to SD) the male flower becomes fertile in the first formed inflorescences.[390] Thus aging in noninductive conditions appears to be a prerequisite for male fertility in these species.

D. Chasmogamy and Cleistogamy

A number of species show a seasonal production of cleistogamous and chasmogamous flowers suggesting that the change in the floral type may be correlated with a daylength

Table 2
SEX EXPRESSION IN A RECEPTOR SCION[a] OF A MONOECIOUS *CUCUMIS SATIVUS* AS A FUNCTION OF FLOWER INDUCTION OF THE DONOR STOCK, *SICYOS ANGULATUS* (SDP)[b,c]

	Stock	
Scion response	Noninduced	Induced
Node of the first male flower	3.7	2.2
Node of the first female flower	9.3	5.7
Number of nodes with male flower	11.0	6.2
Number of nodes with female flower	1.5	8.0

[a] Cucumber receptor plants are grafted at the cotyledonary stage.
[b] Noninduced stock plants are continuously kept under LD. Induced stock plants receive 10 SD at the four-leaf stage.
[c] The composite plant is kept in LD after grafting.

From Takahashi, H., Saito, T., and Suge, H., *Plant Cell Physiol.*, 23, 1, 1982. With permission.

change. In different *Viola* species normal chasmogamous flowers are produced under SD whereas cleistogamous flowers are formed in LD.[391,392] *Bothriochloa decipiens,* a cleistogamous grass, produces its closed flowers in SD and open flowers in LD.[393] Similarly, Lord reported that in *Lamium amplexicaule,* SD result in a significant reduction in chasmogamous flower production. The number of flowers per plant is not affected by daylength.[394,395]

E. Flower Bud Dormancy

Daylength is one of the most important factors controlling dormancy of (vegetative or reproductive) buds.[321,396,397] Four major response types have been recognized. First, there are species in which LD cause continuous growth and SD induce dormancy. In the second group, dormancy is accelerated by SD and delayed but not prevented by LD. The difference between these two groups is that in the first one the response to SD is qualitative whereas it is quantitative in the second one. In the third group, SD stop growth and LD induce a series of three to four flushes of growth interspersed with periods of dormancy which are, in some cases, shortened to such a degree that growth seems to be continuous. Finally there are species, including some common fruit trees, in which the onset of dormancy appears to be independent of photoperiod. Hence, as stressed by Wareing and Vince-Prue, among species which are daylength sensitive there seems to be no exception to the rule that dormancy is hastened by SD and delayed or prevented by LD.[321,397] Without extending the discussion here the reader is referred to the reviews of this field that have already been published.[321,396,397]

F. Time Dependence of Photoperiodic Effects

Generally, the number of favorable photoperiods required for flower development is less than the number of days to anthesis indicating that the late stages of flower development are independent of daylength.[59,242,324,327,329-331,359,387] In the SDP *Lemna paucicostata* 6746, however, SD are required during the complete period of flower development.[326]

The minimum number of favorable photoperiods necessary may be modified by various environmental factors such as light intensity and temperature which will be discussed subsequently. In *Impatiens balsamina* (SDP) the minimum number of SD is affected by the length of the light period. Four 8-hr SD suffice, but if the day is lengthened to 15 hr (a daylength close to the critical) at least 28 cycles are required.[398]

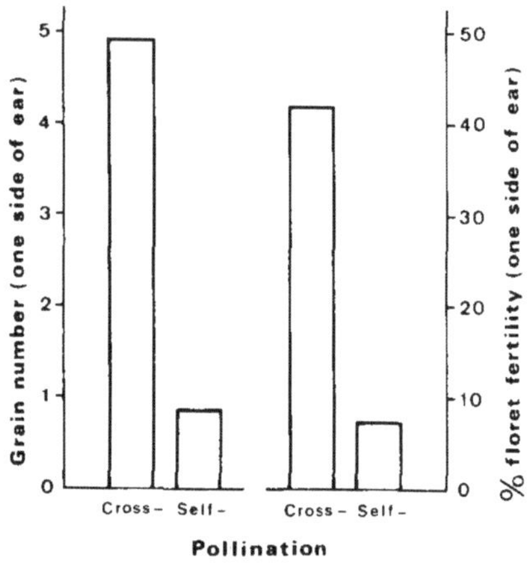

FIGURE 8. Effect of daylength on fertility in the LDP *Hordeum vulgare* cv. 'Proctor'. Florets on one side of the main shoot ears of plants which receive fourteen 10-hr SD from awn initial stage are cross-pollinated with pollen from plants grown throughout in LD. Florets on the other side of these ears are not cross-pollinated so that any grains which set are the result of self pollination. (Adapted from Batch, J. J. and Morgan, D. G., *Nature (London)*, 250, 165, 1974.)

In some species the favorable cycles need not be given consecutively. In *Lemna paucicostata* 6746 floral development is unaffected when a single LD cycle is inserted in a sequence of SD. Abortion occurs, however, when 2 or more LD are interspersed.[326] In *Caryopteris* anthesis occurs about 25 days after the beginning of an effective SD treatment, which must consist of a minimum of 11 cycles. More than 11 SD does not materially hasten anthesis.[343] When plants are subjected to 7 SD followed by 30 days of continuous light before returning to SD, anthesis occurs 18 to 20 days after the end of the period of continuous illumination. The plants thus flower after a total of about 26 SD despite the intervention of as many as 30 days of continuous light. Since (1) during the continuous illumination treatment the plants differentiated five or six additional nodes on the main axis, and (2) the flowers which eventually reached anthesis were confined to the terminal 5 nodes, it appears that all the flower buds which were present on the stem at the time of the initial inductive cycles aborted. It is unfortunate that morphological studies were not included in this work because, without them, it is impossible to determine the stages of development affected by the various daylength treatments.

The stage at which floral development is inhibited in noninductive conditions is dependent upon previous experimental conditions. There is also considerable variation according to the species and the position of the flower bud on the plant. In P47 *Phaseolus* grown in unfavorable LD the first initiated flower bud is inhibited at pollen formation,[235] whereas flowers initiated later are inhibited at an earlier stage.[399] The developmental stage of the flower bud at senescence also varies within the inflorescence of *Caryopteris* plants grown under LD conditions. Many buds contain only very small, or no, floral organ primordia; in others, anthers have completed wall formation, but their development is halted soon after the production of a few premeiotic sporogenous cells.[400] Buds which drop in *Vigna* species in LD are also halted at the premeiotic stage.[401] In *Sinapis* the stage at which flower buds abort under SD depends on light intensity.[402] At low light there is no elongation of the raceme axis and the flower primordia remain undifferentiated. At high light degenerated pollen grains are present in the anthers suggesting that microsporogenesis has been impaired. The latest stage at which buds of 'Indian Summer' *Chrysanthemum* are arrested in LD is that of ovule formation.[153] In *Bougainvillea* the development of the inflorescence usually stops at an early stage of development in LD, most often before perianth differentiation in florets,[231] but also at later stages depending upon plant size and environmental conditions.

The experimental evidence thus suggests that, depending on conditions, flower development may be stopped at different stages preceding a critical phase, after which daylength has no further influence. The existence of critical phases during which an appropriate daylength is required if flower and/or inflorescence development is to proceed normally has been demonstrated in various species. It was found that in *Chrysanthemum* (SDP) the inhibitory effect of an inserted 20 LD treatment is maximum around the macroscopic appearance of the capitulum and declines rapidly thereafter.[403] In *Hordeum* cv. 'Proctor' (LDP) reductions in floret fertility by a 2-week SD treatment are not distributed evenly among the ears.[388] When SD are imposed from the awn initial stage on the main shoot, it is the ear of this shoot which becomes almost wholly sterile while the ear in the axils of leaves 1 and 2 is unaffected and the ear in the axil of leaf 3 shows a small increase in fertility. A SD treatment, which starts when the awns of the main shoot apex are longer than the spike and is without influence on this ear, causes almost total sterility of the axillary ears. Since the general development of the lateral ears is later than that of the main shoot ear, it is likely that the development of pollen in the florets of the former lags behind that in the latter. Then the reduction in floret fertility in the lateral ears by late SD treatments, but not by early ones, supports the idea that pollen development is more sensitive to SD at some early stage of development; this sensitive stage is probably the same for all ears. In contrast, the teratological effects of SD on stamen and ovary formation in all ears of *Triticum aestivum* cv. 'Opal' seem to be dependent on the stage of development reached by the main shoot inflorescence, irrespective of the stage of the laterals.[363] In another spring wheat (cv. 'Niva'), however, different inflorescence abnormalities are induced as a function of the time of unfavorable SD treatment.[351] When SD is given at an early stage of the transition to flowering, i.e., when the shoot apex is elongating, branched spikes occur. During the stages of initiation of spikelets and florets, SD induce the development of bracts; if SD are given after stamen initiation there is elongation of some of the proximal internodes of the rachis. In maize different parts of the apical male panicle are susceptible to modification at specific periods during development in response to daylength.[349,373]

G. Nature of Daylength-Controlled Reactions

In this section we will not consider the features of the photoperiodic control of bud dormancy in woody plants which were already reviewed by Vince-Prue.[321] We will focus essentially on more general aspects of flower development, such as rate of development and abortion. Despite the limited number of studies devoted to the nature of the photoperiodic control of these aspects, the mechanisms involved do not appear to differ from those implied in the control of flower initiation (see Volume I, Chapter 4).

1. Essentiality of the Daily Dark and Light Periods

Experiments with night breaks indicate clearly that the daily dark period is of great importance. In several species requiring SD for flower development, a light interruption of the long nights suffices to inhibit the SD treatment. In *Caryopteris,* anthesis is inhibited by 1 min of incandescent light at 200 lx given near the middle of 14-hr nights. With a 3-hr light interruption near the middle of each dark period, anthesis of stock plants can be prevented for more than 5 years.[343] Chrysanthemums are also highly sensitive to light interruptions given near the middle of 12-hr nights; intermittent lighting for short times (cyclic lighting) is however more effective than a single, continuous light break.[404] In daylength sensitive varieties of *Phaseolus*, a 6-min light break is effective in inhibiting the development of the first-initiated flower buds and in reducing the production of open flowers when imposed two thirds of the way through a 13-hr dark period.[399]

Light interruptions of long dark periods have been shown by Reid to be sufficient to give the LD response in two LD lines of pea: the first open flower, in plants irradiated with red light for 20 min in the middle of 16-hr nights, occurs at a lower node than in plants subjected to uninterrupted nights. The node at which the first flower is initiated is not affected by photoperiod in these lines.[243]

A night-break treatment also influences the development of the tassel in maize and enhances female tendency in *Dahlia* cv. 'Newby'.[349,382]

The essential role of the light period is obvious in plants that require LD for flower development and that develop their flowers until anthesis under continuous light e.g., *Sinapis*, 'Marquis' wheat, *Salvia splendens* 'Feu de la Saint Jean', 'Wintex' barley, etc.[10,230,329,405] Frequently the light period is more effective when given at a high photon flux density and this was demonstrated not only for plants having a LD requirement but also for species which need SD during the period of flower development, e.g., *Bougainvillea* and *Chrysanthemum*.[147,385] Light quality also influences the rate of growth of flowers. In 'Prior' and 'Wintex' barley, *Bougainvillea* and 'Marquis' wheat, inflorescence development is relatively slow in fluorescent light but rapid when a combination of fluorescent and incandescent light is used.[334,405-407] These results further substantiate the idea that, beyond photosynthesis, the daily light period is important in the control of flower development.

2. Perception of Daylength

The site of daylength perception in relation to flower development has been studied by Zehni and associates in the P47 daylength-sensitive variety of *Phaseolus*.[408] In one experiment all leaves other than the two primaries and the first trifoliate leaf are removed as soon as they become visible. In SD these plants develop numerous flowers to anthesis on both terminal and axillary inflorescences. Exposure of the trifoliate leaf only to SD, with the primaries in LD, also results in production of a great number of open flowers. Subjecting the first trifoliate leaf to long photoperiods in plants otherwise given SD reduces the number of open flowers almost to the number produced in plants kept entirely in LD. When several leaves are retained, exposure of the lowest leaf to LD still leads to inhibition of flower bud development, even though there are leaves in SD between the receptor buds and the leaf held in LD. Conversely exposure of the lowest trifoliate leaf to SD in plants otherwise in LD allows a few flower buds to develop to anthesis. Thus in *Phaseolus* the leaves respond to either short or long photoperiods, producing probably inhibitors as well as promoters of flower development.

3. Involvement of Phytochrome

Phytochrome appears to be the photoreceptor involved in the photoperiodic control of flower development in the few species in which this aspect has been studied. As already reported *Caryopteris* requires SD to develop flowers until anthesis and the promotive action of long dark periods is totally nullified by a very short night break.[343] Piringer and associates found that the inhibitory effect of 1 min of R given in the middle of 14 hr nights is completely reversed by a 5 min FR irradiation immediately following the R. Anthesis is repeatedly reversible by short R and FR interruptions (Table 3). Reversibility is largely lost if an interval of darkness of 30 min or more is inserted between the R and FR irradiations. Furthermore, a prolonged FR interruption of 60 min largely prevents anthesis in plants that had no prior R treatment.[343] Clearly, as in the inhibition of flower initiation in SDP, the inhibition of flower development in *Caryopteris* by light breaks is the result of maintaining the level of phytochrome in the Pfr active form above some particular threshold for enough time.

Table 3
EFFECT OF SUCCESSIVE R AND FR IRRADIATION TREATMENTS GIVEN IN THE MIDDLE OF LONG NIGHTS ON FLOWER DEVELOPMENT OF THE SDP *CARYOPTERIS* × *CLANDONENSIS* AND THE LDP *PISUM SATIVUM* (LINE 60)

Treatment[a]	Number of plants developing flowers until anthesis in *Caryopteris*[b]	Mean number of nodes between the node of first open flower and the node of first-initiated flower in *Pisum*
Dark control	5	4.76
R	0	0.5
R,FR	3	3.33
R,FR,R	0	—
R,FR,R,FR	5	—
R,FR,R,FR,R	0	—
R,FR,R,FR,R,FR	5	—
FR	—	3.66
Continuous light	—	0

[a] For *Caryopteris*, R for 1 min, FR for 5 min; seventeen 14-hr dark periods. For *Pisum*, both R and FR of 20 min duration; from sowing.
[b] Five plants per experimental batch.

Data for *Caryopteris*, from Piringer, A. A., Downs, R. J., and Borthwick, H. A., *Am. J. Bot.*, 50, 86, 1963; data for *Pisum*, adapted from Reid, J. B., *Z. Pflanzenphysiol.*, 93, 297, 1979. With permission.

In two LD lines of *Pisum* a 20 min R light interruption of 16 hr dark periods promotes flower development (Table 3).[243] This effect is nullified by a subsequent 20 min FR irradiation indicating that phytochrome is also involved. The high effectiveness of rather short light interruptions should be noted since the promotion of flower initiation in LDP generally requires long light exposure (Volume I, Chapter 4).

As reported above, fluorescent lamps with a low proportion of energy emitted in the FR region are less effective for promoting the development of the inflorescence of some species than a mixture of fluorescent and incandescent lamps, which has more FR energy.[334,405-407] This result again implies phytochrome and the high irradiance response (see Volume I, Chapter 4) and suggests that the proportion of Pfr to total phytochrome is critical in regulating flower development. It is of interest that in 'Clipper' barley and 'Marquis' wheat the photoperiodic efficiency of incandescent light used to extend an 8-hr high light period changes during the course of a daily cycle. It is greater when the supplementary incandescent light is given after, rather than before, the main light period.[335,407] Deitzer and associates showed that seedlings of 'Wintex' barley, induced to flower with a single 24-hr daylight fluorescent photoperiod supplemented with sufficient FR energy to produce a relative R/FR ratio of 0.5, undergo floral initiation about a week after they are returned to 12-hr daylight fluorescent photoperiods (R/FR ratio = 5.5), but floral development does not begin for an additional 2 weeks.[405] Addition of FR light to a subsequent 12-hr photoperiod decreases the lag period between initiation and development by 10 days without affecting the rate of development. Extending the photoperiod to 24 hr has the same effect on the lag period but increases the rate of development. FR present during the first half of this 24-hr photoperiod further decreases the lag period by 2 to 3 days with no further effect on the rate of development. When FR is added throughout the 24-hr photoperiod, there is no further effect on the lag period but further increase in the rate of development.[405] Thus, the presence of FR during the first half of the photoperiod affects the time of onset of floral development, while its presence during the second half affects the rate of this development.

CONCLUSIONS

Daylength is a major factor controlling flower development in many plants. It may act independently but also in conjunction with irradiance since both contribute to the daily light integral (see next section).

In photoperiodic species the optimum daylength for flower development is not necessarily the same as for initiation, but when both physiological processes have a photoperiodic requirement for their completion, all combinations are possible. Photoperiod not only affects the rate of development and the abortion of the reproductive structures, two phenomena which are tightly linked, it also has numerous morphogenetic effects. Daylength is a potent factor determining sex expression in various monoecious and dioecious species. Most often, femaleness is favored by SD but this is not the general rule. Unfavorable photoperiods induce varied reproductive abnormalities in different species, revealing the great plasticity of the reproductive structures. In several woody plants daylength affects the onset of bud dormancy. It is generally accelerated by SD. Frequently, an appropriate photoperiod is required for a limited time, for the completion of some precise stage, suggesting that sequences of determinations are involved in the ontogeny of flowers and inflorescences.

Daylength perception is a property of the leaves, and is attributed, as for flower initiation, to the pigment phytochrome, acting in some species in the high irradiance response.

III. IRRADIANCE AND THE LIGHT INTEGRAL

Inflorescence development is frequently a function of light integral and/or irradiance, both of which vary concurrently in greenhouse and field conditions. Early studies in temperate zones showed that flower production in greenhouses is considerably reduced during the winter months; hence it was generally supposed that low light integral during this period was the major environmental factor responsible. Several studies have indeed shown the importance of light integral in development. In recent years it has been possible to separate the irradiance and light integral factors and to specify whether photosynthesis is the primary physiological system affected and directly responsible for the flower and inflorescence responses observed.

A. Rate of Development and Abortion

1. Time to Anthesis

Progress to anthesis is generally delayed by low light levels. 'Baccara' rose plants grown at 21°C with continuous illumination at 12,000 lx flower about 3 weeks earlier than plants grown at 1,500 or 3,000 lx.[409] The time interval from the macroscopic appearance of the tomato inflorescence to first anthesis is also shorter when the daily light supply is increased. This effect is independent of daylength.[410] In two azalea cvs. the rate of early floral development is markedly increased when irradiance is doubled.[239]

Thorough investigations by Cockshull and Hughes of inflorescence initiation and early development as a function of light supply with 'Bright Golden Anne' chrysanthemum provide the basic data for a photosynthetic-based model of floral initiation and development.[31,233,385,411] The data in Figure 9 show that at 31 J/cm²/day compared to 250 J/cm²/day initiation is delayed by 2 to 3 weeks, leaf number below the capitulum is doubled, and the slope of the curve for inflorescence development as a function of time (i.e., slope is equal

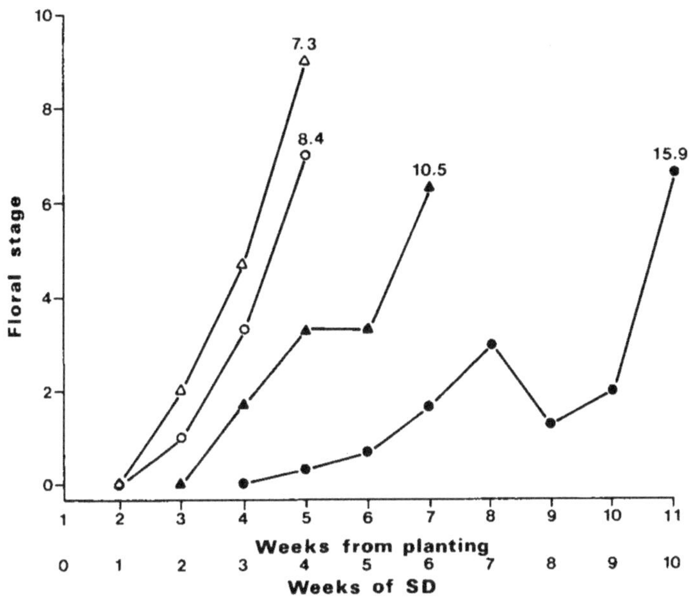

FIGURE 9. Effect of light level on rate of apical development (lateral meristems) in 'Bright Golden Anne' *Chrysanthemum morifolium*. The figures above the curves are number of leaves per lateral branch. — △ — 250 J/cm²/day; — ○ — 125 J/cm²/day; — ▲ — 63 J/cm²/day; — ● — 31 J/cm²/day. Stage 2 = receptable initiation, Stage 5 = floret initiation. (From Cockshull, K. E., *Crop Processes in Controlled Environments*, Rees, A. R., Cockshull, K. E., Hand, D. W., and Hurd, R. G., Eds., Academic Press, London, 1972, 235. With permission.)

to rate) is reduced by 75%. By inserting 2-week intervals of low light (65 J/cm²/day) into a moderate light regime (125 J/cm²/day) the receptacle initiation stage during the first 2 weeks of SD and the floret initiation stages of the third and fourth weeks of SD are found to be very sensitive to reduced light.[385] There is little beneficial carryover from previous high light periods, and, once inflorescence development proceeds beyond floret initiation, further development is not affected by light. The flowering response is determined by the daily total and not directly by irradiance.[411] The plants even respond to the light level integrated over 2 days.[233]

2. *Flower Abortion*

Abortion of flowers·or inflorescences seems to be one of the main light-affected problem in several species including roses, tomatoes, *Bougainvillea,* and some bulbous crops. In hybrid roses the annual light integral and flower production curves are nearly congruent;[412] the prime impacts of light are upon the number of flowers completing development to anthesis and the rate of development.[232,409,413,414]

The minimum light flux for development to anthesis has not been determined. Zieslin and Halevy investigated the effects of different degrees of shade on blindness in 'Baccara' roses (Figure 10).[413] They found that in the two upper buds, which resume growth after pruning, the percentage of flowers reaching anthesis progressively decreases and the percent of blind shoots progressively increases with the increase in shade. Apparently the light flux parameter interacts strongly with cv. and cultural conditions. Also, as we shall learn, light quality is extremely important.

Grapevine "anlagen" (Volume II, Chapter 3, Section III. C.4) require some minimum photon flux to develop into inflorescences rather than tendrils. In the cv. 'Sultana', whole

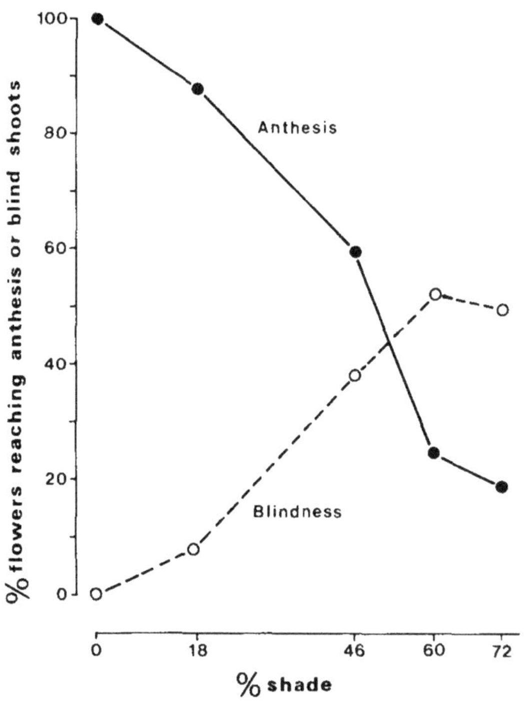

FIGURE 10. Effect of shading rose plants (*Rosa hybrida* cv. 'Baccara'), from pruning, on flower development of the two upper shoots. (From Zieslin, N. and Halevy, A. H., *Sci. Hortic.*, 3, 383, 1975. With permission.)

vine shading (73% reduction of sunlight), for at least 4 weeks in the late spring, when inflorescence primordia of the next season are initiated, reduces the number of shoots with inflorescences (percent fruitful shoots). With the reduced berry weight reported, it seems probable that ovary size, too, is reduced by the late spring shading.[415] Buttrose's data show a strong cv. dependence on photon flux for fruitfulness.[416] 'Shira' is almost without inflorescences after 3 months exposure to 16-hr days of 9000 lx but is fruitful at intensities in excess of 18,000 lx for the same photoperiod; 'Sultana' bears essentially no primordia even when irradiance is at 36,000 lx for 16 hr.

Clear recognition of the importance of photon flux for inflorescence development in tomato was shown by Calvert and Cooper.[417-419] Using plants grown in 4 different light regimes (8-hr SD and 16-hr LD combined with 2 light irradiances: 9 and 18 W/m^2) it was found that increasing daily light integrals greatly reduces inflorescence abortion. In similar light integrals development is far better in SD than in LD.[410]

Inflorescence development does not proceed beyond floret initiation in *Bougainvillea* at low photon flux, something less than 25,000 lx for 8 hr.[147] With photon flux of 4000 lx for 8 hr inflorescence development may cease at bracteole initiation.[420]

Shading of *Gladiolus*, which reduces light intensity to 25% of sunlight reduces the percentage of flowering plants (presumably plants bearing an inflorescence with at least one floret reaching anthesis) as well as the number of florets per spike (presumably number of florets reaching anthesis) in the sensitive cv. 'Sans Souci'.[421] In the less sensitive cv. 'Spic and Span' low light intensity only decreases floret number. Daylength extension is effective in reducing blasting.[422,423] Premature flower abortion at low light levels also appears to be common in *Iris, Lilium, Hydrangea*, etc.[424-428]

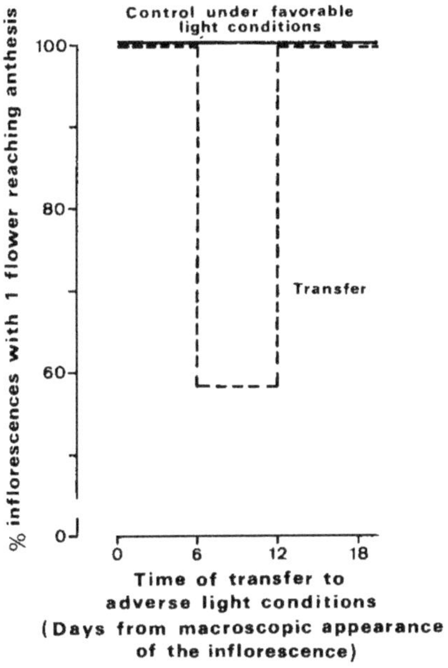

FIGURE 11. Effect of a 6-day low light treatment given at various times from macroscopic appearance of the first inflorescence on development of this inflorescence in tomato plants (*Lycopersicon esculentum* cv. 'King plus') grown in favorable light conditions. Low light conditions: 8-hr photoperiods and a light energy of 12 W/m^2 at the top of the canopy. Favorable light conditions : 16-hr photoperiods and a light energy of 20 W/m^2. (From Kinet, J. M. and Léonard, M., *Acta Hortic.*, 134, 117, 1983. With permission.)

B. Time Dependence of Light Intensity Effects

A relatively narrow time dependence for low light-induced abortion has been found in several species by transferring plants to low light or darkness for a limited period at different stages of flower development. The period of greatest stem elongation, when the flower bud just becomes visible, coincides exactly with the period during which the incidence of bud blast in *Iris* is stimulated by poor light conditions.[424] In tomato one critical stage in inflorescence development is after macroscopic appearance of the inflorescence bud.[410,429] At that time some reduction in the percent of inflorescences which develop to anthesis occurs with only 2 days exposure to low light flux indicating the extreme sensitivity of the inflorescences to reduced photosynthesis. The reverse transfer (from low to high light conditions for 5 to 20 days) greatly promotes inflorescence development.[410] With the 'King plus' cv., Kinet and Léonard further defined the period of greatest sensitivity to low light flux; they found that between the 5th to 6th and 10th to 12th days after macroscopic appearance of the inflorescence, development is severely inhibited by a 5 to 6-day low light treatment (Figure 11).[430,431]

A 26-day, 95% shade, treatment inhibits inflorescence development in 'Sans Souci' *Gladiolus*. Shillo and Halevy showed that plants at the 4 to 6 leaf stage, in which inflorescences are elongating, floret initiation is in full progress, and some florets have all floral organs formed, are most sensitive to light conditions; low irradiance during this period reduces both the percentage of flowering plants and floret number. Shading at a later stage causes only abortion of individual florets.[421]

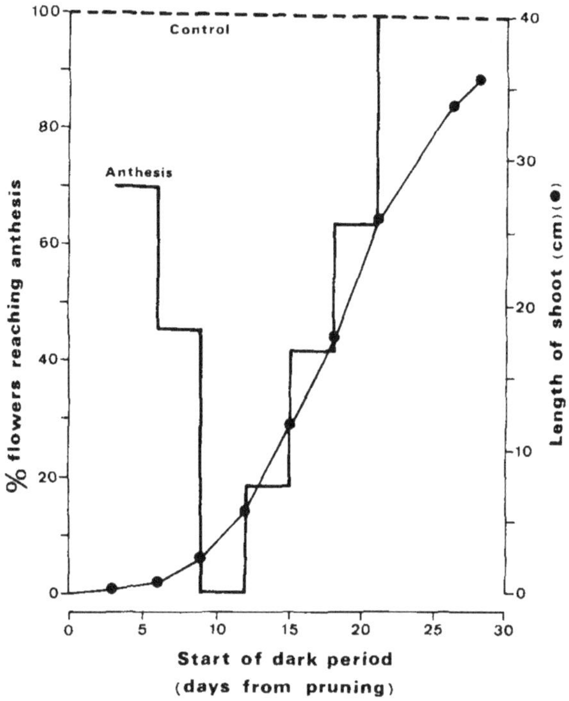

FIGURE 12. Effect of time of darkening rose shoots (*Rosa hybrida* cv. 'Marimba') on flower development and shoot elongation. Branches were pruned to second five-leaflet leaf from top, and the shoot emerging from the upper bud was darkened for 12 days at various times after pruning. Percentage of flowers reaching anthesis and shoot length were measured 34 days after pruning. Growth curve of control plants is shown. Experimental points are entered at the beginning of the dark period. (From Mor, Y. and Halevy, A. H., *Plant Physiol.*, 66, 990, 1980. With permission.)

In *Lilium longiflorum* blasting occurs most frequently in buds ranging from 2 mm to 5 cm in length;[425] in 'Enchantment' lily, too, blasting may occur at almost any time during flower development,[427] but, as shown by Kamerbeek and Durieux, flower buds are most likely to undergo abscission at a bud length of 1 to 3 cm, at meiosis in the stamens.[426,427] In tomato, microsporogenesis also appears to be a critical stage when low light is likely to cause abortion.[432,433] In plants continuously grown under unfavorable light conditions flower failure occurs at the pollen mother cell stage, whereas in plants grown under favorable light conditions and subjected at different times to a 5- to 6-day low light treatment (Figure 11) greatest sensitivity occurs during meiosis in the anthers.[433]

Low light induced blindness in roses is also time dependent; this case will be expanded upon in the following section.

C. Sink vs. Source Activation by High Irradiance

In some plants at least a part of the light requirement cannot be explained by increased photosynthesis, although there is little question that higher photosynthetic activity in the source leaves is probably a major contributing factor to high light-induced promotion of flower development.

A direct action of light (in this case irradiance rather than light integral) on the developing rose bud, with a consequent promotion of sink activity, is clearly described by Halevy and colleagues and appears to be causal in reduction of blind shoots.[413,434,435] Darkening shoots

Table 4
PERCENTAGE OF FRUITFUL BUDS AND WEIGHT OF INFLORESCENCE PRIMORDIA FORMED IN 'SULTANA' GRAPEVINE (*VITIS VINIFERA*) WHEN BUDS ONLY ARE SHADED DURING EARLY DIFFERENTIATION OF THE INFLORESCENCE

	Fruitful buds (%)			Inflorescence primordium weight (μg)		
Season	Not shaded	Transparent cover	Heavy shade	Not shaded	Transparent cover	Heavy shade
1958	96	Not applied	67	108	Not applied	46
1959	60	64	25	53	53	65
1960	92	83	65	115	68	65

From May, P., *Aust. J. Biol. Sci.*, 18, 463, 1965. With permission.

for 12 days, subsequent to forcing axillary bud growth by pruning, reduces flowering in most stages of shoot development, but it is particularly detrimental starting between days 6 and 15 after pruning, at the beginning of exponential stem elongation (Figure 12). Starting at 9 days, darkening completely inhibits flower development; at this stage all sepal and some petal primordia are already present.[434] Light applied directly to the shoot tip promotes flower development even when the tip is treated with DCMU or is held in a CO_2-free atmosphere;[434] apparently photosynthesis of the green shoot tip does not contribute to this action of light. Photosynthetic assimilates come from the mature leaves; a more detailed analysis of this aspect with roses will be made in Chapter 9, covering assimilate transport. Fluorescent light only is not as effective as mixtures of fluorescent plus incandescent light in promoting sink activity in roses.[435] Thus, earlier studies on flower production and abortion in roses as a function of light flux, where fluorescent tubes only provided *all* of high irradiance requirements for flower development must be evaluated in this perspective. High irradiance requirements for flower development in 'Wedgwood' iris are fulfilled at very low CO_2 levels suggesting that in this case, too, light is promoting assimilate transport to the developing flower rather than merely increasing photosyntesis.[436]

Key experiments with 'Sultana' grapevine, with late spring shading of individual axillary buds using aluminium shields, prove that light has direct action on the buds themselves (Table 4).[437] Shading whole vines during a period of 6 weeks reduces the proportion of fruitful shoots, but the critical period for individual buds is considerably shorter. The high-light flux requirement is of about 2 weeks duration, but there is also a further reduction in fruitful bud number if shading is prolonged to 6 weeks. It is clear that, in grapevines as in roses, high irradiance requirements serve in part for sink enhancement of the developing inflorescences.

In contrast, there is no evidence for irradiance-induced sink activation in *Bougainvillea,* tomato, and *Chrysanthemum.*[420,433,438] Shading the inflorescence bud of tomato, from macroscopic appearance through anthesis does not prevent, not even reduce the rate of development.[433] When 'San Diego Red' *Bougainvillea* and 'Bright Golden Anne' *Chrysanthemum* shoot tips, containing developing inflorescences, are shaded there is no reduction in development if the leaves are maintained in high light. Illuminating the shoot tip with high irradiance while mature leaves are held at low irradiance results in greatly reduced rates of inflorescence development and in *Bougainvillea* the inflorescence meristem may abort before floret initiation.[420,438]

D. Interaction with Daylength

Interactions between daylength and light intensity have been recorded in several photoperiodic species. We already reported above that light flux markedly affects inflorescence development in the SDP *Chrysanthemum* 'Bright Golden Anne' and *Bougainvillea* 'San Diego Red'. In the latter, Sachs and Hackett also demonstrated that high light not only promotes inflorescence development, but even may override the photoperiodic signal so that this plant behaves then as a DNP.[439] In the SDP *Kalanchoe*, lowering the light intensity after the macroscopic appearance of the inflorescence causes the development of vegetative inflorescences. This effect is recorded when the leaves, not the inflorescence, are subjected to low light.[440] In contrast, low light favors flower development in another SDP, *Lemna paucicostata* 6746. It prevents or delays bud abortion under noninductive photoperiodic cycles.[441,442]

Changes in photon flux density have a striking effect in several LDP; generally increasing the light level favors floral development. In *Saxifraga rivularis* plants grown under continuous incandescent plus fluorescent light, decreasing the irradiance reduces the number of plants reaching anthesis.[443] In *Sinapis*, grown in 16-hr LD, the rate of production of flower buds by the terminal inflorescence is 3.75 per day with a light intensity of 6000 lx and 1.75 at 1500 lx.[370] As shown by Bagnard, lowering the light level to 300 to 500 lx, after flower initiation occurred at 1500 lx, may halt the development of the flower primordia at an early stage and cause the formation of inflorescence reversions.[402] The anomalous structures are however more effectively produced when the low irradiance treatment is given under noninductive SD.[354] At low light intensities spikelet production occurs at a low rate and stops at an early stage of ear development in wheat, producing ultimately a shorter ear and lower spikelet number at anthesis.[444] In two azalea cvs. Bodson found that early inflorescence bud abortion in SD is prevented by increasing the irradiance.[239]

E. Sex Expression and Fertility

It is commonly acknowledged that reducing light intensity shifts sex expression towards femaleness.[116,320] In cucumber, however, low light flux generally promotes the male tendency.[274,445] In normally gynoecious lines, Kooistra found that plants exposed to natural daylight in wintertime develop up to 30% male flowers whereas in plants receiving supplemental lighting during the light period, only female flowers are formed (Table 5A).[445] In the field, with the long daylength of summer, some gynoecious lines develop male flowers, but the number is augmented if the plants are shaded (Table 5B). Additional studies with gynoecious lines of pickling cucumbers reveal a complex interaction of light flux, daylength, temperature and genotype, but they generally indicate that female flowering increases with increased light intensity.[378] In the 'Bright Golden Anne' *Chrysanthemum*, too, female (ray) floret number is increased when a high daily light total is given at the onset of floret initiation; it is unaffected by either earlier or later treatments.[385] High light also enhances female flower production in the orchids *Cycnoches warscewiczii* and *Catasetum expansum*.[446]

In contrast, profuse tassel branching and functional male florets are observed in the terminal panicle of a day-neutral maize inbred grown under SD given at high light intensity. When irradiance is decreased, only ovules are observed in the apical inflorescence which remains in the leaf whorl. In this line sex reversal from male to female thus appears to be primarily the result of a reduced light intensity.[246] In hermaphroditic species such as tomato, *Kalanchoe*, and strawberry low light conditions induce male sterility.[257,447-449] Howlett related this phenomenon in tomato to early degeneration of microspores with little pollen production as well as to poor germinability of pollen grains.[257] In the 'Sivetta' strawberry mature stamen number is reduced by 80% with 3 days of 75% shading (Figure 13). A similar shading treatment is without effect on stamen development in the cv. 'Karina' indicating that there is a strong cv. dependence on light intensity for male fertility.[449]

Table 5
EFFECT OF LIGHT INTENSITY ON THE PERCENTAGE OF AXILS WITH MALE FLOWERS (FIRST 10 AXILS PER PLANT COUNTED) IN VARIOUS GYNOECIOUS LINES OF *CUCUMIS SATIVUS*

A. Winter Experiment

Line	Treatment[a]	Axils with male flowers (%)
119	Control	17
	Illuminated	0
123	Control	28
	Illuminated	0
001	Control	3
	Illuminated	0

B. Summer Experiment

Line	Treatment[b]	Axils with male flowers (%)
014	Control	48.8
	Shading	64.0
142	Control	0.0
	Shading	18.1
144	Control	0.6
	Shading	24.5

[a] Control plants exposed to natural daylight (light intensity of about 4000 lx; daylength of 8 to 9 hr). Illuminated plants exposed to natural daylight supplemented for 8 hr during the day period with light from 400-W HPL lamps giving a total light intensity of 8000 lx.

[b] Control plants exposed to natural daylight. Shading treatment by covering the plants with double cheese cloth.

From Kooistra, E., *Euphytica*, 16, 1, 1967. With permission.

CONCLUSIONS

All studies reviewed here show that low irradiance and/or low light integral reduce inflorescence development. There is evidence that in most species inflorescences will be altered in form or abort prematurely at low light integral. How low is low? This differs with the species and cv. and other environmental parameters, but the intensities are well above the compensation point and clearly sufficient to support vegetative growth. In some cases the high irradiance requirement appears to promote sink strength in the developing flower or inflorescence, but there is little question that photosynthesis in the source leaves is promoted, too, and is probably a major contributing factor to high light-induced promotion of inflorescence development.

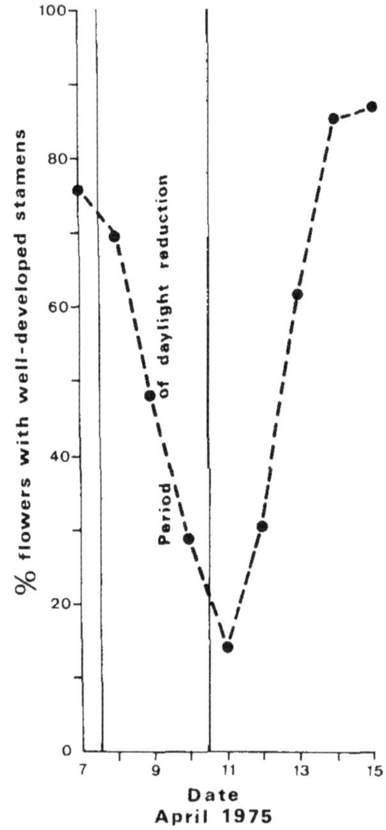

FIGURE 13. Effect of light intensity on stamen development in 'Sivetta' strawberry (*Fragaria*). The number of flowers with well-developed stamens is given as a percentage of the number of flowers that opened on the date indicated. Shading to reduce the intensity of natural daylight to 25% was from April 8 to April 10. (From Smeets, L., *Sci. Hortic.*, 12, 343, 1980. With permission.)

There is a relatively narrow time dependence for low-light induced effects. In several species the most sensitive stage of development appears to be microsporogenesis.

Sex expression is clearly affected by the light intensity; low light flux generally promotes the male tendency in cucumbers. Different responses have been found in other species. Again, microsporogenesis seems to be adversely affected by low light; when the male organs fail, there is an increase in femaleness.

Chapter 5

CONTROL BY TEMPERATURE AND WATER STRESS

TABLE OF CONTENTS

I. TEMPERATURE

As with every other aspect of plant growth, temperature is of paramount importance for flower and inflorescence development with both quantitative and qualitative effects. Besides its action on the rate of development, temperature is critical in certain species, in determining normal development until anthesis or abortion of the bud. It also affects sex expression in monoecious and dioecious plants and controls dormancy breaking of flower buds in various bulbous and woody perennials and in several alpine plants.

A. Rate of Development and Abortion

1. Time to Anthesis

Increase in temperature usually tends to speed development towards anthesis, the rate of morphological development of reproductive structures being more rapid at higher temperatures. This effect has been recorded in many different species, including grasses,[230,242,450] herbaceous dicots,[201,451-453] and woody species.[232,343,409,454] However, in various bulbous plants and woody perennials where flower buds enter dormancy after initiation, the situation is not that simple. Different temperature optima are required for successive stages of flower development. After floral initiation which occurs usually at warm temperature, there is a low temperature optimum for breaking flower bud dormancy (see Section I.G). However, when the cold requirement is satisfied, the situation is not different from that reported for other plants; i.e., higher temperatures reduce the time to anthesis as found in tulips and *Iris hollandica* (Figure 1).[163,424,455]

Lake reported that day temperatures play a greater role than night temperatures in determining the rate of flower development in tomato, but this may have been simply because in his experiments the days were longer than the nights.[452] In fact, the relative magnitude of the effects of day and night temperatures was in approximately the same ratio as the duration of the light and dark periods. In rose it was found that night temperature has a larger effect than day temperature.[232] In this case, too, it may have been because the night was longer than the day. By growing tomato plants either under a constant 21°C or a differential day/night regime with a mean of 21°C, Morgan and Clarke found that there is no significant difference in time to anthesis of the first and second trusses between the two treatments.[456] In *Iris hollandica* day and night temperatures affect time to anthesis similarly showing again that the average daily temperature is the major factor determining the rate of flower development.[424]

The differential effect of air and soil temperature has been investigated in a few species. Generally, increases in both reduce the number of days to anthesis; the effect of air temperature is, however, much greater in *Iris* and *Gerbera*.[424,457] By varying independently the temperature of the flower bud and other parts of the carnation plant, Harris and Scott found that anthesis occurs 4 weeks earlier as a result of heating the bud, but is not significantly hastened by heating the lower part of the shoot.[201] In this case the temperature-dependent processes are in the flower bud rather than in the stems or leaves.

2. Flower Abortion

High temperature prevents flowers from reaching anthesis in numerous species, including tulip, *Iris, Lilium,* tomato, *Citrus,* etc.[162,163,424,426,429,458,459] Generally the higher the temperature, the higher the percent abortion (Figure 1). In *Iris,* the most sensitive stage coincides with the period of most rapid stem elongation, at the time the bud emerges from the leaf sheath, 1 or 2 weeks before anthesis.[424] In tomato, too, high temperature is most deleterious at the time of macroscopic appearance of the truss.[429]

Increasing day as well as night temperature promotes abortion in *Iris,* but increased day temperature is usually more detrimental than increased night temperature.[424] The effect of air temperature is also greater than that of soil temperature in *Iris* and *Citrus*.[424,459]

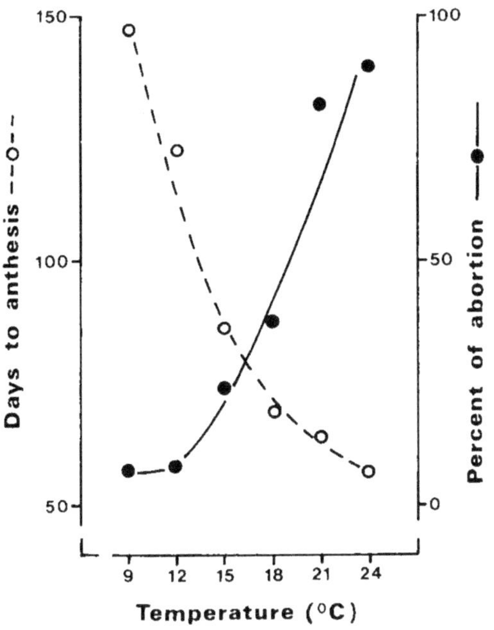

FIGURE 1. The influence of temperature on the number of days from planting to 50% anthesis and the percent of abortion of the second flower bud of the inflorescence of *Iris hollandica*. Each point is the average of different photoperiodic treatments with bulbs of three different cvs.: 'Dominator', 'Prof. Blaauw', and 'Imperator'. Before planting, the bulbs were stored 2 weeks at 35°C + 2 weeks at 17°C + 6 weeks at 9°C. (Adapted from Fortanier, E. J. and Zevenbergen, A., *Neth. J. Agric. Sci.*, 21, 145, 1973.)

In a few species low temperature treatment causes flower bud abortion. It is true in rose where low-air or low-soil temperature promotes blind shoot formation.[232,413,460] These perturbations in rose are more pronounced in early rather than late stages of shoot development. Moe has observed that maintaining plants at 18°C for more than 16 days after pruning, which causes lateral shoot growth and flower differentiation, is sufficient to reduce blind shoot formation induced by 12°C treatments.[460] By 21 days the critical period for low temperature-induced abortion is past. At this time all shoots have initiated stamen and pistil primordia (Table 1).

Bud blasting in Easter lilies is induced by a 10- to 15-day soil chilling treatment when the buds are 0.6- to 1.3-cm long.[425] Cooling the soil during the night only does not induce blasting, and air temperatures at or above 4.4°C are also without effect. Since, in these experiments, wilting is observed during the first 24 hr of the 4.4°C soil temperature treatment, water stress could have contributed to flower bud abortion. Low temperature (1 to 4°C) also increases flower bud blasting in *Gladiolus*. Cold nights combined with poor light conditions during the day are most detrimental.[461,462]

B. Flower and Inflorescence Size

High temperatures which promote flower development frequently result in poor quality flowers. This is well known in bulbous plants.[161] In carnation differences in temperature have relatively little effect on the growth in dry weight of the shoot system as a whole, but at high temperatures the flower accounts for a much higher proportion of the total increase

Table 1
BLIND SHOOT AND BULLHEAD FORMATION IN *ROSA HYBRIDA* CV. 'BACCARA' AS A FUNCTION OF TEMPERATURE

Days at 18°C[a]	No. of blind shoots (per 8 plants)	Percent of bullheads	Percent of shoots at stage[b] (6 shoots): 3	4	5,6	7
0	17	40				
12	17	41.2	66.7	33.3		
16	16	33.3	33.3	66.7		
19	10	32.1		33.3	66.7	
21	4	13.0			83.3	16.7
Constant	3	6.9				

[a] After pruning, plants are grown in a greenhouse at 18 ± 1°C for the designated interval and then moved to 12 ± 0.5°C.

[b] The flowering stages are 3 = sepal primordia; 4 = petal primordia; 5,6 = stamen and pistil primordia; and 7 = anthers visible.

From Moe, R., *Physiol. Plant.*, 24, 291, 1971. With permission.

in dry weight than at lower temperatures.[201] However, the absolute dry weight and the final size of the flowers grown at high temperature are reduced because the duration of growth is shorter. Similar results are obtained when the bud alone is heated. In hybrid geranium (*Pelargonium* × *hortorum*) cv. 'Sooner Red' flower diameter is greatest at 15°C and decreases as temperature is raised to 32°C.[453] The number of flowers per inflorescence is also reduced at 32°C.

Detrimental effects of high temperature have been recorded in cereals. In 'Marquis' wheat, e.g., Friend has found that the length of the developing inflorescence and of the ear and the number of spikelets are highest at low temperature (Figure 2).[444] This result is related to differential effects of temperature on the rates and duration of apical elongation, morphological development of the ear, and spikelet formation. Low temperature retards morphological development of the ear to a greater extent than that of the whole plant, prolonging the duration of ear development and spikelet initiation which results in large ears at anthesis.[444] Differences among cvs. of wheat in their responses to temperature have been reported. According to Rawson and Bagga, spikelet number in different Australian cvs. is influenced mainly by the temperature prevailing before double ridge formation.[450] However, low temperatures during preanthesis growth result in an increased number of grains set per ear suggesting that the proportion of fertile florets on each spikelet is increased.[450,463] These temperature responses, particularly elaborated upon in Australia and New Zealand,[450,463,464] provide the base data for interpreting climatic influences on wheat yield (see Chapter 11).

Branched inflorescences, with appreciably more flowers than on unbranched trusses, develop on tomato plants subjected to a low temperature treatment.[465,466] The critical period for chilling coincides with inflorescence initiation.[433,467] Temperature also has a large influence on the type of inflorescence formed in sweet orange. At low temperatures the many-flowered leafless type predominates, but at high temperature the many-flowered leafy type is the commonest.[454]

C. Number of Floral Parts

In many species the number of floral parts is affected by temperature. Blaauw and coworkers found that in most tulip cvs. there is a tendency to increase the number of floral

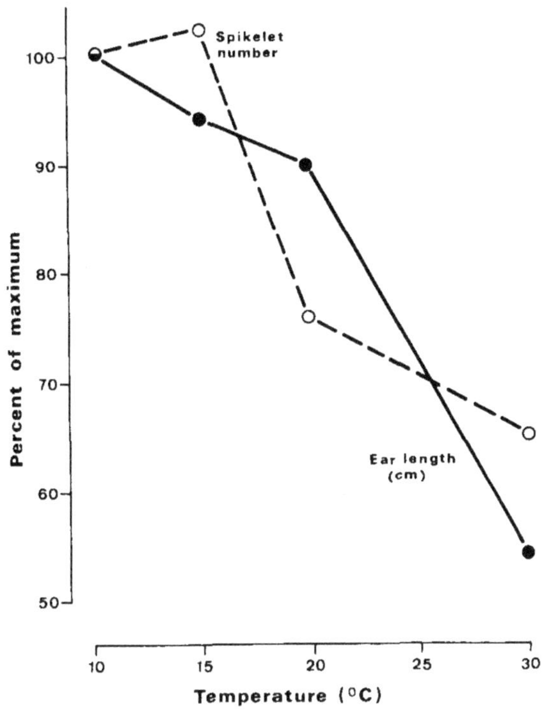

FIGURE 2. Ear length and spikelet number at the time of anthesis in plants of *Triticum aestivum* cv. 'Marquis' grown at different temperatures. (Reproduced by permission of the National Research Council of Canada from Friend, D. J. C., *Can. J. Bot.*, 43, 345, 1965.)

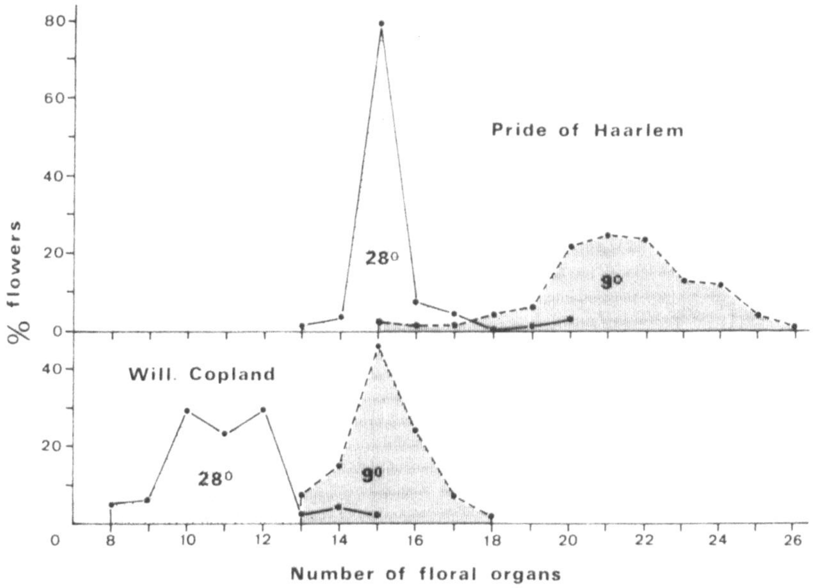

FIGURE 3. Number of floral parts of tulips (*Tulipa gesneriana*) cvs. 'Pride of Haarlem' and 'Will. Copland' formed at 9° and 28°C. (From Blaauw, A. H., Luyten, I., and Hartsema, A. M., *Proc. K. Ned. Akad. Wet.*, 35, 485, 1932. With permission.)

parts in bulbs stored at low temperature after lifting (Figure 3).[468] This number, normally 15 (see Chapter 1, Figure 1), is also a function of the cv. (Figure 3).

Temperature of 15°C or less is also a major factor contributing to malformation in 'Baccara' roses producing so-called "bullheads" (see Chapter 1, Sections III.D,G).[460,469] Bullheads are reduced when the shoots have passed the stage when flower abortion is no longer induced by low temperature (Table 1).[460]

Exposure to low temperatures of 4 to 10°C, at night or continuously over long periods, increases petal number in carnation.[61,470] This increase is attributable to additional petals developing from secondary growth centers formed around the gynoecium,[61,470,471] whereas the number of primary petals arising directly from the receptacle is unaffected.[61] Work by Kohl showed that the most sensitive stage for formation of the secondary centers is when the flower bud is 3 to 6.5 mm in diameter.[471] Garrod and Harris have found that low temperature regimes before the flower bud is visible increase the mean number of secondary centers per flower about 15 fold;[61] no further increase occurs when the low temperature treatment is extended after the appearance of the flower bud. The longer the chilling treatment (up to 3 weeks) during the early stages the higher the number of growth centers.[472]

Hollow carnation flowers produced in the summer have low petal counts. The factor(s) responsible is (are) unknown, but it is not simply high temperature since experiments by Abou Dahab and Garrod and Harris have shown that high temperature treatments, even localized at the shoot tip, result in increased petal number.[61,338] This effect is exerted, once again, in the early stages of flower development but, unlike low temperature, high temperature increases the number of primary petals while the numbers of secondary centers and of petals developing from them are not changed significantly.[61]

Aberrant numbers of floral parts in *Silene coeli-rosa* appear more frequently at high (27°C) and low (13°C) than at intermediate (20°C) temperatures.[70] The staminoid whorls are most often affected and at both high and low temperatures parts are missing rather than in excess (Chapter 1, Table 1).

D. Sex Expression and Fertility

The first unambiguous demonstration of the importance of temperature in sex expression was by Nitsch and associates with 'Acorn' squash.[115] This study, performed in the rigidly controlled environments of a phytotron, clearly established that decreasing the temperature accelerates the appearance of the first female flower and enhances the index of feminization in SD as well as in LD (Table 2). Day and night temperatures do not seem to be of equal value; night temperature appears to be the crucial factor. However, since the duration of the night was twice that of the day, it may be argued that the critical factor is the duration of the exposure to a given temperature. Heslop-Harrison also found that low night temperature promotes femaleness in maize by inducing the appearance of female flowers in the tassel (Table 4.1).[373]

Nitsch and co-workers extended their observations to cucumber and gherkin and found that these plants behave in the same way as squash.[115] In cucumber Cantliffe found that both male and female flower numbers are higher at 26 and 30°C than at lower temperatures, but the increase of male flowers at high temperature exceeds that of female ones.[378] Response to temperature is, however, cv. dependent. Galun observed that the node number at which the first female flower is produced may be lowered when a cold treatment is applied to imbibed seeds.[473] Low temperature can thus affect sex expression in an inductive manner, as it does for other aspects of the flowering response (see Volume I, Chapter 5, Section I).

The production of female flowers is favored by low temperatures in many other monoecious and dioecious species.[111,116] In contrast, in *Begonia* × *cheimantha* low temperatures enhance maleness.[383] The pattern of sex distribution in this plant may, however, profoundly influence the results and mask the effect of the treatments (Chapter 2, Section II. B. 1).

Table 2
EFFECT OF TEMPERATURE ON SEX EXPRESSION IN THE SQUASH (*CUCURBITA PEPO*) CV. 'ACORN'[a]

Temperature (°C)		Node of the first female flower	Index of feminization[b]
Day	Night		
30	30	58	8
26	20	23	53
23	17	20	76
20	10	16	80

[a] All plants were grown under 8-hr SD.
[b] Explanations in Chapter 2, Section II.B.1.

From Nitsch, J. P., Kurtz, E. B., Jr., Liverman, J. L., and Went, F. W., *Am. J. Bot.*, 39, 32, 1952. With permission.

In various cereals, such as *Sorghum bicolor* and rice, low temperature induces male sterility. Sensitivity to chilling is greatest at the premeiotic stage in the former,[474] whereas in the latter it is mainly at microspore release from the tetrads and, to a lesser degree, at the premeiotic period.[475,476] Low temperature also induces stamen abortion in strawberry.[477]

High temperatures, too, were found to adversely affect male fertility in various species including maize, *Tradescantia*, tomato, *Viscaria* (= *Silene*) *candida*, and wheat.[211,478-481] Once again, premeiotic and/or meiotic stages are particularly sensitive. In wheat, the proportion of florets containing abnormal anthers is increased and pollen viability in normal anthers is reduced by exposing plants, during the sensitive stage, to 30°C for 1 day only.[481] Female fertility is also reduced by high temperature.

E. Chasmogamy and Cleistogamy

Temperature is a major factor controlling the development of either chasmogamous or cleistogamous flowers in *Viola* species. In *V. palustris* a low temperature treatment (4°C for 8 weeks) gives a short burst of chasmogamous flowers when the plants are moved to a 20/14°C (day/night) temperature regime in either SD or LD.[392] Under constant 20/14°C, only cleistogamous flowers are produced in LD, whereas a few plants may develop an occasional chasmogamous flower under SD. *V. silvestris* behaves in a similar way.[391] Chasmogamous flowers in *V. hirta* are only produced in SD after a cold treatment.[391]

F. Interaction of Temperature and Other Environmental Factors

1. Daylength

Temperature and daylength interact in their effects on flower development. *Chrysanthemum* was investigated in this respect by Cathey who found that in some cvs. the photoperiod required for the development of the capitulum is shortened by raising the temperature (Table 3).[482] Decreasing temperature also favors capitulum development in the summer-flowering cv. 'Golden Stardust' when grown under LD.[483] This chrysanthemum initiates inflorescences under continuous light at temperatures from 10 to 28°C, but development to anthesis occurs only on plants held between 10 and 16°C. At 22 and 28°C development is arrested at the receptacle stage.

In *Lemna paucicostata* 6746, induced by a single SD cycle at 25°C, abortion occurs with 2 days of continuous light at 25°C but, at 15 and 20°C, 3 to 5 days of continuous light are required before flower abortion occurs.[442] Flower abortion under noninductive daylength conditions can even be prevented entirely by further decreasing temperature. The critical

Table 3
CRITICAL DAYLENGTH FOR CAPITULUM DEVELOPMENT OF THREE DIFFERENT CVS. OF *CHRYSANTHEMUM MORIFOLIUM* AS INFLUENCED BY TEMPERATURE

Cv.	Minimum night temperature[a] (°C)	Critical daylength (hr)
'White Wonder'	10	13.75
	15.6	13.75
	26.7	12
'Encore'	10	13.75
	15.6	13
	26.7	12
'Snow'	10	12
	15.6	10
	26.7	9

[a] The day temperature was 2.8 to 5.6°C above the night temperature.

From Cathey, H. M., *Proc. Am. Soc. Hortic. Sci.*, 69, 485, 1957. With permssion.

temperature is a function of light intensity; at very low light level and low temperature conditions, new flower buds are initiated in continuous light.[441] In the SDP *Bougainvillea*, too, decreasing either night or day temperature favors flower development under LD as measured by the node number of first inflorescence. In SD, temperature has less influence.[147]

Low temperature can thus substitute, at least partly, for SD in chrysanthemum, *Lemna* and *Bougainvillea*. Similar interactions of daylength and temperature have been described for the control of sex expression in various cucurbits.[320]

2. Light Intensity

Temperature interactions with light intensity in flower development are shown clearly in tomato where buds readily develop into flowers over a very wide range of temperatures in summer but in winter, when light level is reduced, flower development is restricted to a comparatively narrow temperature range.[429] Lower temperatures are then required to prevent abortion, especially at the time of appearance of visible buds.[430] Data for *Iris* also show that the daily light flux requirement for normal flower development increases markedly with increased daily temperatures (Figure 4).[424] In *Lilium longiflorum*, too, reducing the temperature during a period of complete darkness reduces the percent of aborted flower buds.[425] In the LDP *Sinapis* low light under noninductive photoperiodic cycles may stop development of flower primordia at an early stage and induce inflorescence reversions (Volume III, Chapter 4, Section III,D). Bagnard has found that lowering the temperature prevents this effect of the low light.[370]

Low temperatures are generally detrimental to flower development of rose during early stages (see Section I.A.2 above). However, when the experiments are started early in the winter under poor light conditions, high temperatures increase rather than decrease the percentage of blind shoots. Experiments begun later in the spring, when daily light flux is greater, show the "normal" response, namely, a significant decrease in blind shoot formation with increased temperature.[460]

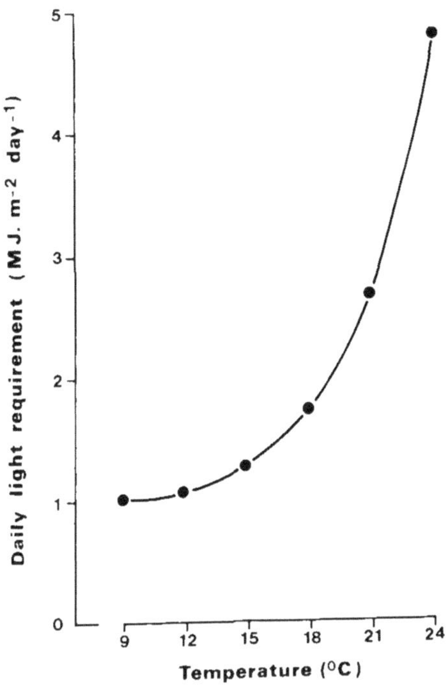

FIGURE 4. Daily light requirement for 50% anthesis of the first bud of *Iris hollandica* as a function of average daily temperature. Data averaged for several experiments in different years and with bulbs of different cvs. (From Fortanier, E. J. and Zevenbergen, A., *Neth. J. Agric. Sci.*, 21, 145, 1973. With permission.)

G. Breaking Flower Bud Dormancy

To produce rapid growth under favorable conditions, bud dormancy should be broken by exposure to cold treatment, though chilling is not always an absolute requirement for development to anthesis. Flowering in peach and tulip can proceed at a slow rate without exposure to temperature lower than 15°C.[171,484,485] In peach, however, resting flower buds whose growth has been interrupted during a protracted period of chilling (2.5 months at −1 to 0°C) ultimately overtake the development of buds kept above 15°C whose growth has never ceased.[171]

1. Effective Temperatures and Durations of Treatment

The threshold value below which temperature has a marked effect with respect to breaking flower bud dormancy varies among species and cvs. It is around 12°C for almond and below 9°C for apple buds.[165] In 'Desert Gold' and some other peach cvs. 6°C is most effective in breaking rest of flower buds, about twice as efficient as a 10°C treatment.[486,487] In 'Sungold' nectarine, a relatively low chilling cv., Gilreath and Buchanan have observed that the rate of resumption of bud growth is almost the same at 10 as at 7°C (Figure 5).[488] In *Azalea* 9 to 12°C is more efficient than 3 to 6°C.[489] The optimum temperature is around 5°C in the alpine plant, *Soldanella minima;* at 10°C flowering is poor and delayed.[170]

In bulbous plants the optimal temperature for maximum efficiency is known to differ according to species.[4,162,163] For tulip, Hartsema and co-workers found the optimum for early anthesis to be near 5°C, but flower damage resulted and 9°C was suggested.[4] In some studies the optimum is near 2°C,[490] and with the 'Apeldoorn' cv., cooling of dry bulbs to −1°C may

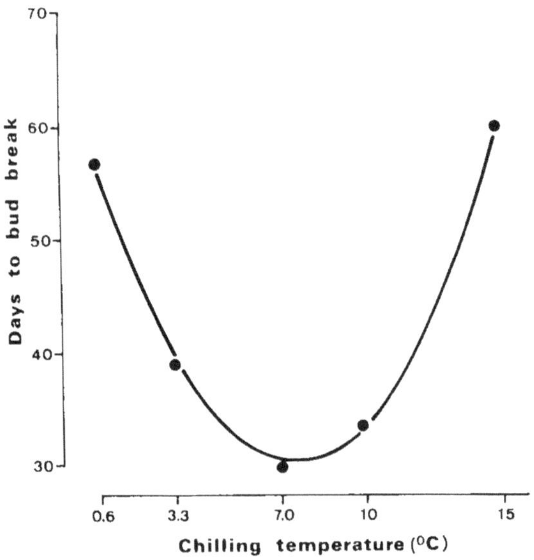

FIGURE 5. Dormancy breaking of reproductive buds of 'Sungold' nectarine (*Prunus persica*) as a function of chilling temperature. The duration of the chilling treatment is 750 hr. (From Gilreath, P. R. and Buchanan, D. W., *J. Am. Soc. Hortic. Sci.*, 106, 426, 1981. With permission.)

be advantageous.[491] With some other cvs., however, there is a tendency for higher percentages of blasting after cooling at very low temperatures. Hence, recommendations are a compromise between earliness and flower quality; the former, but not the latter, is improved by lower temperatures.[492] A technique known as "5°C forcing", "direct forcing",[162] or "special precooling"[163] has evolved mainly for cut tulip production. The main difference between standard cooling and special precooling is that in the first method all or part of the low temperature treatment is given after planting, while with special precooling the whole of the low temperature treatment is applied to the bulb before planting. Information for other species can be found elsewhere.[4,161,162,163]

The optimal length of the cold treatment required to terminate rest is termed the chilling requirement or cold requirement (CR). In woody plants it is usually expressed as the number of hours below a given base temperature, generally 7.2°C (45°F), necessary to complete rest. CR has been determined for a large number of species and cvs. It varies from 200 to more than 1000 hr in various peach cvs.,[165,493] as well as in various species of *Vaccinium*.[494,495] CR to break rest of both 'Bartlett' pear and 'Delicious' apple is considered to be about 1500 hr, but it is known to vary from year to year and from place to place.[496,497]

In bulbs the length of the cold treatment needed for most rapid growth has not been reported in detail except for tulips where the optimal length was shown to vary with each cv. For the earliest flowering of 'Apeldoorn', 'Paul Richter', and 'Rose Copland' cvs., CR is 159, 128, and 140 days at 9°C, respectively.[498] The minimum requirement for hyacinths for early forcing is 56 to 70 days and that for regular hyacinths 91 to 105 days.[163]

The number of days from completion of chilling to bud break is reduced in relation to the length of the chilling period.[488,499] In *Azalea* a cooling period beyond 6 weeks reduces the postchilling time (forcing period) required to reach full bloom, but the sum of the cooling and forcing periods is increased.[489] Cooling for 6 weeks or more also gives more flowers and more uniform anthesis than a 3 week chilling treatment. In 'Paul Richter' tulip the number of days from the end of cold to anthesis can be markedly reduced by increasing the duration of chilling (Figure 6).[500] However, Rees found that the median flowering date

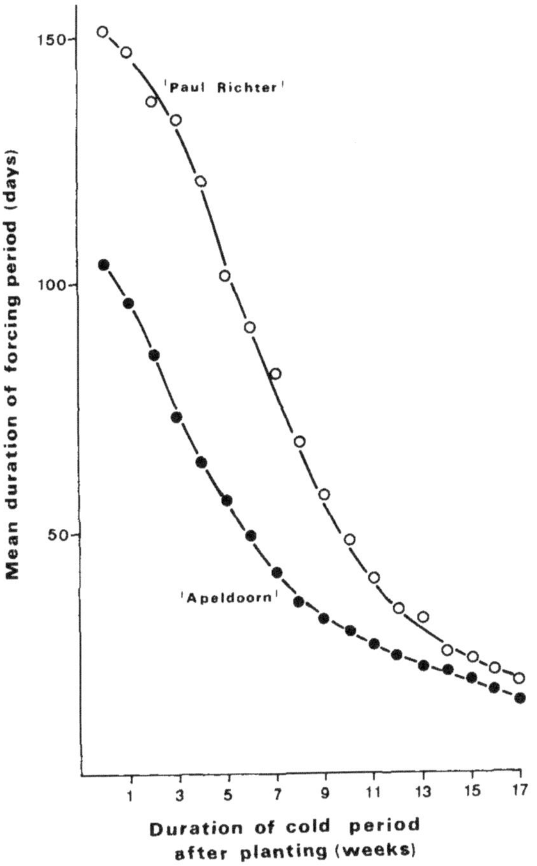

FIGURE 6. Duration of the forcing period in an 18°C glasshouse as a function of length of previous cold storage treatment in two tulip (*Tulipa gesneriana*) cvs. Before planting, bulbs are kept at 9°C for 6 weeks. After planting, they are kept outdoors during winter. The forcing period is the number of days from the end of cold treatment to anthesis. (From Rees, A. R., *J. Hortic. Sci.*, 44, 27, 1969. With permission.)

(reflecting the sum of the cooling and subsequent forcing periods) is earliest with 12 weeks of low temperature.[500] Variation in days to anthesis is reduced with longer cold periods, because during the reduced development at low temperature the slower developing buds tend to catch up.

Insufficient chilling adversely affects flower bud development in both bulbs and woody plants. It results in damage to the perianth of tulips; e.g., in 'Paul Richter' the percentage of harvested flowers with green-tipped tepals falls from 92 to 6 with increased chilling from 6 to 12 weeks.[500] White-tipped perianth in 'Apeldoorn' almost disappears when the cold period exceeds 14 weeks. Still, in 'Apeldoorn' up to half the bulbs fail to flower when given brief cold periods.[500] Failure of floral bud development also results from insufficient chilling in deciduous fruit trees. Abscission of buds or floral parts is frequent after warm winters in various stone and pome fruits including almond, apple, apricot, cherry, nectarine, peach, pear, and plum.[501,502]

In tulip the influence of increased duration of low temperatures on total stem length is also evident.[500] The longer the cold treatment, the longer the stem in 'Paul Richter' and 'Apeldoorn'. The internode between the highest leaf and the base of the flower appears relatively insensitive; in contrast, the lower internodes, and especially the middle portion of the stem, increase with increased chilling.

2. Perception of Low Temperature

We have not been able to find experiments in which only the dormant floral buds of woody plants are chilled. There is, however, some information suggesting that it is likely that the buds themselves are the sensitive organs. In *Syringa*, when only one of a pair of coterminal buds is treated with vapors of ethylene chlorhydrin, the treated bud, alone, resumes growth while the other bud remains dormant.[503] This is not to say that dormancy resides in the bud apical meristem; some experiments indicate a primary role of bud scales and perhaps other leaf primordia.[504]

3. Interactions of Chilling and Other Environmental Factors

The response to a chilling treatment may be affected by other climatic factors and these interactions have important practical consequences.

a. High Temperature

In mild winter locations chilling temperatures may be interrupted by unseasonably high temperatures. Also, bulbs that are given a low temperature treatment before planting are often subjected to higher temperatures during transport (''heating in transit'') or when offered for sale.

Rees demonstrated that in 'Apeldoorn' tulip, when a 150-day treatment at 9°C is interrupted after 6 weeks by a 30°C treatment for 1, 2, or 3 weeks, all flowers are killed.[505] Three weeks at 25°C damaged the anthers. After 1 week at 15, 20, or 25° C, anthesis is delayed by 0.5 to 3.0 days after 2 weeks by 4.5 to 8.0 days and after 3 weeks by 12.5 to 14.5 days. Since transfer to the glasshouse was delayed by 7, 14, or 21 days, depending on the duration of the high temperature interruption, the results indicate that warm storage does not nullify the effects of precooling, rather it reduces the duration of the period from the end of the chilling treatment to anthesis.

In peach floral bud break is retarded by brief periods of high temperatures during winter.[506] 'Sungold' nectarine plants which are given a 14-day interruption at 30°C in the middle of chilling periods varying from 450 to 750 hr also show some chilling negation, since they never reach a level of bud activity comparable to that of uninterrupted controls.[488] However, this effect is only evident in a fraction of the buds. When the interruption is given in the middle of a 350-hr chilling period, there is no reaction of the floral bud to the cold treatment. This result would substantiate the theory developed by Erez and Lavee for peach vegetative buds suggesting that a minimum amount of chilling must be received before a ''temperature-dependent fixation process'' can occur.[507] Thus, in 'Sungold' nectarine a 175-hr chilling period seems insufficient for the theorized fixation to occur.

b. Light Intensity

Generally, low light intensity during dormancy promotes subsequent floral bud growth in various woody plants. The value of reduced light intensity has been clearly demonstrated in evergreen azaleas where naturally cooled plants flower earliest when the light is reduced to 5% during cooling. When plants are artificially chilled at 11 and 15°C, light intensity of 20,000 lx prolongs rest and inhibits flower opening.[508]

Freeman and Martin stressed that light may operate in two different ways in affecting bud rest.[487] Light itself may be effective in a photochemical reaction but it could also act by radiant heating of the bud; temperature of dormant buds exposed to intense sunshine in California may be 11°C higher than that of the surrounding air. In peaches where bud temperatures were constantly monitored and, as needed, adjusted to within ± 1°C, high light intensity during dormancy reduces bud break indicating that light is operating by itself.[487] High light intensity is also known to inhibit floral bud opening.[509]

c. Water Availability

In 'Sungold' nectarine overhead sprinkling hastens bloom. The advance is due to water applied to the canopy and not to increased water in the root zone since sprinkled trees reach 50% bloom about 11 to 12 days earlier than control trees and water applied to the root does not affect bloom date.[510] The cooling effect of water applied to the canopy was believed to be responsible for the advance in bloom in nectarine. However, differences in bloom dates are greater than predicted by accumulation of chill units suggesting that other factors are involved in addition to temperature. Data collected with 'Desert Gold' peach support this view since flower bud growth in this cv. is promoted by mist during dormancy though temperature at bud level is constantly controlled and, as needed, adjusted to within ± 1°C.[487] Hence, misting must have an influence on bud rest beyond its effect on temperature.

In both 'Bartlett' pear and 'Starkrimson' apple, development of buds exposed to winter rain is enhanced in comparison to buds covered to keep any rainfall from contacting them.[511] Immersing branches in water for 1 or 2 days after a cold treatment insufficient to fulfill the CR also results in increased bud development (Figure 7). In various other species bud dormancy is released by immersing the buds or by dipping cut branches in a water bath at 30 to 40°C. This treatment, known as Molisch's method, forces the flower buds to reach anthesis earlier than normal,[164,165] and is of practical significance in the floricultural industry. Clearly the low temperature and reduced water availability (water stress) may both contribute to the breaking of dormancy; i.e., during periods of low temperature there is often water loss from shoot tissues that cannot be readily replaced by water absorbed by the root system (which is also at low temperature and consequently absorption is impaired). We suggest then that dormancy breaking, at least with the studies on hand, may be linked to some of the physiological processes that take place during stress hardening, where both chilling and water stress contribute significantly. Future research in the area of bud dormancy should attempt separation, if possible, of the low temperature and water potential factors.

II. WATER STRESS

Water status of the soil, though it markedly influences flower development both quantitatively and qualitatively, has not received much attention; so far, as expected owing to practical implications of water supply in agriculture, most drought studies have been made on crop plants.

In cereals a period of water deficit generally affects both the production of new spikelets by the meristem and the differentiation of the spikelets already initiated. Sensitivity of the apical meristem is particularly high and cessation of its activity is not uncommon after an episode of water stress (Figure 8A).[28,512] This inhibition of the meristematic activity may lead to a gap in the row of grains on the mature ear of barley.[28] The differentiation of existing primordia is usually less affected. Water stress reduces the rate of floral organogenesis of the lateral spikelet primordia of barley, but a slow persistent differentiation of these spikelets continues, even during severe stress (Figure 8B).[512] Relief of the stress leads to rapid development of the spikelets and results in an immediate, but transient, resumption of apical activity (Figure 8A, B).

In maize water stress imposed during tassel initiation and early development reduces both the growth rate and the mature size of that structure.[227] It induces a marked degree of sterility in the spikelets on the lateral branches although spikelets on the main axis are generally fertile. Male gametogenesis in barley and wheat is also particularly sensitive to water status of the soil.[513,514] Withholding water during the premeiotic and meiotic stages and immediately after meiosis induces pollen sterility in wheat, whereas female fertility is unaffected.[514]

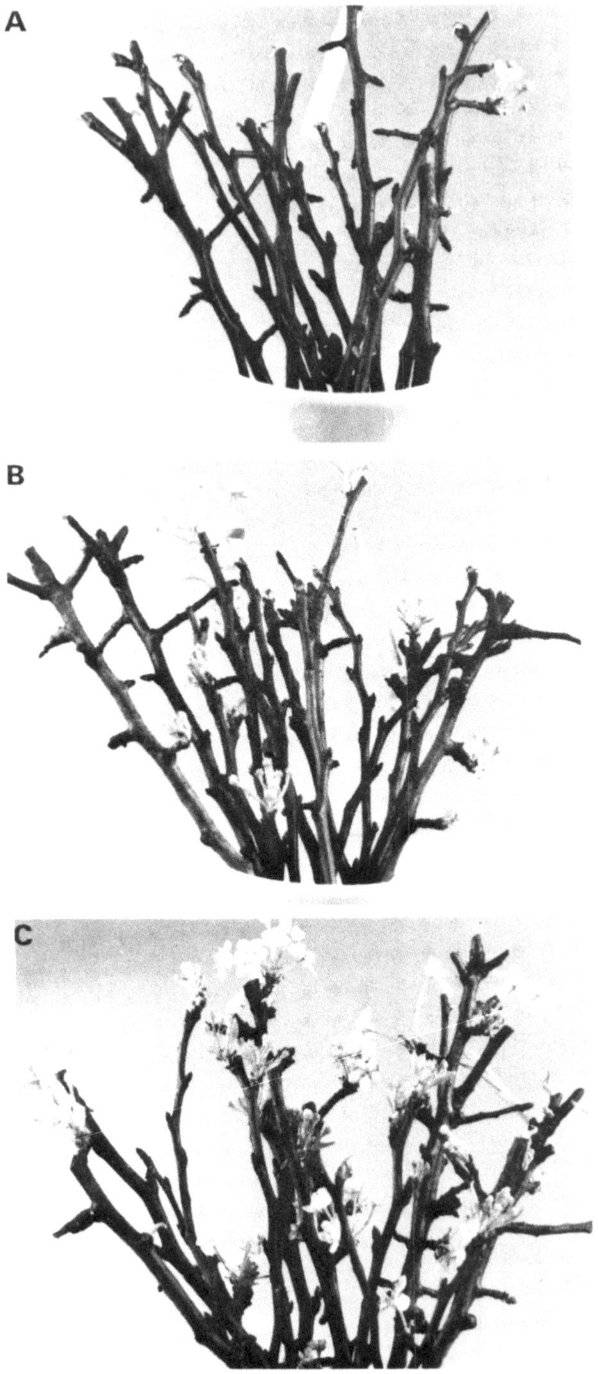

FIGURE 7. Effect of 1- or 2-days water soak of 'Bartlett' pear (*Pyrus communis*) cuttings after 650 hr chilling on reproductive bud break. Photographs after 30 days forcing at 20°C. (A) Control. (B,C) 1- and 2-days water soak, respectively. (From Westwood, M. N. and Bjornstad, H. O., *J. Am. Soc. Hortic. Sci.*, 103, 142, 1978. With permission.)

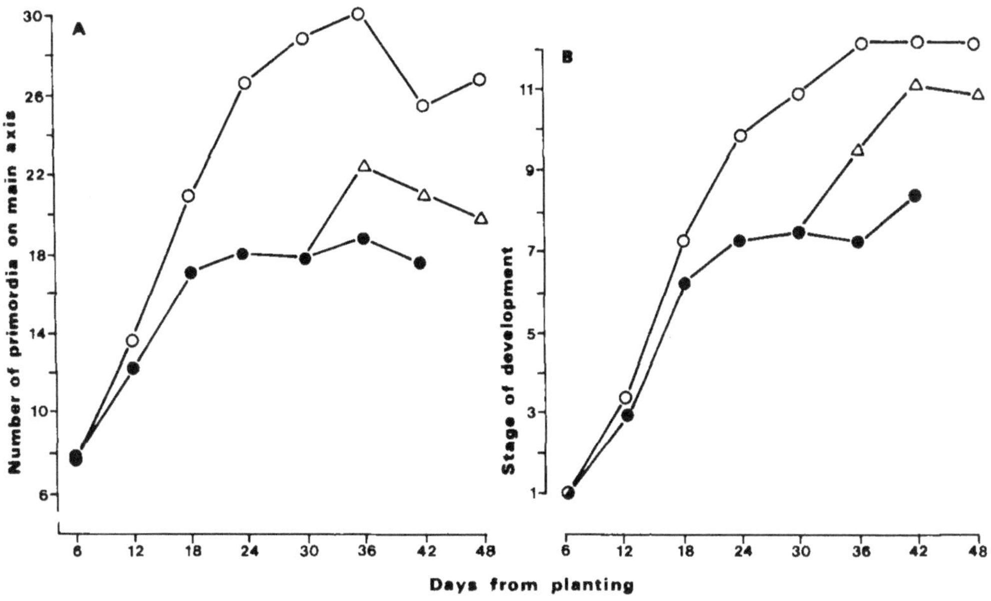

FIGURE 8. Effect of a prolonged water stress on (A) total number of primordia and (B) spikelet differentiation on main apical meristem of *Hordeum vulgare* cv. 'Prior A'. The developmental stages of the most advanced spikelet are assessed using the scale of Aspinall and Paleg (Chapter 2, Table 2). (○) Plants watered throughout growth. (●) Plants not watered after sowing (day 0). (△) Plants not watered between days 0 and 30. (From Husain, I. and Aspinall, D., *Ann. Bot. (London)*, 34, 393, 1970. With permission.)

In general high available soil moisture tends to promote femaleness,[111] and according to Moss and Downey male organs develop even within the axillary female inflorescences (cobs) of maize plants exposed to a water deficit 59 days after germination.[515] This stimulation of the growth of male flowers after a period of water stress imposed at a similar phase of growth does not appear in studies by Damptey and Aspinall with another maize cv.[227]

Water status of the soil may influence the balance between chasmogamous and cleistogamous flowers. In the grass *Stipa leucotricha,* the lower the soil water, the higher the percentage of cleistogamy.[516]

Much evidence points to the importance of soil moisture for flower bud dormancy release in several tropical plants. In the orchid *Dendrobium crumenatum* as in various species of *Coffea* and of the geophytic bulbs *Zephyranthes* and *Pancratium*, gregarious anthesis usually occurs a few days after rain.[517-519] In these plants flower buds generally develop to a definite stage before stopping growth, and rainfall starts the resting buds growing again at the same time so that the plants (and branches on plants) flower simultaneously.

In *Coffea arabica* flower buds stop growing at a premeiotic stage, when the corolla has reached a length of about 4 mm.[172] In *Pancratium* the critical stage corresponds with pollen formation; only flowers with pollen emerge while the remaining unemerged flower buds have only pollen mother cells in their anthers.[519]

Flower development in *Pancratium* is inhibited if the soil is kept continuously wet by regular watering.[519] Similarly, flowers of regularly watered 'Arabica' and 'Robusta' coffee plants develop to premeiosis, but fail to open and senesce 4 to 6 months later.[518,520,521] In some cvs., however, flowers develop and open but they are small and sterile:[520] these are the so-called "starflowers".

Normal flower development to anthesis in coffee plants is stimulated by alternate periods of drying and watering.[521,522] Rain and submergence of the buds in water are also efficient

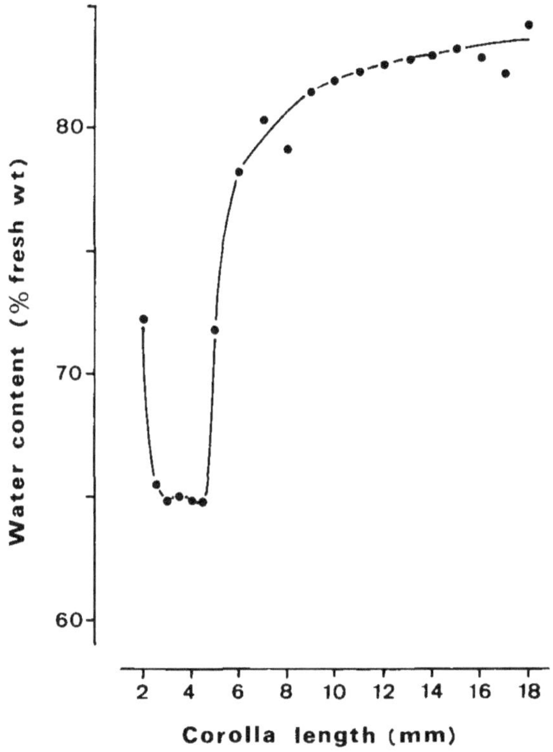

FIGURE 9. Water content of developing flower buds of *Coffea arabica*. Flower buds of different sizes were picked over a period of about a week from a tree which was growing at 30/24°C (day/night); the top of the tree was submerged in water in order to break flower bud dormancy. (Adapted from Mes, M. G., *Port. Acta Biol. Sér. A*, 5, 25, 1958.)

ways of breaking dormancy.[523,524] Hence, it is not only the availability of large amounts of water which permits normal flower development, but a prior water stress also appears to be a critical factor. This idea is supported by the observation that coffee cuttings, before or after the dormancy period, open normal flowers some time after being placed in the multiplication bed.[520] It is well known that cuttings are always subjected to a water stress until new roots are formed. In this respect, the finding of Mes that the water content of coffee flower buds fluctuates during development (Figure 9) is also of interest.[172] On trees whose top has been submerged in water in order to break flower bud dormancy, buds which are 4.5-mm long — the stage at which growth normally stops — have the lowest water content. There is a sudden increase in water content of buds which are 5- to 7-mm long; this increase is coincident with meiosis and microspore formation.

Went has suggested that it is not the rain, but the drop in temperature during and after the rain that may break dormancy in the flower buds of coffee plants.[518] Flowering occurs indeed within 13 days in trees regularly watered but transferred from alternating temperatures of 30/24 to 23/17°C.[518,523] However, branches of stressed coffee trees submerged for 1 hr in water at 35°C flower better than branches treated at 25°C and immersion in water at 15°C has almost no effect.[524] In *Pancratium* and *Zephyranthes* the incidence of flowering after a dry period is unaffected by the temperature used (21 and 34°C) for watering the plants.[519]

The factor involved in dormancy release could thus vary depending on growing conditions. Recall that in temperate climates a drop in the water content of the buds is associated with low temperatures that are able to break dormancy and that rainfall or water soaking by immersing branches in warm water stimulates flower bud development. Thus low temperature and water status of the plant contribute to breaking flower bud dormancy in both temperate and tropical species, but the interaction is not well understood.

CONCLUSIONS

Temperature has both qualitative and quantitative effects upon flower development. High temperatures generally increase rate of flower development resulting in earlier anthesis but smaller reproductive structures. They also induce flower abortion in several species. The mean temperature, rather than alternation of day and night temperature, seems most important. The effect of air temperature is usually greater than that of soil temperature. Low temperature also induces abortion in some species. Here, both air and soil temperatures are generally active.

The number of floral parts and sex expression are markedly affected by temperature. Most often low temperature promotes femaleness in dioecious and monoecious species. In hermaphroditic plants male fertility is frequently impaired by extremely low or high temperatures.

Temperature strongly interacts with light conditions. In some photoperiodic plants a low temperature treatment may override the daylength requirement for flower development. Low temperatures also counteract the detrimental effects of low light flux in various species suggesting that assimilate supply, redistribution, and utilization may be essential factors in mediating temperature effects.

Water stress reduces growth rate of florets and inflorescences and induces male sterility in cereals.

Both low temperature and reduced water availability play a major role in breaking flower bud dormancy in temperate and tropical species. Since low temperatures are usually accompanied by a drop in the water content of the tissue, water stress could be viewed as the critical factor. Further studies are required to solve this problem.

Chapter 6

BASIC ROLE OF CORRELATIVE INFLUENCES

TABLE OF CONTENTS

I. INTRODUCTION

Correlative influences have been shown to affect markedly transition from vegetative to floral condition (Volume II, Chapter 1). We cannot imagine that their action ceases during reproductive development. In fact, their complexity during this developmental process should be amplified since, besides influences of roots, leaves, stems, axillary shoots, apical dominance, etc., strong interactions should exist among the different flowers of an inflorescence and the various floral organs of a flower. In the first part of this chapter we will attempt to isolate specific studies showing these varied influences on flower and inflorescence development. For the most part, they depend upon removal of a given organ or varying quantitatively the ratio between various organs. Root pruning, decapitation, defoliation, vegetative axillary bud removal, etc., are techniques that have been used.

In an attempt to determine whether or not correlative influences are basic to reproductive development and to simplify the experimental systems, studies with isolated inflorescences, flowers and floral organs were performed. These in vitro investigations, reviewed in the second part of this chapter, have frequently been combined with microsurgery techniques to further explore the morphogenetic features within the reproductive meristems.

II. CORRELATIVE INFLUENCES IN INTACT PLANTS

A. Role of the Root System

The presence of the roots does not seem to be an absolute prerequisite for reproductive development in some species. Chailakhyan and associates and Viana found that removal of the root system in *Bryophyllum daigremontianum* has a slight promotive effect on the rate of inflorescence development with no apparent effect upon inflorescence structure.[525,526] Near-normal inflorescence development also occurs in derooted *Perilla* and *Phaseolus* plants and in derooted explants of *Chenopodium* and *Scrophularia* grown in vitro.[527,528] In *Brassica crenata*, inflorescence development is merely delayed a few days in the absence of roots.[527]

In other experimental systems, however, roots were shown to play some role in flower or inflorescence development. In *Carex* continuous removal of roots from shoots grown in vitro results in a reduction in inflorescence branching (fewer lateral spikes) which is accompanied by an increase in length of the terminal spike.[529] However, the major influence of root removal in this species is on reduced inflorescence initiation, not just development alone. In woody cuttings of grapevine growth of inflorescences usually ceases after bud burst, and they shrivel and die. Failure of inflorescence growth appears to be related to the absence of a root system at planting since, in prerooted cuttings, inflorescences are retained and grow normally.[219,530,531] Short roots (2 mm) are as effective as long roots (5 cm) in preventing atrophy of inflorescences.[531]

Mullins has suggested that some factors originating in the roots are required for normal inflorescence development.[531] A similar idea arises from the results of Chailakhyan and Khryanin who observed that root removal in the dioecious species, hemp and spinach, leads to a significant increase in the ratio of male to female plants (Table 1).[532,533] In grapevine, as in hemp and spinach, cytokinins substitute for roots.[531-533]

Another result which is suggestive of a role for roots in reproductive development was obtained with tomato where flower number in the inflorescence is increased by reducing root temperature. As demonstrated by Phatak and associates, this temperature effect is graft transmissible.[534] Also, in tomato, the effect on inflorescence development of a transfer from small pots to large pots under low light conditions is time dependent.[535] A transfer at the time of, or shortly after, inflorescence initiation is particularly detrimental suggesting that root formation and/or growth and early inflorescence development are antagonistic events in light conditions reducing assimilate supply; later transfers promote development. Cooper

Table 1
THE ROLE OF THE ROOT SYSTEM IN SEX EXPRESSION OF TWO DIOECIOUS SPECIES, *CANNABIS SATIVA* AND *SPINACIA OLERACEA*

	Cannabis sativa		*Spinacia oleracea*	
Treatment	**Male (%)**	**Female (%)**	**Male (%)**	**Female (%)**
Control with secondary roots	10.5	89.5	15	85
Secondary roots continuously excised	87.5	12.5	85	15

Note: All plants are derooted at the seedling stage. In control plants, secondary roots are allowed to develop. In the minus root batches, secondary roots are continuously removed. All plants are grown on a nutrient solution.

Adapted from Chailakhyan, M. Kh. and Khryanin, V. N., *Planta*, 138, 185, 1978; 142, 207, 1978. With permission.

and Hurd have reported, too, that delaying planting in tomato beyond the stage of visible inflorescence primordium reduces the abortion of this inflorescence.[536] The root restriction imposed by a small pot and the consequent reduction of the competition, within the plant, for the available assimilates is suggested as being responsible for this favorable effect. Under high light conditions, which promote inflorescence development, the dry weight accumulation in the root system relative to total plant dry weight is, however, far greater than under unfavorable light conditions which lead to abortion.[202] Thus, the relationship between root and inflorescence development is not simple and is affected by the environment.

B. Role of Leaves

The influence of leaves upon reproductive development is complex and strongly dependent on the physiological age of these organs. Mature leaves are usually required and their removal inhibits development,[537-539] a result expected if photosynthetic assimilate is a limiting factor (see Chapter 9).

Excision of young, expanding leaves promotes inflorescence growth in some species. This is clearly shown in Mullins' experiments with grapevine where, in leafy cuttings, the flowers fail to emerge from their enclosing bracts and defoliation promotes rapid growth.[540] Defoliation also greatly promotes early stages of floret development in *Bougainvillea* and, in tomato, it prevents abortion induced by low light conditions and increases the number of flowers in the inflorescence.[541,542] In the latter species, removal of leaves produced just before the inflorescence is most effective, i.e., leaves which develop at the same time as the inflorescence appear most inhibitory.[542]

Apparently in *Bougainvillea* the presence of young leaves accounts for the high FR light requirement for inflorescence development. When the young leaves are removed inflorescence development proceeds in high irradiance regime provided by mainly fluorescent light whereas, when the young leaves are present, incandescent (FR) irradiance must be increased giving a much higher FR/R ratio.[406]

The young leaves have been shown by Sachs and colleagues to act as competing sinks for assimilates, but at the same time they appear to serve as centers of growth substance production. In *Bougainvillea* there is a clear reduction in stem elongation (concomitant with advanced inflorescence development) with removal of leaves less than 0.1 fully expanded. Auxin applications do not restore stem elongation nor inhibit inflorescence development in the defoliated plants, whereas GA_3 treatments do both.[420] In tomato the increase in flower number is not observed in defoliated plants treated with GA_3,[543] suggesting that, in this species as in *Bougainvillea*, the inhibitory effect of young developing leaves may be, at least partly, through production of endogenous GAs.

An interesting case is that of 'Baccara' rose where Zieslin and Halevy have investigated potential source and sink roles for mature and young leaves in flower development.[538] Removal of mature leaves may increase blindness nearly 4-fold (10 to 15% up to 50%). Excision of young expanding leaves only has little effect on blindness compared to controls. However, if both the young and mature leaves are removed, blindness increases to above 90%. The authors suggest that in this case the young leaves are required to promote sinkness in the developing terminal bud.

In cucumber mature leaf removal enhances female sex expression but young leaf excision stimulates maleness,[544] indicating once again that leaf influence upon floral development varies with physiological age.

C. Role of Apical Dominance and Axillary Shoots

Several studies suggest that the apical bud controls development of axillary flowers or inflorescences in the same fashion that it controls development of axillary shoots. As a result apical dominance may have not only direct, but also indirect effects upon reproductive development since axillary shoots frequently compete for available assimilates with reproductive structures.

1. Apical Dominance

In various species, decapitation promotes flower or inflorescence development. For example, removal of the main flowering shoot of *Impatiens* enables inhibited axillary flower buds to develop to anthesis especially when SD induction is not optimal.[545] Similarly, decapitation slightly stimulates inflorescence development in tomato plants transferred under low light conditions from the 6th to the 12th day after the macroscopic appearance of the inflorescence.[546] In defoliated cuttings of grapevine, apical bud removal promotes inflorescence development, too.[540] However, when the leaves are present, the inflorescence is strongly suppressed whether or not the apical bud is removed. In this case, the primary role resides in the leaves and the apical bud appears to be a weak competitor. Studies by Yeung and Peterson demonstrate that in *Hieracium* the terminal meristem and buds in the axils of the upper five or six bracts always develop into inflorescences.[547] The buds in the axils of the next few bracts are usually determined as reproductive buds, but remain in an undeveloped prefloral state except when the upper developing inflorescences are removed. This case bears some resemblance to that described for *Geum* in Volume II, Chapter 1, Section II.A, and clearly shows that apical dominance may control reproductive development of axillary meristems at very dissimilar stages.

2. Axillary Shoots

Continuous removal of all lateral or secondary shoots on a *Chrysanthemum* plant allows the terminal capitulum of the main shoot to complete its development under unfavorable LD.[153] Similarly, in the LDP *Sinapis* degeneration of the apical meristem of the raceme under SD given at a low light intensity may be prevented by excising the lateral shoots.[402] Studies by Ramina and co-workers show that inflorescence development in *Bougainvillea*, whether expressed as node to first flower or number of inflorescences developing, is an inverse function of branch number.[539] They also found that percent dry weight and soluble

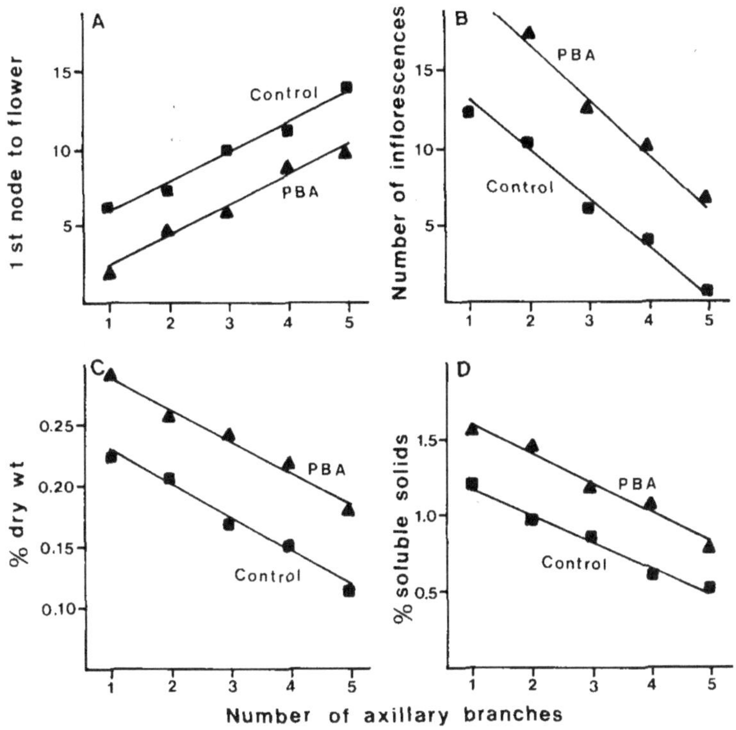

FIGURE 1. Regression lines for relationships between number of axillary branches and inflorescence development in *Bougainvillea* 'San Diego Red'. Results are expressed as first node to flower (A) or number of inflorescences developing (B), dry weight gain (C), and soluble solids (D) in the shoot apical region. PBA at 500 mg/ℓ was applied every 4 days to shoot tips in 10-μℓ droplets. (From Ramina, A., Hackett, W. P., and Sachs, R. M., *Plant Physiol.*, 64, 810, 1979. With permission.)

solids increase concomitantly with increased inflorescence development and thus in inverse ratio to branch number (Figure 1). A PBA treatment on shoot tips reduces the detrimental effect of the axillary branches. The dry weight of carnation flowers at anthesis is also increased by removing floral lateral shoots.[201] Competition for assimilates thus possibly accounts for these observations. The greater the number of axillary shoots competing with the reproductive structures the greater the impact on development.

D. Evidence Coming from Grafting Experiments

Although studies with *Mercurialis, Cannabis,* and *Melandrium* fail to show graft-induced modifications of sex expression,[111] graft experiments with muskmelon and cucumber repeatedly show sexual alteration of scions grafted on stocks of a different sexual type. Thus, grafting of gynomonoecious muskmelon scions to andromonoecious seedlings of the same species or of pumpkin greatly increases the number of male flowers in the scion (Table 2). These flowers occur at an earlier node as compared to the spontaneous occurrence of male flowers in ungrafted control plants.[548] Gynoecious muskmelon scions grafted on monoecious rootstocks were also found to develop hermaphroditic flowers instead of the normal female one.[549] In cucumber the stock, especially when it is shootless, causes a considerable change in the flowering pattern of the scion; gynoecious stocks increase the female tendency of monoecious scions whereas monoecious stocks raise the node number of the first female flower and/or cause the development of male flowers in gynoecious scions.[222]

Obviously, unknown sex regulating substances flow from the stock to the scion and the

Table 2
EFFECT OF GRAFTING GYNOMONOECIOUS MUSKMELON (*CUCUMIS MELO*) SCIONS TO ANDROMONOECIOUS SEEDLINGS OF THE SAME SPECIES ON THE INDUCTION OF MALE FLOWERS IN THE SCIONS

Treatment	Male flowers (%)	Position of the first node bearing a male flower
Ungrafted control	0.7	40
Control grafted to gynomonoecious muskmelon	0.5	38
Grafted to andromonoecious muskmelon	31	18

Adapted from Mockaitis, J. M. and Kivilaan, A., *Nature (London)*, 202, 216, 1964.

fate of a given flower primordium is dependent on messages coming from the root as well as other plant parts.

E. Flowering Gradients

Flowering gradients, i.e., changes in the expression of the reproductive condition along the axis of a plant, have mainly held the attention of workers interested in flower initiation (Volume II, Chapter 1, Sections II.B.5 and III.B.2). However, as far as flower development is concerned, the occurrence of similar gradients can also be demonstrated. In several cucurbits the nodal sequence of flower types shown in Chapter 1, Figure 14A reveals clearly that the ability of a bisexual flower primordium to develop either stamens or a pistil changes as a function of its position in the vine. Nonrandom distribution of sexes in numerous other monoecious species probably also reflects the existence of some sexual gradient. Such a gradient has been found in maize by Raman and associates.[550] They showed that different buds along the stem possess different capacities to develop into normal plants when rooted and transplanted. Lowermost buds produce, for the most part, plants with normal male tassels and lateral female inflorescences. Buds from higher nodes produce, in culture, plants in which the terminal inflorescence contains both male and female flowers. Uppermost nodes (no axillary buds were derived from nodes above the tenth one) produce a totally female terminal inflorescence.

In *Bougainvillea*, flower initiation occurs at each node, but the inflorescences of the lower nodes generally fail to develop to anthesis.[231] Similarly, bud position in rose strongly affects flower development as shown by varying pruning position.[460,551,552] Flowers formed on higher shoots on the parent stem generally develop more rapidly and are larger than flowers arising from lower buds. Severe pruning may also result in blindness.

Lord demonstrated that cleistogamy in *Lamium amplexicaule* must be explained in the context of the whole inflorescence development since there are floral variations as a function of floral position in the developing inflorescence.[119] From lower to upper flowering nodes there is a change from cleistogamous to pseudocleistogamous and then chasmogamous flowers.

The occurrence of these flowering gradients is probably one expression of the varied correlative influences operating in a plant which determine gradients of different chemicals (growth regulators, assimilates, etc.).

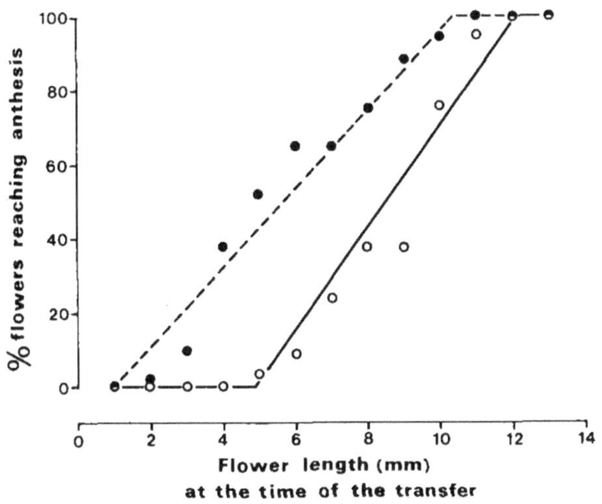

FIGURE 2. The effect of the first flower reaching anthesis in the first inflorescence of *Lycopersicon esculentum* 'King plus' on the development of the other flowers of this inflorescence under low light conditions. At different times after the macroscopic appearance of the first inflorescence, tomato plants are transferred from favorable light conditions (16-hr LD, 20 W/m^2) to low light conditions (8-hr SD, 12 W/m^2). All flowers are measured at the time of the transfer. At the end of the experiment, they are distributed among two groups: (○—○) flowers which are not preceded by one which reaches anthesis (including the first flower of the inflorescence) and (●---●) flowers which are preceded by one which reach anthesis. In both groups, the percent of flowers which reach anthesis, as a function of their size at the time of transfer to low light conditions, is recorded.

F. Correlations in the Reproductive Structures

1. In the Inflorescence

In *Cleome spinosa* inflorescence fruited zones alternate with sterile zones which result from the suppression of pistil development.[293] Removal of young fruits decreases female abortion.[273,537] De Jong and Bruinsma showed that the inhibitory factor of the fruits originates from the seeds rather than from the carpels. Removal of ovaries or ovules at anthesis prevents female abortion almost completely. These effects are commonly attributed to a reduced internal competition for assimilates or growth substances between reproductive sinks.

Experiments of Kinet with tomato also clearly indicate that there are strong interactions between flowers during the development of the inflorescence.[433] By transferring plants from high light to low light conditions at different times after the macroscopic appearance of the inflorescence, it appears that the normal development of a flower (regardless of its position in the inflorescence) is not only dependent upon its size (i.e., developmental stage) at the time of the transfer, but also on the fate of the preceding flowers in the inflorescence (Figure 2). Flowers, which are not preceded in the inflorescence by a flower that reaches anthesis, will not reach anthesis if they are not at least 5 mm long at the time of the transfer. In contrast, flowers, which are preceded by one that develops to anthesis, have a greater chance to reach anthesis, even when less than 5 mm long at the time of the transfer. Allometric comparisons of petals, stamens (anther + filament), and style to sepals in tomato inflorescences, before and after anthesis of the first flower, also demonstrate the strong influence of the first flower upon the development of the other flowers of the inflorescence. As shown in Figure 3, there is a strong linear relationship between sepal length and petal, stamen, and style length. However, the relative growth relationships of the three latter organs to that of

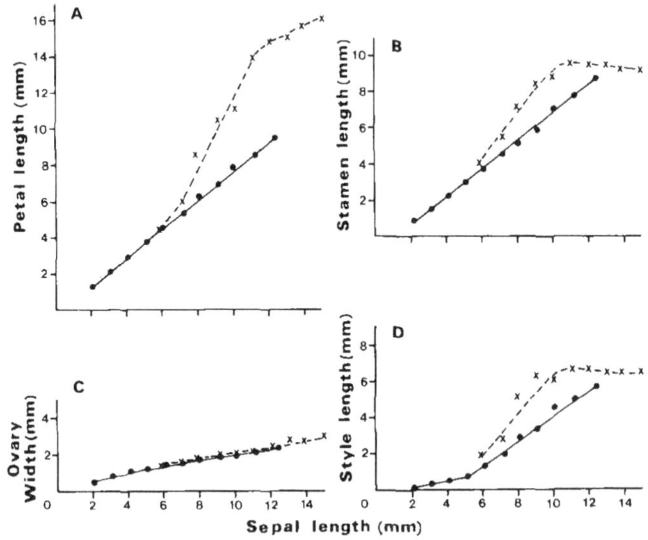

FIGURE 3. Comparative allometric growth plots of petal, stamen, ovary, style, and sepal dimensions in flowers of *Lycopersicon esculentum* 'King plus' before (●—●) and after (x—x) first anthesis. (A) Petal length vs. sepal length. (B) Stamen length vs. sepal length. (C) Ovary width vs. sepal length. (D) Style length vs. sepal length.

the sepals change in all developing flowers of the inflorescence after the first flower has reached anthesis. After the first anthesis, all subsequent flowers develop more rapidly and a tendency towards synchronous floral development becomes obvious (see discussion in Chapter 1, Section IV.B).

The mechanism involved in the stimulation of flower development in tomato at first anthesis is still unknown. Two hypotheses may account for this observation. First, at anthesis, a flower becomes a better sink which drains nutritive factors and/or growth regulators in the inflorescence in such quantities that all other flowers which develop in the same inflorescence benefit by this increased supply. Second, the open flower releases promotive factors which stimulate the development of the other flowers.

Another interesting case is that of cereals such as wheat and barley where first floret initiation correlates with a number of events (reported in Chapter 1, Section IV.C). In barley cessation of spikelet initiation occurs concurrently with stamen initiation in the most advanced spikelet.[28,335] Kirby and Faris and Nicholls believe that this cessation of initiation could be caused by a lack of nutrients resulting from increased demand for assimilates by the growing spikelets.[145,553] An alternative interpretation was proposed by Cottrell and associates who suggest that at the time of stamen initiation in the most advanced spikelet an inhibitor is released which passes up the spike where it prevents initiation.[554] This hypothesis is supported by the observation that GA_3 application to barley plants grown under 8-hr SD accelerates spikelet development resulting in earlier stamen initiation and cessation of spikelet formation. However, meristem growth continues indicating that metabolite supply to the apical dome cannot be completely restricted. Later on, renewed spikelet production may occur during an additional phase of initiation, distant from the previously formed spikelets, an observation which is consistent with escape from an inhibitor emanating from below.[554]

In *Begonia*, where female flowers differentiate only after production of a variable number of male flowers (Chapter 1, Figure 14B to D), Berghoef and Bruinsma have also found that female flower buds are not produced directly on a callus formed on inflorescence pedicels cultured in vitro.[244,555] Male flowers are produced first, whereas female flowers always originate in the axils of bracts of male flower pedicels and never start to differentiate before initiation of stamen primordia.

All the above results thus indicate clearly that there are varied interactions between the various parts of an inflorescence. Both promotive and inhibitory influences have been detected. Their nature remains, however, to be elucidated.

2. *In the Flower*

Strong correlations between the development of certain flower parts have been reported repeatedly. Allometric studies reveal how growth processes of the various floral organs are precisely integrated to form a complete and characteristic flower.[134] They also allow the detection of critical points during development, when differences between the relative growth rates of these organs arise. In *Nigella*, e.g., Greyson and Sawhney focus attention upon initiation of floral organs.[129] They show that petals and stamens grow at different absolute growth rates and that this difference is apparent very early, suggesting that petal primordia (50 to 100 cell estimate) possess some organ-specific attributes not shared by stamens. The difference between the stamens of a normal tomato plant and a stamenless mutant also relates to some difference at initiation.[129] In contrast, cleistogamous and chasmogamous flowers of *Lamium amplexicaule* are indistinguishable at inception and during early organogenesis. Divergence does not occur until the late stages of anther differentiation prior to meiosis.[556]

The allometric plots of the length of the various floral organs of tomato versus sepal length presented in Figure 3 reveal that, at the time of anthesis of the first flower of the inflorescence, the relative growth relationships of petals, stamens, and styles (all deciduous organs with determinate growth) to sepals change, in a similar way, in the other flowers of the truss. On the contrary, the relative growth rate of the ovary to the sepals remains constant. This result strongly supports the idea that there are some particular correlations between sepals and ovary, on the one hand, and between petals, stamens, and style on the other hand.

Nitsch reported in 1965 that in some species ovary development is positively related to sepal development, whereas petal development is linked to stamen development.[320] He also stressed that factors which favor one pair of organs tend to depress the growth of the other pair. For example, J. and Y. Heslop-Harrison found that, in *Silene pendula ruberrima*, LD and NAA applications cause male sterility and reduce corolla development, whereas ovary growth is strongly stimulated.[389,557]

In *Bryophyllum daigremontianum*, low light intensity causes stamen abortion and reduces petal development, with no concurrent increase of gynoecium size (which is also reduced in extreme cases).[558] Hence, there is no negative correlation between gynoecium and androecium in this species, but Resende suggested that there is a floral gradient with progressive regression of the flower from a "floral" to a "vegetative" type after appropriate environmental or chemical treatments.[154,558] The floral organs are suppressed in the following sequence: stamens, petals, ovary, and finally the sepals which are frequently retained, with or without change in their morphology. A similar gradient was found in *Hyoscyamus* and in *Bryophyllum crenatum*.[559,560] It is noteworthy, however, that in the latter species a kinetin treatment which inhibits petal and stamen development may increase ovule number in the carpels, suggesting some antagonism between carpels and petals + stamens.[560]

In *Citrus* the rates of fresh and dry matter accumulation in the different floral organs suggest that the ovary, on the one hand, and the petals and stamens, on the other hand, represent two quite different types of growth.[561] Wounding a single petal or, to a lesser degree, application of NAA to one petal was shown by Goldschmidt and Huberman to change the normal pattern of [^{14}C] sucrose distribution in the flower (Table 3).[561] In response to either treatment, radioactivity increases in all the petals indicating that they are somehow connected with each other. The radioactivity concurrently decreases in the stamens (with little effect upon ovaries). These results suggest that there is an intimate relationship between

Table 3
DISTRIBUTION OF LABEL FROM [^{14}C] SUCROSE IN FLOWER PARTS OF CONTROL AND WOUNDED OR NAA-TREATED FLOWERS OF *CITRUS SINENSIS* 'SHAMOUTI' (dpm × 10^{-3}/0.1 g FRESH WT)

Treatment	Wounded[a] or NAA-treated[b] petal	Nontreated petals	Stamens	Ovary
Intact control	—	4.3	16.6	6.4
Wounded	44.0	8.1	5.2	4.8
Lanolin control	3.2	3.3	8.5	3.9
NAA	6.2	5.6	5.7	5.5

[a] Wounding was performed by crosswise scratching with a scalpel.
[b] Lanolin paste containing 1% NAA.

From Goldschmidt, E. E. and Huberman, M., *J. Exp. Bot.*, 25, 534, 1974. With permission.

petals and stamens since they seem to draw upon a common pool and show similarity in growth behavior.

Lord reported evidence for developmental coordination between the stamens and corolla in numerous other species, where reduction of corolla size accompanies male sterility and/or androecium suppression.[118] Emasculation also results in an inhibition of corolla elongation in *Glechoma hederacea*.[562]

For Lloyd and Webb, the most frequently mentioned differences between male and female flowers (secondary sex differences) are in petal size: the corolla is usually smaller in female than in hermaphroditic or male flowers.[117] This is not a universal rule since numerous exceptions are known.

Ontogenetic studies reviewed in Chapter 1 do not shed much light on this matter. Recall, however, that in some species, common primordia occur (Chapter 1, Figure 12), each differentiating into an adaxial stamen primordium and an abaxial petal primordium suggesting, once again, that these types of floral organs may share some specific features.

CONCLUSIONS

The development of a reproductive structure is under the influence of signals coming from all plant parts and interacting in a complex manner. Besides apical dominance and influences of the vegetative organs of the plant (root system, leaves, axillary shoots), strong correlations within inflorescences and flowers have been observed. Flowers and fruits developing in an inflorescence may markedly alter the fate of a younger flower. Interactions between the various floral organs also occur. Thus there are long-distance interactions between reproductive structures and vegetative organs. They are revealed in several species by a change in floral expression along the stem (flowering gradient) and by grafting experiments. There are also short-distance interactions especially inside the inflorescence and flower primordia.

Both inhibitory and stimulatory influences have been described. They may be interpreted in terms of either competition between various organs for a similar metabolite pool or production of inhibitors or promoters by some organs. Assimilate partitioning and growth substance distribution are generally viewed as essential factors governing the establishment of the complex interactions existing in a plant. It is not easy to separate the respective roles of these two processes which largely overlap (discussed in Chapter 9).

III. CORRELATIVE INFLUENCES IN CULTURED PLANT PARTS

A. Culture of Excised Reproductive Primordia

In vitro culture of young excised reproductive primordia in the last two decades has addressed the question of dependence of these isolated structures on other plant parts for their normal development. Three types of primordia have mainly been investigated: (1) young inflorescences, (2) flower buds, and (3) floral organs.

1. Inflorescences

Inflorescences of various species have been successfully grown in vitro to anthesis, usually only when the growing medium is supplemented with some growth substances (GAs or cytokinins).[563-566] This result suggests that these inflorescences are unable to produce their own growth substance supply and are thus dependent on other plant organs for this supply.

2. Flowers

The first reports of successful culture of flower bud primordia in vitro appeared in the early 1960s, Galun and co-workers grew cucumber flower buds collected at the bisexual stage (i.e., at a rather advanced stage when all floral organs were initiated) to anthesis,[109,207] and Tepfer and associates obtained some development with *Aquilegia formosa*, but buds, taken at an early stage, do not reach anthesis.[208,209,567] Furthermore, buds do not grow unless sepals are removed. Thus, in vivo the inhibitory influence of the sepals is negated by some as yet unknown mechanism. Later, several other species including *Viscaria candida* and *V. cardinalis*, *Anagallis*, muskmelon, tobacco, *Nigella*, *Cleome iberidella*, and *Browallia* were grown in culture.[210-216,297,568,569] Several unsuccessful attempts have also been reported.[570-573]

At the outset, distinction should be made between initiation and development of floral organs, two processes which appear to have different requirements in vitro.

The normal sequence of initiation of all floral organs generally occurs in young flower buds excised at the sepal or sepal plus petal primordium stage. In tobacco and in a *Nigella* genotype with "single" flowers these initiation processes are independent of exogenous growth substances suggesting that the floral apices may well be autonomous in this regard.[215,216] However, the required growth substances may have been present endogenously before excision and derived from other plant parts.[574] In a *Nigella* genotype with "double" flowers floral organ initiation fails to occur on a growth substance free medium indicating that the postulated autonomy may not be of universal occurrence.[216]

Development of floral organs to maturity has more complex requirements than initiation. Completion of development, culminating in meiosis and gamete maturation, is reported in only a few studies. Blake and Porath and Galun have obtained pollen formation in *Viscaria* and muskmelon flower buds excised when the anthers were premeiotic;[211,214] here, the stage at excision seems important. Meiosis occurs in tobacco flowers cultured from the sepal stage, and, as suggested by Hicks and Sussex, could have been triggered as a result of some influence from another part of the tobacco flower itself (such as sepals or petals).[215] Callus tissue which develops on the base of the pedicel of excised flower buds could also contribute stimulatory factors. It is noteworthy that the amount of callus is greatest at kinetin concentrations which lead to best flower development (Figure 4).

Other studies show that development is promoted when the flower bud is excised with some other plant tissues attached. Thus, floral organs in *Viscaria* flower buds, bearing only sepals initially, develop completely only when a pair of leaves is included on each explant.[210] In *Anagallis*, development to maturity is obtained when the flower bud, on which sepal primordia are present, is excised with a small stump of subjacent shoot tissue attached.[213] Furthermore, roots develop early at the base of this explant probably as a result of the presence of IAA in the culture medium.

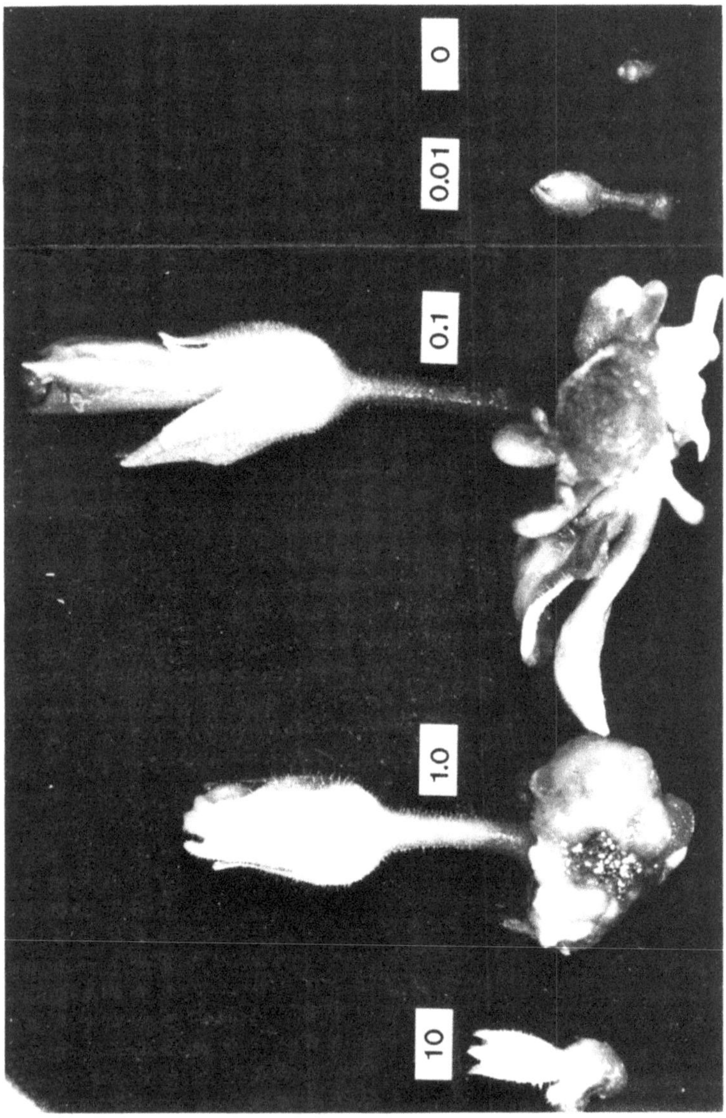

FIGURE 4. Cultured flowers of *Nicotiana tabacum* 'W 38' grown on basal medium with kinetin. Numerals indicate concentration of kinetin in mg/ℓ. Note that callus amount parallels flower development. (Reproduced by permission of the National Research Council of Canada, from Hicks, G. S. and Sussex, I. M., *Can. J. Bot.*, 48, 133, 1970.)

In several species complete development seems easiest to obtain when the flower bud is kept on the inflorescence which is cultured from an early stage. In such conditions, meiosis was found to occur in *Begonia franconis* and maize.[564,566]

Bilderback and associates have long emphasized the importance of the culture medium.[567,575] Most normal flower buds of cucumber, *Viscaria*, muskmelon, and *Aquilegia* are produced on media containing coconut milk and growth regulators.[109,207,210,211,214,567,575] Flower buds of other species were usually found to grow on chemically defined medium when supplemented with growth substances. For tobacco, kinetin appears to be absolutely required (Figure 4). At 0.1 ppm, pedicel, calyx, and corolla are well developed and pollen grains are formed in the anthers.[215]

Available evidence thus supports the idea that flower development, as is true for inflorescence development, depends on growth regulators coming from other plant parts.

3. Floral Organs

There is still little information on the mechanisms regulating development of floral organs isolated from the flower buds. Peterson has determined the nutrient and growth regulator requirements for ovule formation and pistil development of *Nigella sativa*.[296,576] Some studies with tobacco ovaries or placental tissue indicate that alternative developmental sequences may be expressed. For example, ovules, instead of maturing normally, develop into carpelloid structures,[577,578] suggesting that normal development is prevented by isolation from other plant parts. However, development of apparently normal ovules from isolated preovular placental tissue and pistils has been reported more recently on growth substance-free media.[579,580] McHughen showed that young pistil primordia excised from two different tobacco species behave differently in vitro. *Nicotiana rustica* hybrid pistils consistently develop a mature morphology and grow to the size of pistils developing in vivo, but *N. tabacum* cv. 'W 38' pistils do not.[580] Hence, the former appears to be almost entirely autonomous for regulating its own development from an early stage, whereas the latter seems to be dependent on regulatory information originating from other plant parts. A series of experiments was designed by McHughen to test the possible influence of other floral organs on *N. tabacum* pistil maturation.[580] Pistils were excised and cultured with other floral organs attached. As shown in Table 4, better pistil development occurs on explants with petals and stamens and greater inhibition on explants containing all floral organs or where the sepals or petals are retained. Thus, no one type of floral organ contributes exclusively to the development or inhibition of the pistil.

Although developmental regulation of *N. tabacum* pistils is through complex interactions with other plant parts, petals are autonomous from the moment of inception. They develop normally in isolation and do not require exogenous growth substances from the time of initiation to the early stages of maturity.[581] Sepals would have the same autonomy as petals.[580,581]

Stamen development in vitro contrasts substantially with that of cultured pistils and petals. They generate a variety of stigmatoid outgrowths.[582,583] Since kinetin was present in the growing medium and its level appeared to be a determining factor in outgrowth formation on stamens, Hicks may have confirmed in tobacco the well-known feminizing action of the cytokinins (Chapter 7, Section V.C). Little can be concluded from these results regarding the possible autonomy of stamen primordia.

B. Microsurgery Combined with In Vitro Culture

A variety of surgical manipulations have been performed to test the dependence of one region of the reproductive primordium on the formation and development of another. For this purpose, in vitro culture of the excised reproductive primordium has most often been used in combination because it offers greater facility of manipulation than when primordia are left attached to the plant.

Table 4
THE DEVELOPMENT OF THE PISTIL OF *NICOTIANA TABACUM* 'W 38' AS AFFECTED BY THE OTHER FLORAL ORGANS CULTURED ATTACHED FOR 4 WEEKS

Organs present on the explant	Maturity of the pistil[a]
Pistil	+ +
Pistil + calyx	+
Pistil + corolla	+
Pistil + stamens	+ +
Pistil + calyx + corolla	+
Pistil + calyx + stamens	+ +
Pistil + corolla + stamens	+ + +
Whole bud	+

[a] + = little development; + + = development similar to that of the isolated pistil alone; + + + = considerable development including the production of apparently normal ovules.

From McHughen, A., *J. Exp. Bot.*, 33, 162, 1982. With permission.

1. Bisection of Flower Buds and Floral Organs

Cusick initiated experiments in which flower buds of *Primula bulleyana* were surgically halved by a vertical incision, i.e., bisection, at various developmental stages.[584] Similar investigations were later performed on *Portulaca grandiflora,* tobacco, and *Aquilegia formosa.*[585-587] The main objective was to determine when, in ontogeny, the specific pattern of organ primordia is fixed in the young developing flower meristem and secondarily to provide some information on interactions which may occur between different organ primordia on the flower meristem during the organogenetic phase.

Although two of these studies were with bisected flower buds attached to the plant and two others with cultured buds, the regeneration data are basically similar for all species. In order to easily understand the description of the results of Hicks and Sussex with tobacco (Figure 5),[586] it is essential to focus attention on the floral organs which are initiated next to the incision, i.e., in a position which would have been the center of the flower if it had not been bisected. In progressively older buds median distal tissues of the meristem gradually lose their potential for organ regeneration. From the presepal, up to and including the sepal primordium stage, petal and stamen primordia usually regenerate at least in one half and pistil primordia in both halves, but sepal primordia do not regenerate (Figure 5B). At the petal and stamen primordium stages, neither petal nor stamen primordia regenerate; only pistil regeneration occurs (Figure 5C, D). Finally, all regenerative capacity is lost at the carpel primordium stage; when the plane of the incision coincides with that of the median septum of the gynoecium, the two carpels each form a half pistil (Figure 5E). The flower meristem thus appears to pass through a sequence of physiological phases which governs the sequential appearance of the floral organs. Only three organ-forming phases are clearly resolved: a separate sepal-forming phase, a separate carpel-forming phase, and a common petal- and stamen-forming phase. If separate petal- and stamen-forming phases do exist, but follow quickly, the surgical analysis might simply fail to resolve them. Parallel progression

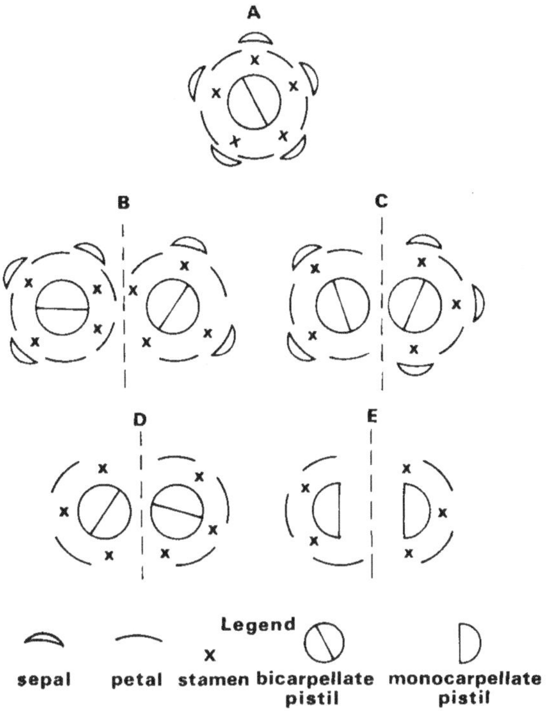

FIGURE 5. Selected floral diagrams of normal (A) and bisected (B to E) flower buds of *Nicotiana tabacum* 'W 38'. After the bisection, the halves were separated. (A) Normal, (B) bisected at the 5-sepal, (C) bisected at the 5-petal, (D) bisected at the 5-stamen, and (E) bisected at the 2-carpel primordium stage. In (D) and (E), sepals were removed. (From Hicks, G. S. and Sussex, I. M., *Bot. Gaz. (Chicago)*, 132, 350, 1971. Reprinted by permission of The University of Chicago Press. Copyright 1971 by the University of Chicago.)

through the sequence of organ initiation and the sequence of physiological phases governing the appearance of the floral organs, as well as the observation that the normal sequence of initiation cannot be reversed or disrupted readily by bisection, suggest that the previously initiated primordia influence the whole flower bud and determine the next type of organ to appear. Hicks and Sussex emphasized the great lability of the center of the meristem, which, after bisection and depending on the stage of the bud at bisection, can regenerate varied floral organs.[586] They suggested that phase changes in this labile region govern the formation of the lateral flower appendages.

Later studies, reported below, further support the idea that both the center of the flower meristem and the existing floral organ primordia are involved in the control of the initiation of the next floral organs to appear.

Bisection of floral organs left attached to cultured flower meristems of the 'W 38' tobacco shows that this surgical manipulation does not alter normal determination of sepals, petals, and stamens,[588-590] i.e., there is no tendency of the half organs to change into another kind of floral organ or into a foliar structure. The youngest incised primordia usually regenerate two complete primordia of the same class suggesting that they normally pass through a labile phase. In older stages they only regenerate two half organs, indicating that the ultimate fate of floral organs become fixed as they develop. Similar conclusions can be drawn for the pistil from studies with bisected flower buds.[586]

Table 5
ORGANOGENESIS IN *NICOTIANA TABACUM* 'W 38' FLOWER BUDS GROWN IN CULTURE WITH OR WITHOUT THE APICAL REGION OF THE FLOWER MERISTEM

	Percent of explants continuing development	
Explant	**With organogenesis**	**Without organogenesis**[a]
Buds with intact or injured apical region	100	0
Prepetal stage buds minus apical region	12.5	87.5
Prestamen stage buds minus apical region	11.5	88.2
Precarpel stage buds minus apical region	0	100

[a] Development as evidenced by continued growth of primordia already present and by healing of the apical region, but no further organ initiation.

Adapted from McHughen, A., *Ann. Bot. (London)*, 46, 633, 1980.

2. Excision of Floral Primordium Parts

Experiments by McHughen indicate that the apical region of the flower meristem of tobacco plays an integral but so far undefined role in controlling the initiation of floral organs.[591] The portion of the apical dome distal to sepal, petal, or stamen primordia initiated is carefully excised and the remainder of the flower bud cultured on a growth substance-free medium. Care is taken to ensure that a sufficient annulus of tissue is left, distal to the last initiated whorl, to provide the physical platform necessary to support organogenesis of the following whorl. Floral organ initiation occurs infrequently on those buds lacking an apical region (Table 5), and this effect is attributable only to the apical region and not to injury or nutritional factors inhibiting growth since all control buds support the production of the next whorl of organs although some are with an injured apical region and, in the manipulated buds, growth of organs already present continue while no new organs are produced.

Previously formed organs in tobacco also regulate initiation of next whorl.[592] Thus, the carpels emerge rarely when the stamen primordia are excised. Removal of the sepal and/or the petal primordia has no such effect. The initiation of the stamens depends on the regulatory influence of the sepal and petal primordia; the initiation of the petals requires the presence of the sepals and finally, the initiation of the sepals is controlled by the lower plant body.

CONCLUSIONS

Development of excised reproductive primordia is under the influence of a network of interactions between various plant organs. Initiation of the floral appendages of a cultured flower bud is controlled by both the central region of the flower meristem, which exhibits great lability, and the organ primordia already existing. In some species it seems to be independent of influences outside of the flower.

Development of floral organs to maturity has more complex requirements than initiation. Maturation of sexual organs, culminating with meiosis, occurs uncommonly in vitro. In most studies, development is promoted on media supplemented with growth regulators which can mimic endogenous signals normally translocated in the intact plant from other organs. This conclusion is supported by results showing normal development when shoot tissue or young leaves are left attached to the explant.

Sepals and petals, isolated from the flower bud, appear to be rather autonomous, but in vitro development of the pistil appears to be dependent on more than one of the other classes of floral organs.

Chapter 7

EXOGENOUS SUBSTANCES

TABLE OF CONTENTS

I. GENERAL CONSIDERATIONS

Much of what we said in introducing Chapter 6 of Volume II on exogenous substances in relation to flower initiation applies also in reference to development. Interpretation of results is difficult because the observed responses are generally the consequence of complex interactions (1) at different levels of organization (organismal, tissue, cellular, etc.) and (2) in time sequences not always clearly defined or even considered. The "PESIGS" rules and their limitations are rarely analyzed, so little more can be said about this approach to growth regulator studies. But some new general ideas have come to our attention that are pertinent at this point.

We have now more literature concerning metabolism, storage, and transport (or lack of transport) of growth regulators. This new information affects questions concerning amount and form of the applied compound, whether it is readily absorbed, whether it is transported from its site of application, and, if it is transported, which tissues are responding. Unfortunately, we do not have many studies on hand that relate specifically to flower development. In fact there are fewer studies in this category than there were concerning flower initiation.

Timing and duration of treatment problems are related not only to stage of development at the time of application, but also to persistence of the compound or an active derivative. Studies on metabolism are extremely significant here. If habituation of tissues in vivo to growth regulators follows the epigenetic model developed by Meins and Binns for cell division factors in vitro,[593] we should expect to find continuous variations in response to exogenous growth substances which, again, impinge on matters related to dose and timing. Finally, the site of action may not be the same as the site of application of the applied compound; work on transport should shed light on this problem.

Applications are often made weeks before observations on flower development are recorded. There are unquestionably more immediate responses that determine subsequent events. Perhaps our overall knowledge is improved, and certainly practical applications can be developed from the long-term, delayed observations, but from a physiologist's point of view, they are insufficient. Hypotheses on mechanism of action, and possible consequent generalizations, depend upon data for short-term, immediate responses and on sequences of responses at several levels of organization.

II. AUXINS

A. Rate of Development and Abortion

Auxin-induced inhibition of flower development has been frequently observed in several early studies. For example, IAA-lanolin pastes inhibit development of the flower buds of *Lycopus* and the inflorescences of *Chrysanthemum*.[153,594]

Lang has stressed that auxin effects are strongly dependent on concentration (dose) and that inhibition of flower development usually results when large amounts are applied; in contrast, with low doses promotion of flower development is often observed.[595] de Zeeuw found increased length and fresh weight of the tomato inflorescence after the application of NAA at 0.5 mg/g of lanolin, but inhibition of development at 500 mg/g.[596] A similar response is obtained with tomato plants grown under a constant low light regime.[433] Under these environmental conditions the inflorescence usually aborts (Chapter 4, Section III) except when a mixture of a cytokinin and a GA is applied to the inflorescence primordium (see further sections in this chapter). Development of the inflorescence is further stimulated by the addition to the mixture of IAA at 1 or 2 mg/ℓ, while concentrations of 10 mg/ℓ and above inhibit.

Auxin action is also influenced by the timing of the treatment (see Volume II, Chapter 6, Section II.A). In some photoperiodic species, e.g., *Xanthium, Chenopodium* and *Ci-*

Table 1
EFFECT OF β-INDOLYLBUTYRIC ACID (IBA) ON THE IN VITRO FLOWERING RESPONSE OF *CICHORIUM INTYBUS* BUDS ISOLATED FROM ROOT EXPLANTS CULTURED FOR 2 WEEKS UNDER INDUCTIVE LD[a]

Medium composition		Buds producing
Before bud isolation	After bud isolation	an inflorescence (%)
BM	BM	70
BM	IBA 5.10^{-6} *M*	100
IBA 5.10^{-6} *M*	BM	25

[a] After 2 weeks under LD photoperiodic induction is achieved. The photoperiod is 24 hr throughout. BM = basal medium (sucrose 45 g/ℓ + agar 12 g/ℓ).

From Margara, J., *C. R. Acad. Sci.*, 277, 497, 1973. With permission.

chorium, treatments before or during induction inhibit, whereas after the inductive period they promote flowering (Table 1).[597-599] Auxin also prevents floral initiation and stimulates inflorescence development in an orchid hybrid.[600] Seidlová and Khatoon showed that the stimulation in *Chenopodium* is restricted to the terminal flower, whereas the lateral buds are inhibited.[601] This suggests that in *Chenopodium* the effect of applied IAA may be explained by increased apical dominance and indicates that the site of application, as well as measurement of the flowering response itself, may be decisive in evaluating the effect of auxin.

All available evidence indicates clearly that discrimination between floral initiation and development is essential. Several studies where stimulation of flowering by auxin is reported do not make clear which stage of flowering is concerned. Since promotion was usually detected during inductive photoperiods, it is likely, from the above results, that these effects are not upon inductive processes, but on those related to flower development.

B. Reproductive Axis Growth

In 1926, Söding observed that scape growth in *Bellis perennis* is suppressed by removal of the inflorescence, but is restored to near normal by replacing the capitulum on the stalk using gelatin to seal the union.[602] This result indicates clearly that the inflorescence contributes substances important for the development of the stalk. Later work has established that auxin can substitute for the flower or inflorescence bud in various species including *Fritillaria, Gerbera,* tulip, narcissus, *Chrysanthemum,* etc.[143,603-608] Stem elongation, however, is by no means a simple phenomenon and its control involves varied mechanisms as revealed by studies with tulip. In this species, reduced stem length after decapitation is largely due to decreased growth of the top internode which is largely restored by application of IAA to the cut surface (Figure 1).[606] Growth of the lower internode is almost unaffected by disbudding and seems to be under the partial control of GAs, perhaps produced in response to cold. Leaves are also involved in the growth of the tulip stem since their excision inhibits internode elongation. Administration of auxin after leaf removal restores growth, mainly below the site of application, suggesting that these organs contribute auxin mainly for the elongation of the internode immediately below their insertion point on the stem.[608]

Sometimes the effect of flower removal on stem growth is not completely counteracted by auxin and the addition of GA is required for more normal elongation. The role of GAs in stem elongation in a variety of plants, especially those exhibiting the rosette habit of

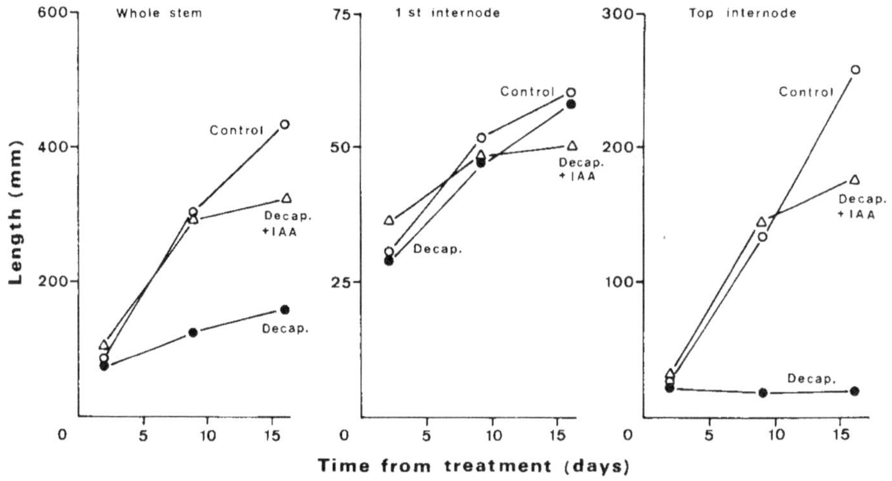

FIGURE 1. Effect of IAA on the growth of the whole stem, the first and the top internodes of decapitated tulips *(Tulipa gesneriana)*. (○) = Intact control. (●) = Decapitated plants: the flower bud is excised as close as possible to the point of insertion of the perianth segments. The cut surface receives plain lanolin. (△) = Decapitated plus IAA in lanolin (0.1% wt/wt). (From Hanks, G. R. and Rees, A. R., *New Phytol.*, 78, 579, 1977. With permission.)

growth, has long been known (Volume II, Chapter 2, Section II and Chapter 6, Section IV). Their effect upon decapitated plants is briefly reviewed in a further section of this chapter.

C. Abnormalities

Auxin-induced abnormalities in reproductive structures have been reported on several occasions. The malformations, especially well studied by Debraux and Astié in *Saponaria officinalis* and *Tropaeolum majus*, are so varied that their detailed description is impossible here.[609] After a 2,4-D treatment, *Saponaria* exhibits transformed inflorescence structures as a consequence of induced fusion and reduction or abortion of lateral axes of the normal inflorescences. Virescent bracts are also produced in response to auxin application in inflorescences of *Saponaria* and *Euphorbia cyparisias*.[610] In wheat a 2,4-D treatment results in various inflorescence abnormalities including branched heads, whorls of spikelets, superimposed pairs of spikelets and paired spikelets borne at the same level.[611] Disturbance due to treatment is temporary.

Fusion of parts in flowers with normally free appendages and the independent growth of organs in flowers in which they are normally concrescent are examples of organ transformations induced by auxin. Thus, flowers of *Digitalis purpurea,* with normally five free sepals and a corolla tube surmounted by five petal lobes, show partially or completely fused sepals and/or a more or less complete separation of five petals after 2,4-D treatment.[610] Also, the normally zygomorphic corolla sometimes becomes actinomorphic. Some flowers are so transformed that they resemble flowers of Solanaceae rather than Scrofulariaceae. In *Saponaria,* a Caryophyllaceae, 2,4-D may cause petal or stamen fusion; on rare occasion, nine of ten stamens fuse producing a structure reminiscent of the staminal tube in Leguminosae. Clearly, some anomalous features in one species may be the normal condition in another. Auxins may also induce virescence of one or several floral members. The virescent "flower" of *Tropaeolum* illustrated in Figure 2, composed exclusively of peltate leaves, results from applications of 2,4-D. Reduction or increase in the number of floral appendages is also frequently observed in response to an auxin treatment.[609]

Auxin-induced proliferation of the floral receptacle is also observed in *Tropaeolum* where

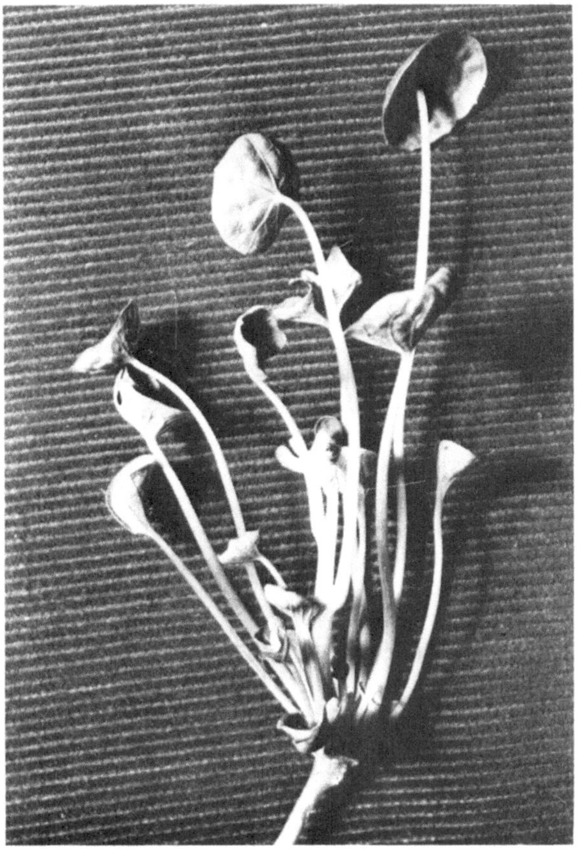

FIGURE 2. Virescent "flower" of *Tropaeolum majus* in which all floral members are transformed into peltate leaves by applications of a saturated solution of 2,4-D. (From Astié, M., *Ann. Sci. Nat. Bot., 12e Sér.*, 3, 619, 1962. With permission.)

in response to a 2,4-D treatment rudimentary flowers occur in the stamen whorls.[610] These studies on reproductive abnormalities induced by 2,4-D unfortunately do not provide much information on the role of auxin in flower development since precise data concerning the method of application and concentration are often lacking. However, the main interest of this work is that it shows that the developmental pattern of the primordia of reproductive structures is not determined at initiation, but remains highly plastic for some time.

D. Sex Expression

Since the pioneer work of Laibach and Kribben showing that treatment of young seedlings with NAA or IAA increases the proportion as well as the number of female flowers in cucumber,[612,613] numerous other studies report similar effects of auxins on sex expression. This subject has already been reviewed on several occasions.[111,116,320,321,614] The summary in Table 2 reveals that there are only a few exceptions to the general rule that auxins enhance the female tendency in monoecious, dioecious, and hermaphroditic species.

In cucurbits auxins usually lower the node to the first female flower and increase the ratio of female to male flowers. Their feminizing effect is also found in other monoecious plants, such as maize, where NAA induces female flowers in the tassel (Figure 3),[373] and in *Carex* that develops female spikelets on sites of potential male flowers after IAA or NAA treatment.[621] Another feminizing effect of auxin treatment in maize is the suppression of branching

Table 2
EFFECTS OF AUXIN APPLICATION ON SEX EXPRESSION IN SELECTED PLANTS

Effect	Species	Sex type	Experimental system	Ref.
Promotion of femaleness	*Cucumis sativus*	Monoecious	Intact plant or excised flower bud	109, 207, 612, 613, 615—618
	Cucumis melo	Andromonoecious	Intact plant	292
	Cucurbita pepo	Monoecious	Intact plant	115, 615
	Luffa acutangula	Monoecious	Intact plant	619
	Citrullus lanatus	Monoecious	Intact plant	620
	Zea mays	Monoecious	Intact plant	373
	Carex sp.	Monoecious	Intact plant	621
	Cannabis sativa	Dioecious	Intact plant	622—624
	Spinacia oleracea	Dioecious	Intact plant	533
	Silene pendula	Hermaphroditic	Intact plant	389, 557
	Lycopersicon esculentum	Stamenless mutant	Intact plant	625
Absence of effect	*Melandrium rubrum*	Dioecious	Intact plant	104
	Mercurialis annua	Dioecious	Intact plant	614, 626
Promotion of maleness	*Mercurialis annua*	Dioecious	Excised node of genetically female plant	627, 628

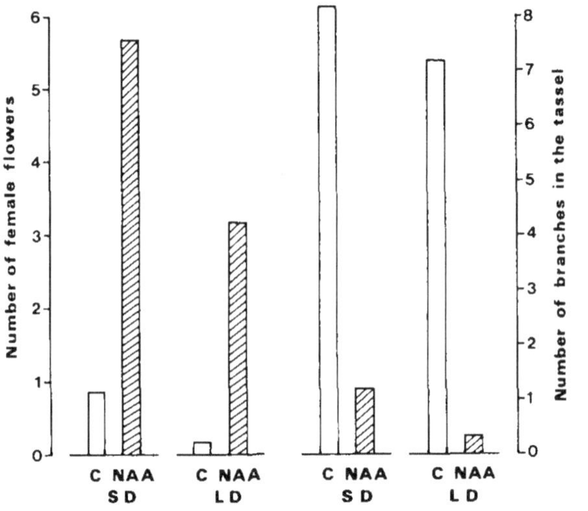

FIGURE 3. Effect of NAA on the formation of female flowers and lateral branches in the tassel of *Zea mays*. C = untreated control; NAA = plants treated with NAA 100 mg/ℓ, 0.15 mℓ injected into the cavity formed by the younger leaves, 4 times over a period of 4 weeks spanning the period of differentiation of the tassel. Two days after the final application plants are transferred either to an 8-hr SD or to 20-hr LD. (Adapted from Heslop-Harrison, J., *Proc. Linn. Soc. Lond.*, 172 Session, 108, 1961.)

in the tassels (Figure 3), many of which have one axis only, a character of the female inflorescence. In *Chrysanthemum,* a gynomonoecious plant, NAA sprays inhibit the development of the hermaphroditic disc florets with little effect upon the number of female ray florets thereby increasing the overall femaleness of the capitulum.[629]

Female sex expression in some dioecious species is also promoted by auxins. In hemp Heslop-Harrison showed that with NAA applications, following photoperiodic induction but

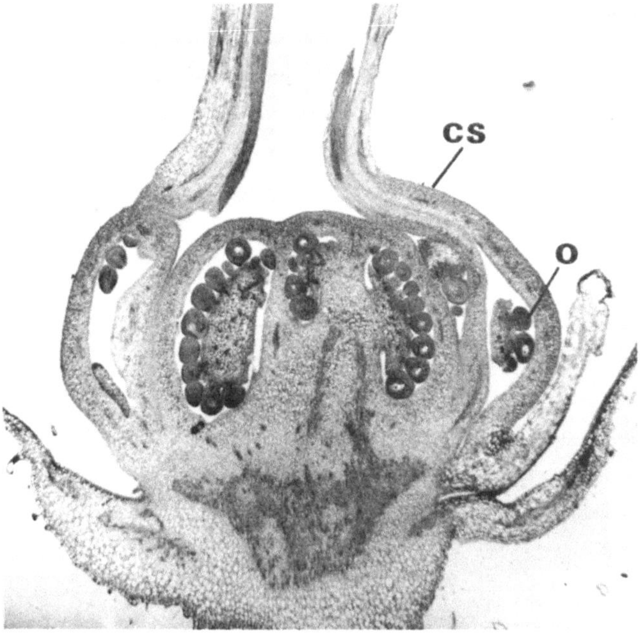

FIGURE 4. Longitudinal section of a flower of the stamenless-2 (sl_2/sl_2) mutant of *Lycopersicon esculentum* after IAA treatment (two applications of 25 μℓ of 10^{-4} *M* in the axil of the leaf close to the shoot apex). Observations are made on the first flowers initiated in the inflorescence subsequent to the treatment. Note the ovules (o) in the swollen basal ends of the carpelloid stamens (cs). (Reproduced by permission of the National Research Council of Canada, from Sawhney, V. K. and Greyson, R. I., *Can. J. Bot.*, 51, 2473, 1973.)

immediately before the differentiation of flowers, male plants can form female flowers in sites which would normally be occupied by males.[622] Applications of IAA at a later stage, when the sex of the plant is readily determined, also lead to formation of female flowers on male plants.[623] Intermediate flower types, with male and female organs more or less developed, are also produced.

Studies with hermaphroditic plants corroborate the generalization of the role of auxins in feminization. In *Silene* the growth of the calyx and gynoecium is promoted after auxin applications while the androecium and corolla are suppressed.[389,557] In a stamenless mutant of tomato IAA applications induce the formation of carpel-like organs in sites normally occupied by stamens. These carpelloid stamens adhere closely to the central pistil and bear ovules inside the basal swollen end (Figure 4).[625]

Heslop-Harrison found that single doses of NAA supplied to hemp plants at the onset of the flowering period are effective for only a limited number of nodes,[622] suggesting that only flower primordia passing through a phase of sensitivity or competence, when auxin levels are high, differentiate carpels instead of stamens. Chailakhyan and Khryanin reported, however, that vegetative hemp seedlings treated with IAA for only a 24-hr period through the root system ultimately develop either female plants or plants with hermaphroditic flowers, but never plants with only male flowers.[624] Spinach seedlings similarly treated with IAA through the root system also develop a higher percentage of plants bearing female flowers than do untreated plants.[533] Moreover, soaking seeds of cucumber and *Luffa acutangula* in NAA solutions also has a slight feminizing effect.[473,619] These results do not support necessarily the view that auxin can regulate sex expression at an early stage, before flower

Table 3
INTERACTION BETWEEN IAA AND BA IN THE CONTROL OF SEX EXPRESSION IN EXCISED NODES OF *MERCURIALIS ANNUA* GROWN IN VITRO

IAA concentration (g/ℓ)	BA concentration (g/ℓ)				
	0	10^{-6}	10^{-4}	10^{-3}	10^{-2}
0		F	F	FF	FFF
10^{-6}			FF	FF	FF
10^{-4}	M	S		F	F
10^{-3}	M			F	F

Note: F, FF, FFF: weak, strong, and very strong feminization, respectively, of nodes excised from male plants. S: male sterility in nodes excised from male plants. M: masculinization of nodes excised from female plants.

From Champault, A., *Bull. Soc. Bot. Fr.*, 120, 87, 1973. With permission.

initiation, since the endogenous hormonal status of plants — treated at the seedling stage or earlier as seeds — at the time of flower differentiation is unknown.

Since sex determination occurs in very young primordia, auxin concentration at the flower meristem or its immediate vicinity was considered to be most significant.[111,622] Experiments with excised flower primordia grown in vitro lend support to this idea. Galun et al. demonstrated that cucumber buds that would have become male flowers *in situ* owing to their location on the plant, develop ovaries and become female flowers when excised at the bisexual stage and cultured in media supplemented with IAA.[109,207] The initiation and development of female organs are also promoted by IAA or NAA in *Aquilegia* and *Cleome iberidella* flower buds grown in vitro,[209,630] and in excised inflorescences of *Lemna gibba* G_1.[631] In contrast, auxins do not affect sex expression in buds, cultured in vitro, from 'Acorn' squash and spinach,[110,632] two species in which they promote femaleness in intact plants. The importance of the medium for in vitro culture must not be underestimated since IAA-induced carpel initiation in excised *Aquilegia* buds is not observed in a medium deprived of coconut milk,[295] and stimulation of pistil growth by NAA in *Cleome* occurs only in the presence of zeatin.[630]

Auxin action is in fact dependent on experimental conditions, which explains why inconsistent results can be obtained, even with the same species. For instance, IAA and NAA have failed to affect sexuality in intact plants of *Mercurialis annua*,[614,626] while male flowers are produced on node explants collected on genetically female plants and grown in vitro in a medium supplemented with IAA or NAA (Table 3).[627,628] In the andromonoecious plant, *Cleome spinosa*, pistil development is reduced by spraying the main inflorescence with NAA,[633] but it is stimulated by low concentrations of NAA in *C. iberidella* flower buds grown in vitro.[630] The inhibition of pistil development in intact *C. spinosa* plants is suggested to be an indirect effect of auxin which would increase sink activity of subjacent developing fruits of the inflorescence, depriving the developing flowers of essential metabolites.[633] Once again the case for *Begonia* is complex. The proportion of female flowers was found to be enhanced by IAA in *B.* × *cheimantha*,[383] but in *B. franconis* an increase in the number of male flowers initiated before the female flowers occurs in intact plants.[112] Finally, auxins are without effect on the in vitro differentiation of flower buds of the latter species.[564] We have already stressed (Chapter 2, Section II.B.1) that the structure of the inflorescence in *Begonia* may strongly affect sex expression obscuring the influence of treatments.

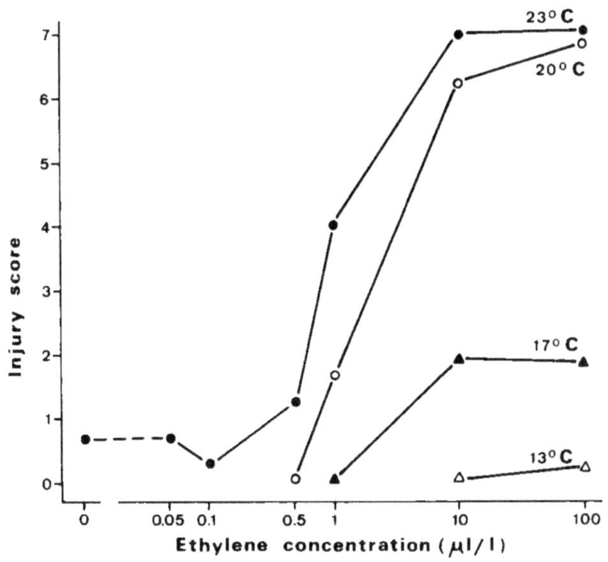

FIGURE 5. Flower bud injury in bulbs of *Tulipa gesneriana* as a function of ethylene concentration and storage temperature. Bulbs are exposed to ethylene for 6 weeks and dissected immediately after treatment for injury scoring. (0 = no injury symptoms, to 7 = wilting of all leaves and floral organs.) (From de Munk, W. J., *Neth. J. Plant Pathol.*, 79, 41, 1973. With permission.)

III. ETHYLENE

A. Rate of Development and Abortion

Ethylene-induced flower bud abortion in many species was first described in the early 1930s by Zimmerman and coworkers who found that young flower buds of roses, lilies, narcissus, and tulips are killed by relatively low concentrations of illuminating gas containing ethylene.[634-636]

More recently de Munk has shown that flower bud blasting in tulip can be induced by exposing bulbs to ethylene during the dry storage period, i.e., before planting.[249] The damage is greater the longer the exposure (from 2 to 6 weeks), the higher the ethylene concentration and the greater the storage temperature. At 13°C ethylene has almost no effect even at high concentration; when the temperature increases from 17 to 23°C, disorders increase and the threshold ethylene concentration for injury decreases (Figure 5). The developmental stage of the flower bud at the time of ethylene treatment strongly influences the response. When bulbs are exposed to ethylene early during dry storage (before the end of August in Holland) blasting — if it occurs — is generally delayed until some days before expected anthesis. With later exposures, particularly those coinciding with microsporogenesis, blasting symptoms are detectable immediately after ethylene application.[249] Treatment of plants in the greenhouse, i.e., after planting, also induces immediate blasting symptoms.[637,638] At that time sensitivity to ethylene is very high since a 3-day exposure is sufficient and the injurious concentration is only half that required for injury during storage. de Munk and co-workers reported that sensitivity to ethylene differs markedly with cv. and increases inversely with the size of the bulb.[249,637,638]

Flower bud blasting and abscission in *Lilium* 'Enchantment' are both induced by ethylene. Blasting can be produced at almost any time during inflorescence development, whereas abscission is evoked only during a critical stage concomitant with the end of meiosis in anthers.[427] 1-Amino-cyclopropane-1-carboxylic acid (ACC), the last product synthesized

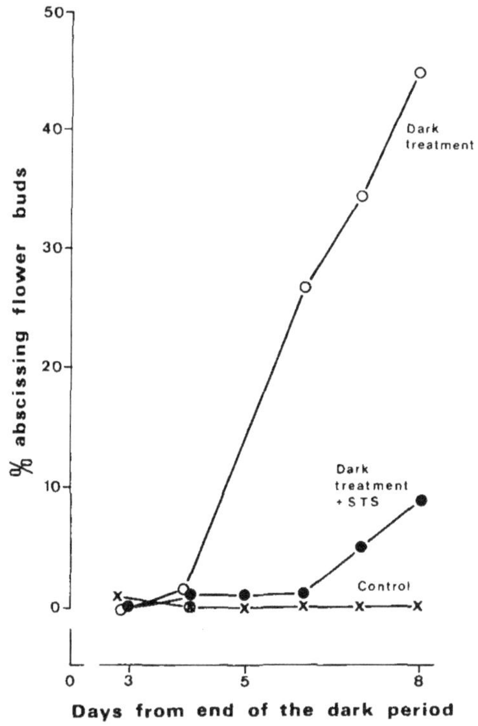

FIGURE 6. Effect of silverthiosulfate (STS) on dark-induced flower bud abscission in *Lilium* 'Enchantment'. (x) Control: no dark treatment, no STS. (○) A 7-day dark period starting when the lower flower buds of the inflorescence are about 3-cm long, no STS. (●) A dark period as above; 2 of the lowest flower buds of each inflorescence are injected with 0.2 *mM* STS (ca. 50 μℓ) at the beginning of the dark period. (From van Meeteren, U. and de Proft, M., *Physiol. Plant.*, 56, 236, 1982. With permission.)

from methionine before the formation of ethylene, is equally effective in inducing bud blasting.[639] Abscission after ethylene treatment also occurs in *Zygocactus*.[640]

Water-soluble ethylene-releasing compounds, such as 2-chloroethylphosphonic acid (CEPA), the active ingredient of ethephon, have been used extensively in recent work as the source of ethylene. As expected treatment with CEPA induces flower bud abortion in many species including roses, *Lilium*, and tomato.[641-643] It also reduces the number of flowers produced in inflorescences of various *Lilium* cvs.[644,645] In the 'Enchantment' lily CEPA suppresses production of secondary flowers normally originating in the axils of the primary bracts.[646] In rose the response to apical applications of CEPA is dependent on the developmental stage of the treated shoot.[641] Flower abortion is greatest when the shoot is 25-cm long, i.e., when initiation of most petals has been completed and differentiation of the stamens begins. Thus, as in tulip and *Lilium* 'Enchantment', the most sensitive stage is during stamen growth and differentiation.

Ethylene action in plant tissues may be prevented by silver and cobaltous ions, high CO_2, low O_2, aminoethoxyvinylglycine (AVG), and a benzothiadiazole (5-methyl-7-chloro-4-ethoxycarbonylmethoxy-2,1,3-benzothiadiazole, MCEB). AVG and Co^{2+} inhibit ethylene production by preventing the conversion of methionine to ethylene,[647,648] while Ag^+ interferes directly with ethylene action at a receptor site.[649,650] Ag^+ is remarkably effective, especially when intact plants are treated with salts, such as $AgNO_3$ or $Ag(S_2O_3)_2^{3-}$ (STS). It also provides prolonged protection from ethylene. The most popular inhibitors, Ag^+ salts and AVG, drastically inhibit bud abscission induced by a dark treatment in *Lilium* 'Enchantment' (Figure 6),[651] but do not prevent blasting of numerous flower buds.[639] In *Zygocactus*, STS reduces abscission of flowers and buds caused by high temperature or low light intensity and provides protection for at least 4 weeks following application.[640] This compound also prevents inflorescence abortion induced in tomato plants by a low light treatment.[433] These results are additional evidence supporting an important role of ethylene in flower bud abortion and abscission.

Perhaps indicative of a positive role of ethylene in flower development is the finding that a single application of AVG on attached branches of rabbiteye blueberries *(Vaccinium ashei)* and peaches, after completion of rest period, delays anthesis by 10 days. The mechanism of flower retardation in these species is unknown. It is of interest that a relationship between completion of the cold requirement for flowering of tulip bulbs and increased ethylene production has recently been suggested (Chapter VIII, Section III.A).

B. Sex Expression

1. Response to Ethylene and Ethylene-Releasing Compounds

Modification of sex expression by chemical treatment was first reported in 1938 by Minima.[654] Cucumber plants exposed to gases released by the imperfect combustion of wood in stoves used for greenhouse heating were found to exhibit an increased female to male flower ratio. This effect was attributed to carbon monoxide, a gas which was shown later to promote femaleness in cucumber, hemp, and *Mercurialis ambigua*,[655-657] but ethylene cannot be excluded as the effective agent in Minima's study.

Since 1968 a great number of publications indicate that ethylene is a major feminizing factor. This work, mainly on cucurbits, shows that CEPA has effects similar to, but more dramatic than, the auxins (Table 4). In cvs. with a more or less pronounced male tendency it may completely inhibit male flowering and promote female flower initiation for an extended period, inducing gynoecious characteristics for a number of flowering nodes.[658,659] In monoecious and andromonoecious muskmelons, it leads to the formation of female flowers on the main stem which normally produces male flowers only.[380,549]

The degree and persistence of conversion depend on the concentration of CEPA, the stage of development at the time of the treatment (Figure 7), the number of applications, and environmental conditions such as temperature or daylength.[380,659,662,663,665,678] Genetic factors are also implicated since, in contrast with findings reported for monoecious and andromonoecious cvs., CEPA is without effect or reduces femaleness in plants with a strong female tendency,[659,663,676] suggesting that sex expression may be a function of regulator level and is determined by an optimum concentration curve.[663]

Reciprocal shoot tip grafts between CEPA-treated and nontreated monoecious cucumbers show that sex expression is not affected if treated scions (shoot tips sprayed) are grafted onto nontreated stocks. In contrast, femaleness is increased in nontreated scions grafted onto treated stocks.[679] This suggests that CEPA induces female flowering through the release of ethylene from the surface of the leaves and not the shoot tips.

In several species other than cucurbits CEPA causes a shift towards femaleness and/or a reduction of male flower development. CEPA promotes pistil growth in *Lemna paucicostata* 6746, *L. gibba* G_1, and *Cleome spinosa,* an andromonoecious species.[631,633] In dioecious plants, such as hemp,[623,670] *Morus nigra,*[671] and *Carica papaya,*[672] it induces the formation of female flowers on male plants or increases the proportion of female plants, whereas in CEPA-treated wheat and barley, two hermaphroditic plants, anther differentiation is suppressed.[674,675] Time of application is critical: male sterility in wheat occurs only when CEPA is applied before or very close to meiosis in pollen-mother cells.[674]

Table 4
EFFECTS OF CEPA APPLICATION ON SEX EXPRESSION IN VARIOUS PLANTS

Effect	Species	Sex type	Experimental system	Ref.
Promotion of femaleness	*Cucumis sativus*	Monoecious	Intact plant or excised apex	658—661
		Androecious	Intact plant	662
		Andromonoecious	Intact plant	663
		Gynoecious heterozygous	Intact plant	663
	Cucumis melo	Monoecious	Intact plant	380, 549, 664
		Andromonoecious	Intact plant	380, 549, 664, 665
		Gynomonoecious	Intact plant	666
		Hermaphroditic	Intact plant	664
	Cucurbita pepo	Monoecious	Intact plant	665, 667
	Cucurbita moschata	Monoecious	Intact plant	667, 668
	Cucurbita mixta	Monoecious	Intact plant	667
	Cucurbita maxima	Monoecious	Intact plant	667
	Citrullus lanatus	Monoecious	Intact plant	620
	Citrullus vulgaris	Monoecious	Intact plant	669
	Cannabis sativa	Dioecious	Intact plant	623, 670
	Morus nigra	Dioecious	Intact plant	671
	Carica papaya	Dioecious	Intact plant	672
	Vitis vinifera	Male line	Intact plant	673
	Triticum aestivum	Hermaphroditic	Intact plant	674
	Hordeum vulgare	Hermaphroditic	Intact plant	675
Absence of effect	*Cucumis sativus*	Gynoecious	Intact plant	659, 676
Promotion of maleness	*Cucumus sativus*	Gynoecious	Intact plant	663
	Zea mays	Monoecious	Intact plant	677

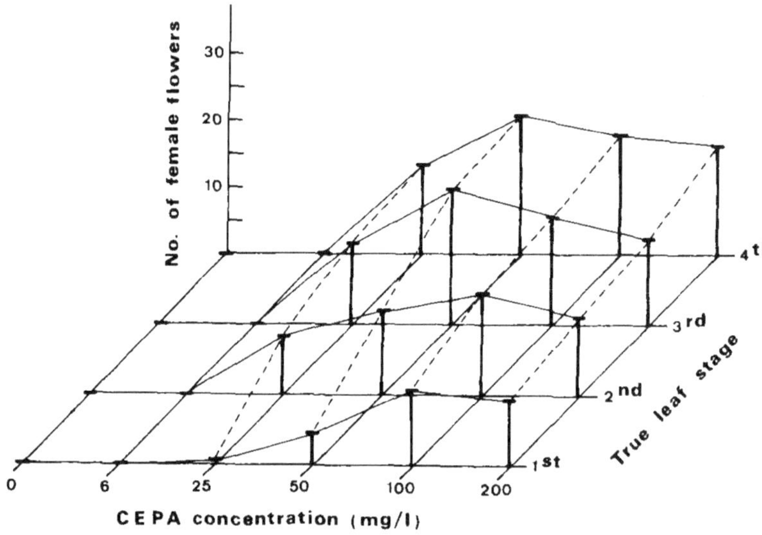

FIGURE 7. Effect of CEPA concentration and the stage of development at the time of the treatment on the number of female flowers induced in androecious *Cucumis sativus* through the 20th node. CEPA is sprayed on the foliage until runoff. (From Augustine, J. J., Baker, L. R., and Sell, H. M., *J. Am. Soc. Hortic. Sci.*, 98, 197, 1973. With permission.)

Table 5
EFFECT OF REDUCED PRESSURE VENTILATION AND ETHYLENE ON SEX EXPRESSION OF GYNOECIOUS *CUCUMIS MELO* OBSERVED THROUGH THE 25th NODE (PLANTS ARE SUBJECTED TO THE TREATMENT WHEN HAVING DEVELOPED FOUR LEAVES)

Pressure (atm)	Gas mixture	Plants with hermaphroditic flowers (%)	Node no. of first hermaphroditic flower	No. of hermaphroditic flowers
1	Air	0	∞	0
0.2	99.85% O_2 0.15% CO_2	100	13	12
0.2	99.85% O_2 0.15% CO_2 + 2 μℓ/ℓ[1] ethylene	0	∞	0
0.2	99.85% O_2 0.15% CO_2 + 0.2 μℓ/ℓ[1] ethylene	75	11	2

From Byers, R. E., Baker, L. R., Sell, H. M., Herner, R. C., and Dilley, D. R., *Proc. Natl. Acad. Sci. U.S.A.*, 69, 717, 1972. With permission.

There is one exception to the rule that CEPA increases femaleness. In maize the female tendency exhibited by the tassel during winter is reduced by exogenous CEPA.[677] It is well to note here that the response of corn to GA, too, is different from that of the majority of species (see Section IV.E.1 below).

Although early studies revealed little or no effect of gaseous ethylene itself,[115,655,660] experiments by Byers and coworkers provide strong evidence that ethylene is the factor responsible for determination of gynoecious sex expression in *Cucumis melo*.[680] They found that gynoecious muskmelon plants, which normally produce only female flowers, develop hermaphroditic flowers when grown under reduced pressure, which facilitates removal of endogenous gases by diffusion and thus reduces internal ethylene concentration (Table 5). The response is affected by the stage of development at the time of the treatment. Plants subjected to hypobaric ventilation at the 3-leaf stage produce hermaphroditic flowers at node 5 while plants treated at the 4- and 5-leaf stages form hermaphroditic flowers at nodes 13 and 17, respectively. When plants are returned to atmospheric pressure or when the ventilating stream is supplemented with ethylene, at a concentration of 2 μℓ/ℓ, gynoecious-like sex expression is restored (Table 5). Ventilation with 0.2 μℓ/ℓ of ethylene is only partially effective in maintaining the gynoecious character.

The same treatments applied to a gynoecious cucumber were without effect.[680] This failure may be related to the fact that the cucumber cv. used produced more than three times as much ethylene as the gynoecious muskmelon. Reduction of the internal ethylene concentration by fivefold apparently is not sufficient to achieve a below-threshold concentration.[680]

Since auxins induce ethylene evolution in many plants, including cucumbers,[647,681] while ethylene lowers auxin level,[647,682] many effects which were attributed to high levels of auxin are now considered to be a result of ethylene production under the influence of auxin. Ethylene would then be a more direct regulator of femaleness than auxin. These aspects will be discussed further in the next chapter.

Table 6
PROMOTION OF MALENESS BY ETHYLENE INHIBITOR APPLICATION IN VARIOUS INTACT PLANTS

Species	Sex type	Compound applied	Ref.
Cucumis sativus	Gynoecious	High CO_2 level, MCEB, $AgNO_3$, STS, AVG	680, 682—687
Cucumis melo	Gynoecious	High CO_2 level, MCEB, $AgNO_3$, AVG	680, 688, 689
	Gynomonoecious	AVG	666
Cannabis sativa	Dioecious	$AgNO_3$, $CoCl_2$	690, 691
Ricinus communis	Female line	$AgNO_3$, $CoCl_2$	692

2. Response to Ethylene Inhibitors

Almost all studies with ethylene inhibitors have used gynoecious cucumbers and gynoecious or gynomonoecious muskmelons (Table 6).[666,680,682-689] In all instances these chemicals increase maleness, inducing the formation of hermaphroditic and/or male flowers in cvs. which normally produce only female or female and hermaphroditic flowers. Anther development in gynoecious muskmelon following $AgNO_3$ or AVG treatments is of special interest because it provides a means of maintaining the gynoecious lines.[688,689] Before the introduction of ethylene inhibitors there was no chemical treatment capable of inducing maleness in gynoecious muskmelon.

Silver and cobaltous ions also promote male sex expression in hemp and castor bean *(Ricinus communis)* by inducing the formation of male flowers in female plants.[690-692] Hermaphroditic flowers are also produced in hemp but not in castor bean.

In conclusion, all results with ethylene inhibitors offer additional evidence for an essential role of ethylene in the control of female sex expression.

IV. GIBBERELLINS

A. Rate of Development and Abortion

1. Time to Anthesis

The role of gibberellins (GAs) on the development of flowers and inflorescences has been extensively investigated and, as with auxins, both promotive and inhibitory effects have been reported. Among the promotive effects of GA, applied after floral initiation, acceleration of development and advanced anthesis have been observed in numerous species, including chrysanthemum,[693] carnation,[694] barley,[695] *Xanthium,*[696] *Lilium longiflorum,*[697] bulbous iris,[698] grapevine,[699] *Bougainvillea,*[420] *Fuchsia,*[700] etc. That GA is a major factor influencing the time of flowering is further supported by the observation that GA inhibitors such as daminozide and Phosfon delay flowering of chrysanthemum by inducing a slower rate of flower bud development.[701] A single dose of GA_3 completely overcomes this retarding effect of daminozide and Phosfon.[693] In the 'Enchantment' lily, ancymidol also delays anthesis. This effect is progressively reduced by increasing levels of GA_3 which, in the absence of ancymidol, does not significantly affect the time to anthesis.[702]

Advanced flowering after GA treatment is not found in all species, and results with a single species are sometimes inconsistent. Time of GA application relative to the stage of development of the flower bud or the inflorescence may strongly influence the response. Spraying *Zygocactus* plants with GA_3 when the flower buds are 1- to 3-mm long advances anthesis by more than 2 weeks, while treatment at an earlier stage delays or inhibits flowering.[703] Early inflorescence development in *Bougainvillea* plants grown under favorable SD is inhibited by GA_3 while CCC causes more rapid growth,[231] but GA promotes development in inflorescences after floret initiation.[420] In tomato GA_{4+7} hastens anthesis only when applied

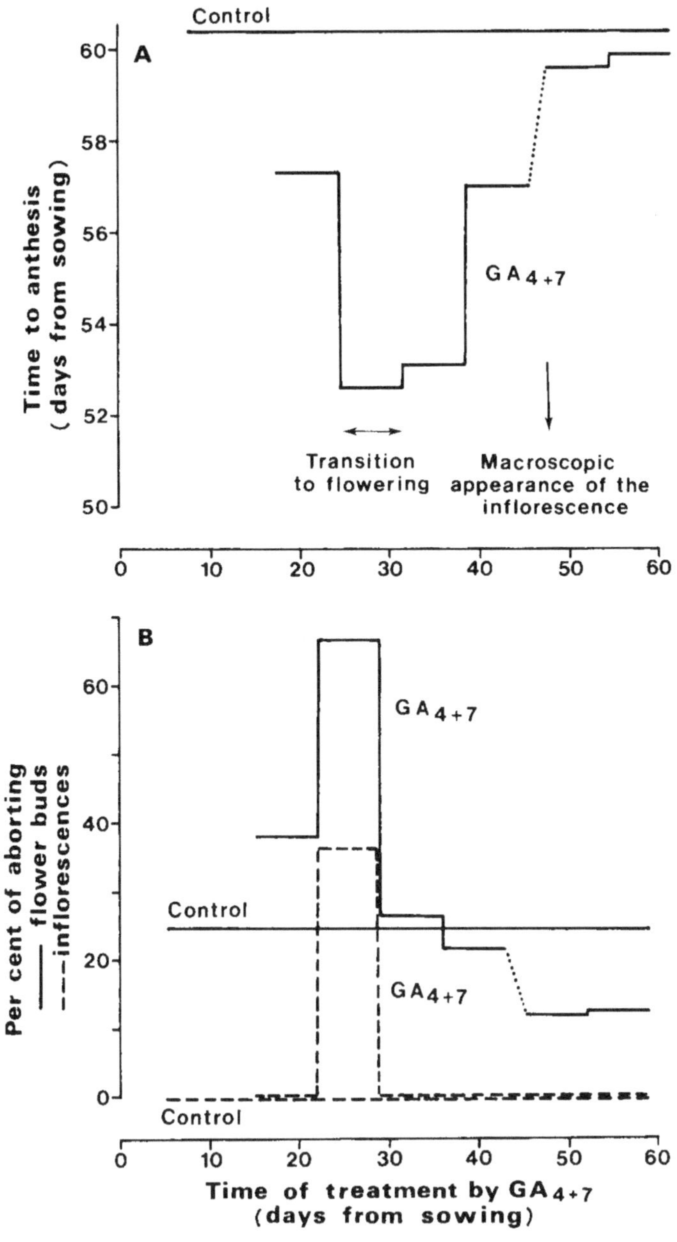

FIGURE 8. Effect of time of application of GA_{4+7} (15 mg/ℓ) on flower and inflorescence development in *Lycopersicon esculentum* cv. 'King plus' as measured by (A) time to anthesis and (B) percent of aborting flower buds and inflorescences. GA_{4+7} is applied to the apex before macroscopic appearance of the inflorescence and directly on the inflorescence after its appearance. Plants are grown in favorable light conditions which promote development to anthesis. There are 4 applications at 2-day intervals.

before macroscopic appearance of the inflorescence; later applications are almost without effect (Figure 8A).

The ability of a GA to hasten flower development is also dependent on its structure and on environmental conditions. GA_1, GA_3, and GA_{4+7} are very active whereas GA_{13} is almost without effect in chrysanthemum.[704] In 'Clipper' barley, acceleration by GA_3 of early development of spike and spikelets is greater in 8-hr SD than in 16-hr LD.[335]

Advanced anthesis in response to GA treatment is frequently obtained in bulbs, alpine, and woody species where flower buds have a rest period. In fact, GA application partially or completely terminates rest in these cases, thus partly or totally substituting for the chilling or water-stress requirement (see Section IV.F in this chapter). After termination of rest these plants generally react to GA in the same way as other species; i.e., anthesis is more or less advanced depending on the timing of the treatment. For example, injecting GA into dry bulbs at the end of chilling, just before planting, accelerates flowering more than injection into the emerging sprout, several weeks after planting.[484,485]

2. Flower Abortion

In some species GAs were found to reduce flower abortion (induced experimentally by manipulation of the environment or by chemical treatments). Flower blasting in tulip induced by ethylene is prevented by a single injection of GA_3 or GA_{4+7} into the flower bud.[250,606,638] Similarly, blindness occurring in shaded rose plants is partially overcome by GA_3 treatment,[641] and abortion induced in tomato by a transfer to low light during the last stages of inflorescence development is also partly prevented by GA_{4+7} applications.[431]

Enhanced abortion after a GA treatment has also been found in several species. GA_3 application to the tip of the shoot of rooted cuttings of *Bougainvillea* greatly increases the number of nodes where inflorescences do not develop to anthesis.[705] These inflorescences cease development at an early stage, before floret initiation, and become lignified thorns. In *Citrus* there are several peaks in flower inhibition by GA_3. One precedes flower initiation, but another coincides with flower differentiation.[706] GA application at that time results in a reduction in the number of leafy inflorescences and a parallel increase in the number of vegetative shoots, suggesting that the apex reverts from the floral to the vegetative state.[706] Applying GA_{4+7} during the early stages of inflorescence differentiation in tomato causes flower bud failure and leads to complete abortion of the truss in almost 40% of the treated plants (Figure 8B). In Easter lily, as in tomato, GA_{4+7} promotes early flower bud abortion.[697] A decline in flower number in 'Enchantment' lily, in response to a GA_3 treatment after the start of flower initiation, is the probable result of early bud abortion, too.[702]

Experiments with GA inhibitors also indicate that a high GA level may inhibit early reproductive development. Abdul and co-workers showed that application of CCC to tomato plants, either before the completion of flower initiation or after flower buds have become visible, reduces the incidence of flower abortion in plants grown at a high temperature with little light.[707] A comparable effect of CCC was observed by Abdalla and Verkerk who found that the shedding of flowers of tomato in response to a particularly high temperature regime (35°C during the day and 25°C during the night) can be reduced by application of the growth retardant as a soil drench.[708] In contrast, the incidence of flower abortion is apparently unaffected by daminozide applied as a spray.[707] In good correspondence with the above results, the yield of diffusible GAs from the shoot tips of tomato plants treated with CCC is greatly reduced, while it remains unaffected after a daminozide treatment.[707] Promotion of floral development up to anthesis by growth retardants, such as CCC and daminozide, is also obtained with *Bougainvillea,*[705] and unexpectedly with rose where, as already reported, GA_3 also reduces flower atrophy.[641] According to Zieslin and Halevy[641] CCC might prevent a decline in GA content of rose as it does in some other species.[709]

From the above results it appears that GAs in general have a dual effect on flower and inflorescence development. Depending on the stage reached by the reproductive structure, they increase or decrease abortion. This dual effect of exogenously applied GA in tomato is also observed in plants continuously grown under a poor light regime which induces the early failure of the whole inflorescence (Figure 9A). Applications of GA directly on the

FIGURE 9. Effect of application of a mixture of BA (10 mg/ℓ) and GA_3 (25 mg/ℓ) on the development of the inflorescence of plants of *Lycopersicon esculentum* cv. 'King plus' grown under a poor light regime. The mixture is applied 5 times at 2-day intervals from the 6th day after the macroscopic appearance of the inflorescence. (A) Control inflorescence (arrow), (B) Treated inflorescence. (From Kinet, J. M., *Sci. Hortic.*, 6, 27, 1977. With permission.)

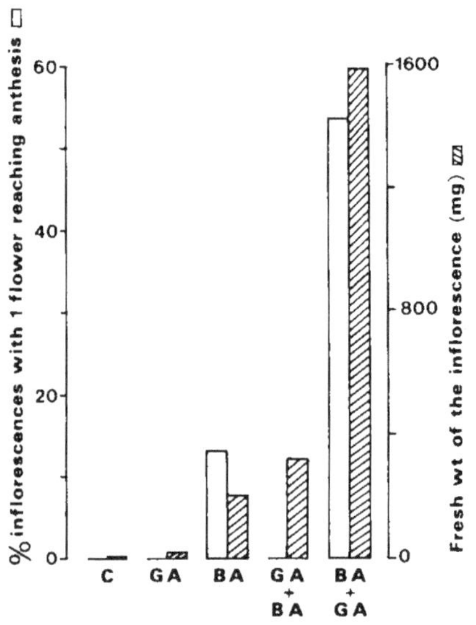

FIGURE 10. Effect of sequential applications of BA (10 mg/ℓ) and GA_{4+7} (25 mg/ℓ) on the development of the inflorescence of plants of *Lycopersicon esculentum* cv. 'King plus' grown under a poor light regime. C = untreated control; GA = 5 applications of GA; BA = 5 applications of BA; GA + BA = 5 applications of GA followed by 5 applications of BA; BA + GA = 5 applications of BA followed by 5 applications of GA. (Adapted from Kinet, J. M., Hurdebise, D., Parmentier, A., and Stainier, R., *J. Am. Soc. Hortic. Sci.*, 103, 724, 1978.)

inflorescence from the time it can be seen by the naked eye do not prevent abortion. There is a slight increase of the fresh weight of some inflorescences, but most GA-treated trusses die more rapidly than untreated ones, suggesting some level of inhibition. However, when a cytokinin is applied along with GA, several flowers develop to anthesis (Figure 9B). When the cytokinin is applied alone this rarely occurs.[542] The two growth regulators have sequential effects in regulating the development of the inflorescence, as shown in an experiment where inflorescences receive various combinations of five applications of BA and five applications of GA_{4+7} (Figure 10).[431] Initial treatments with GA have only a small effect, but better inflorescence development is obtained when BA is applied first. In tomato, therefore, the response of inflorescences to GA is dependent upon some previous action of a cytokinin. That is, cytokinin presumably stimulates the development of the reproductive structure up to a stage where it responds positively to GA.

GA structure influences flower abortion, too. In tomato grown under low light conditions, GA_{4+7} is more effective than GA_3 in stimulating inflorescence development in combination with BA.[202] In Easter lily GA_{4+7}, but not GA_3, induces early flower bud abortion.[697]

Possibly more information on the mode of action on GAs in inflorescence development may be gained by examining effects of GAs (and their antagonists) as a function of their site of application. Stimulation of floral development by GAs is generally observed after treatments restricted to the reproductive structure, e.g., in *Bougainvillea*,[420] *Fuchsia*,[700] tomato,[710] and *Camellia*.[711] In tulip injection of the flower bud is a more reliable application method than injection of the mother bulb or application as a soil drench or by spraying the shoot.[638] In rose, treatments of the differentiating apex and of the leaves of the flowering branch are both effective in preventing blindness.[641] In contrast, inhibition of floral devel-

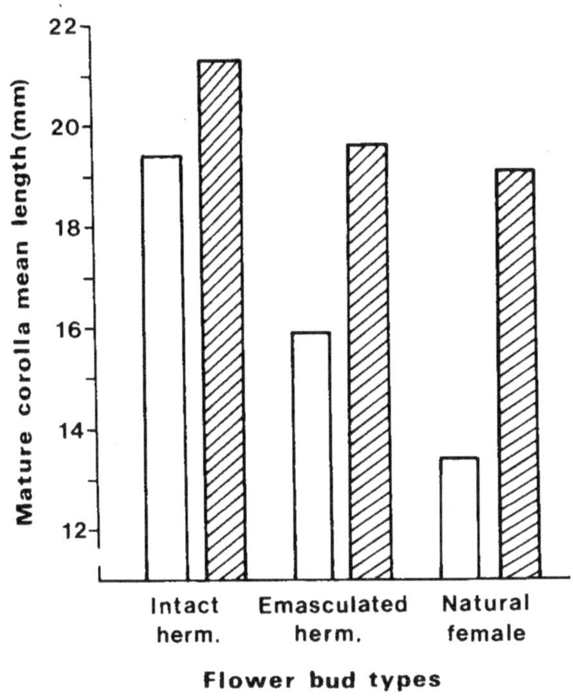

FIGURE 11. Effect of GA on mean length of mature corollas in *Glechoma hederacea*. Distilled water was inserted in control flower buds (open columns) and GA_3 10 mg/ℓ in treated buds (hatched columns). (Adapted from Plack, A., *Nature (London)*, 182, 610, 1958.)

opment by GAs as well as promotion by growth retardants may be obtained, in most instances, with nonlocalized treatments, such as foliar sprays or soil drenches, suggesting that other plant parts and, therefore, correlative influences are involved in the response to GAs. However, in tomato the decrease in shoot growth caused by daminozide is not accompanied by any apparent promotion of flowering,[707] whereas in CCC-treated plants, with reduced shoot growth, inflorescence development is enhanced. Thus, the reduction in growth of vegetative parts induced by CCC cannot be the indirect cause of improved floral development; more likely there is some more direct action of CCC on the inflorescence. Abdul and coworkers suggest that effects of CCC on flowering in tomato are mediated, in part at least, by reduction in levels of endogenous GAs.[707]

B. Flower and Inflorescence Size

GAs promote calyx and/or corolla growth in several species.[211,295,700,711-717] In intact hermaphroditic flower buds of *Glechoma* only a small increase in mean length of the mature corolla results from treatment with GA_3, but when applied to natural female flowers or to emasculated hermaphroditic flower buds, which both display a reduced corolla, GA restores corolla size to that of intact flowers (Figure 11).[713] Increased petal size in rose occurs only in plants grown at low temperature and with application directly to the receptacle; treatments at lower positions on the flowering shoot either have no effect at all or cause elongation of the receptacle only.[715]

An important role for GAs in corolla growth is further supported by the finding that growth retardants reduce and GAs restore petal area in various plants.[718,719] Also, application of GA to potentially cleistogamous flower primordia with normally reduced corolla and anthers causes corolla expansion and production of open flowers which seem chasmogamous.[394,720] In contrast, CCC-treated plants of *Lamium amplexicaule* produce only cleistogamous flowers in environmental conditions in which chasmogamous flowers are produced normally at the upper nodes of the plant. Again CCC-induced inhibition is relieved by GA.[394] Lord and Mayers showed that the GA-induced open flowers of *Lamium* are not morphologically equivalent to the normal open flowers.[721] Pollen counts remain as low as in most closed flowers and corolla cell number remains low in GA-treated flowers. Thus, chasmogamous flowers appear to require other factors responsible for "normal" chasmogamous-like anther and corolla development.

Increased flower size by GA_3 in carnation and *Impatiens* results from the production of additional petals.[61,722] In carnation increased petal number occurs only when applications are made during the early stages of flower development and is attributable to formation of secondary growth centers. The number of primary petals is unaffected by GA.[61]

Increased inflorescence size found in some species after GA treatment usually results from elongation of the reproductive axes: peduncles and pedicels.[433,540,712] In grapevine cuttings, inflorescences do not respond to GA when cytokinin is limiting; when this limitation is removed, either by leaf excision or by adding cytokinin, GA causes very great inflorescence elongation, demonstrating again that the response to one growth regulator is often dependent upon the level, or previous action, of another.[540]

C. Reproductive Axis Growth

Sachs demonstrated that, with the inflorescence removed, intercalary growth can be maintained in the scape of *Gerbera* by application of GA_3.[143] As reported above, auxin, too, is a significant factor in controlling elongation of the *Gerbera* scape. Both growth regulators only partly substitute for the inflorescence (Figure 12), but their combination promotes nearly normal elongation. The response is dependent on GA concentration and the length of the scape at the time of decapitation. In decapitated scapes (receptacle and involucral bracts removed with the inflorescence) GA promotes cell elongation with reduced cell division while in deflowered scapes (receptacle and involucral bracts intact) GA promotes cell division as well as cell elongation so that the pattern of scape elongation is nearly the same as that for intact scapes. Similar results were obtained with IAA. Thus the bracts and receptacle apparently contribute something required for cell division which acts in concert with GA and IAA.

In *Chrysanthemum* the application of GA_3 promotes the elongation of intact shoots and more than compensates for the removal of the inflorescence buds and/or young leaves.[607] Combinations of GA and IAA, or GA and BA, decrease internode length compared to GA alone, suggesting that GAs play the major role in promoting stem elongation. GA_1, GA_3, and GA_{4+7} are very active, while GA_{13} is much less so.[704]

The dependence of stem elongation in various bulbous plants on low temperature is well established, especially in tulip (Chapter 5, Section I.G.1). Attempts to replace the cold requirement by GAs have generally been unsuccessful or only partly successful.[484,485,608,698] Furthermore, in tulip and narcissus stem growth is not restored in cold-treated, decapitated plants by GA application to the cut surface.[604-606] Although these results suggest that GA may not regulate stem elongation in these plants, Hanks and Rees found evidence for their involvement in tulip stem growth using the growth retardant, ancymidol.[606] Applied as a soil drench it leads to significant reduction of tulip stem height by inhibiting elongation of the lower internodes mainly. Simultaneous applications of GA_3 prevent ancymidol-induced inhibition. In some other species, such as chrysanthemum and various *Lilium* cvs., ancymidol and several other growth retardants, including CCC, Phosfon, daminozide, and new quaternary ammonium compounds such as piproctanyl bromide and ADOPB, also reduce stem extension.[162,645,646,701,702,723] The observation that their dwarfing effect may be counteracted, in some cases, by exogenous GAs further supports the idea that the latter are involved in stem growth.[702,704,723]

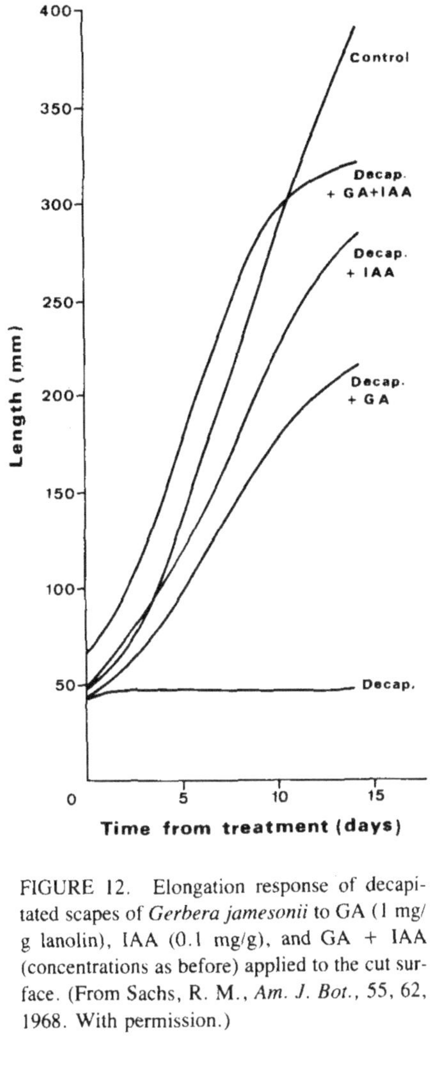

FIGURE 12. Elongation response of decapitated scapes of *Gerbera jamesonii* to GA (1 mg/g lanolin), IAA (0.1 mg/g), and GA + IAA (concentrations as before) applied to the cut surface. (From Sachs, R. M., *Am. J. Bot.*, 55, 62, 1968. With permission.)

D. Abnormalities

GAs induce a variety of reproductive malformations. GA_{4+7} applied at the onset of the differentiation of the first inflorescence of tomato causes the development of leaf-like organs and branches in this inflorescence truss, which is usually unbranched. The additional branches may be vegetative or reproductive.[433] In *Calendula officinalis* GA_3 induces the development of secondary capitula in place of floret primordia on the primary capitulum. Formation of tertiary capitula is also stimulated.[724] In GA_3-treated plants of *Streptocarpus rexii* an additional flower sometimes develops in the inflorescence which is normally uniflowered,[725] suggesting that the normal reproductive structure of this species is an inflorescence reduced to a single flower. In barley elongated rachillae and internodes, rudimentary leaf-like structures in sites normally occupied by spikelets, as well as branched ears, commonly occur in GA-treated plants.[554,695,726] Usually the axillary branches replace the basal spikelets of the spike and bear either rudimentary leaf-like primordia or spikelets.[695,726] As expected environmental conditions such as photoperiod and temperature and the stage at which the barley plants are treated modify the effect of the growth regulator. Generally, abnormalities are absent in plants grown under inductive LD or low temperatures.[695,726] They are induced almost exclusively when GA is applied at the time double ridges appear.[695] Cottrell and associates found that in GA-treated plants grown under SD the main period of spikelet initiation is curtailed with a prolonged period of apical dome extension followed by production of additional spikelets at the tip of the dome.[554]

Virescent petals develop in *Nicotiana* in response to GA_3 and low temperature treatments.[727] In soybean virescence of all floral organs is found when GA_3 is applied in combination with the morphactin, chlorflurenol methylester.[728] In *Streptocarpus rexii* GA_3 induces, in some flowers, an almost complete separation of the petals which are normally concrescent to form a floral tube.[725] A reduction in corolla size may also occur in this species and in eggplant, where its total suppression is sometimes observed.[729]

Finally, GA stimulates production of proliferous flowers in *Anagallis* where vegetative shoots may originate from the apical part of the placenta in GA-treated plants grown under noninductive SD.[730]

These abnormalities induced by GA demonstrate clearly that normal development results by limiting and selecting among a number of potential paths. Unquestionably GAs are required for normal development, but their ability to induce malformations suggests that their production must be regulated in time and with respect to other factors.

E. Sex Expression

1. Response to GAs

GAs affect sex expression and fertility in a large number of species. Recent reviews by Frankel and Galun and Vince-Prue reveal a tangled maze of the accumulated results.[116,321] GAs increase male tendency in some species and femaleness in some others; contradictory reports with the same species are not unusual.

Wittwer and Bukovac first reported in 1958 that GAs increase the proportion of male to female flowers and delay the appearance of the first female flower in cucurbits;[731] numerous others have corroborated their findings (Table 7). In gynoecious cucumbers, GAs induce male flower formation (Figure 13), but are without similar effect on gynoecious muskmelon lines.[549,689,745]

The change in location of the first female flower of monoecious cucumbers results not only from an increase in the number of nodes with male flowers but also from an acropetal shift of male, mixed (male and female), and female flowers; that is, GA, when applied at an early stage, inhibits flowering at the first nodes.[616] Early application of GA to gynoecious cucumbers also inhibits flowering at the lower nodes and results in blind nodes where neither floral nor vegetative buds develop.[685]

Fuchs and collaborators showed that in a gynoecious cucumber line GA treatment inhibits the development of female buds, leading to their abortion, and stimulates the development of adjacent male buds.[735] Combined with the finding that GA_3 does not promote stamen differentiation in flower buds cultured in vitro,[109,207] these results support the idea that GA does not directly affect sex expression in the flower primordium but rather, as Kubicki has suggested, acts indirectly by inhibiting the development of female buds.[746] However, early abortion due to environmental conditions or early removal of female buds in gynoecious cucumber plants does not induce the development of male buds;[735] thus one suspects a more direct role for the GAs.

Increased maleness after GA application is found in other monoecious, dioecious and hermaphroditic species (Table 7). For some of these species, e.g., *Luffa, Xanthium*, and castor bean, there are apparently contradictory reports indicating that GA stimulates femaleness and/or inhibits male expression (Table 8). It is not possible to account for all these opposite responses but frequently experimental conditions differed and different cvs. or mutants of the same species were used. Furthermore, in many species there is a nonrandom distribution of male and female flowers that must be appreciated before results can be interpreted. GAs, especially nonpolar GAs, are also effective in promoting both male and female flowering in some conifer families.[756]

Table 7
SELECTED SPECIES IN WHICH GA APPLICATION PROMOTES MALE SEX EXPRESSION

Species	Sex type	Experimental system	Ref.
Cucumis sativus	Monoecious	Intact plant, or excised apex	616, 661, 663, 732
	Gynoecious	Intact plant	663, 684—686, 732—735
	Hermaphroditic	Intact plant	735
Cucumis melo	Andromonoecious	Intact plant	736
	Hermaphroditic	Intact plant	549
Cucurbita moschata	Monoecious	Intact plant	668
Cucurbita pepo	Monoecious	Intact plant	737
Luffa acutangula	Monoecious	Intact plant	738
Xanthium strumarium	Monoecious	Intact plant	35
Ricinus communis	Monoecious	Intact plant	739
Cleome iberidella	Andromonoecious	Excised flower bud	630
Cleome spinosa	Andromonoecious	Intact plant	633
Cannabis sativa	Dioecious	Intact plant	623, 624, 670, 690
Mercurialis annua	Dioecious	Excised node of genetically female plant	628
Spinacia oleracea	Dioecious	Intact plant	533
Carica papaya	Dioecious	Intact plant	740
Bryophyllum proliferum	Female line	Intact plant	559
Lycopersicon esculentum	Stamenless mutant	Intact plant	625, 741
Cosmos bipinnatus	Stamenless mutant	Intact plant	742
Hordeum vulgare	Stamenless mutant	Intact plant	743
Asparagus officinalis	Female line	Intact plant	744

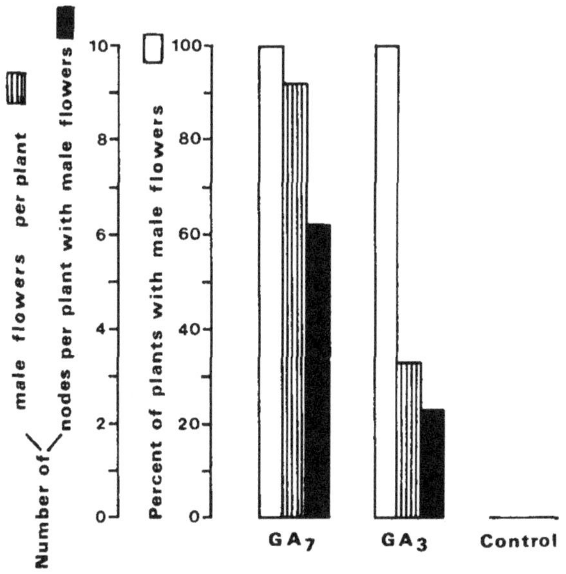

FIGURE 13. Comparative effects of GA_7 and GA_3 in inducing the formation of male flowers in a gynoecious *Cucumis sativus*. All values based on a total of 10 plants per GA treatment and the first 20 nodes of each plant. (Adapted from Wittwer, S. H. and Bukovac, M. J., *Naturwissenschaften*, 49, 305, 1962.)

Table 8
SELECTED SPECIES IN WHICH GA APPLICATION IS WITHOUT EFFECT OR PROMOTES FEMALE SEX EXPRESSION

Effect	Species	Sex type	Experimental system	Ref.
Absence of (or insignificant) effect	*Cucumis melo*	Gynoecious	Intact plant	549, 689, 745
	Cucumis sativus	Monoecious	Excised flower bud	109, 207
	Nicotiana tabacum	Stamenless mutant	Excised stamen primordium	747
Promotion of femaleness[a]	*Luffa acutangula*	Monoecious	Intact plant	619
	Zea mays	Monoecious	Intact plant	677, 748
	Xanthium strumarium	Monoecious	Intact plant	749
	Ricinus communis	Monoecious	Intact plant	750
	Bryophyllum proliferum	Female line	Intact plant	559
	Lycopersicon esculentum	Hermaphroditic	Intact plant	625
	Hyoscyamus niger	Hermaphroditic	Intact plant	559
	Oriza sativa	Hermaphroditic	Intact plant	751
	Capsicum annuum	Hermaphroditic	Intact plant	752
	Lactuca sativa	Hermaphroditic	Intact plant	753
	Allium cepa	Hermaphroditic	Intact plant	754
	Brassica oleracea	Hermaphroditic	Intact plant	755

[a] Promotion of femaleness by stimulating gynoecium development and/or inhibiting androecium.

GA promotes femaleness and/or induces male sterility of normal plants of *Zea mays* (Figure 14).[677,748] In mutants with nonfunctional male flowers, however, sustained GA_3 application restores fertility indicating a promotion of maleness.[757] Enhanced femaleness is also observed since female florets develop in the tassel of treated plants.

GAs usually promote maleness in hermaphroditic systems where abnormalities in the development of male organs are common, whereas in normal plants they frequently stimulate gynoecium growth and/or inhibit androecium development. Application of GA_3 to young plants of a recessive stamenless mutant of tomato produces "phenocopies" of the normal plants.[625] Anther development has also been induced by GA in stamenless mutants of *Cosmos* and barley.[742,743] In contrast GA_3 application to normal tomato plants stimulates femaleness by inducing multilocular ovaries of larger diameter at anthesis than found in untreated controls.[758] Marked style elongation, inducing a heterostylous condition, also occurs in flowers of indeterminate tomato plants treated with GA_3 4 to 6 days before anthesis.[759] Style growth is also stimulated by GA_3 in *Mirabilis*, too.[716] In grapevine, pistil weight is increased after a GA treatment.[699]

Sawhney and Greyson suggested that, in systems where stamen development is inhibited or abnormal, GA_3 application stimulates maleness, presumably by overcoming a deficiency of endogenous GAs, whereas in systems with normal development GAs stimulate femaleness, possibly through an increase in auxin content.[625] An alternative interpretation is that the response to GA is dependent on endogenous GA concentration and therefore is ultimately determined by the sum of existing endogenous plus exogenously derived GAs. If there is an optimum in the dose/response curve then an increase in GA up to the optimum level promotes while a supraoptimal concentration inhibits male expression. This view is supported by the results of Resende and Viana with a female line of *Bryophyllum proliferum* that usually shows female flowers only with undeveloped corollas and very rarely hermaphroditic flowers.[559] In this species GA induces the formation of perfect flowers but continuous treatment at high concentration restores the initial situation where completely female flowers, some with a greater number of carpels than normal, revealing an increased female tendency, are produced. The GA response in cucumber is also dependent on concentration, but there

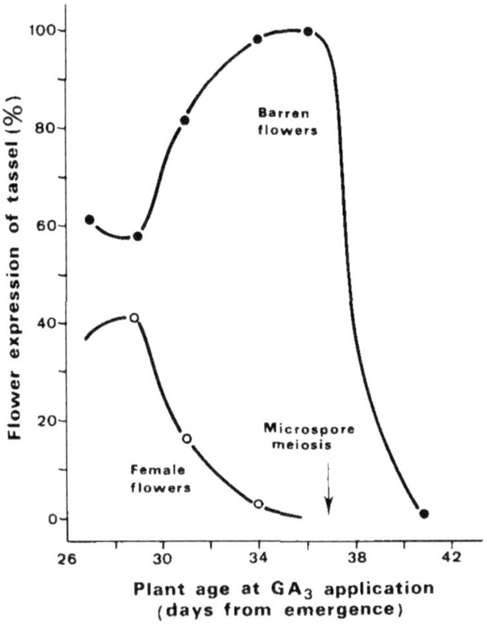

FIGURE 14. GA_3-controlled sex expression of tassels of *Zea mays* as a function of time of GA_3 application. The arrow indicates the start of microspore meiosis. (Reproduced from Hansen, D. J., Bellman, S. K., and Sacher, R. M., *Crop. Sci.*, 16, 371, 1976. By permission of the Crop Science Society of America.)

is no evidence so far for an optimum concentration curve. Usually, increasing the GA concentration, the number of applications, and the duration of the treatment period increases male flower production and the node to the first female flower.[616,732,733]

The activity of GAs is also related to their structure. GA_7 (Figure 13) and GA_{4+7} are very effective in inducing the development of male flowers on gynoecious cucumbers.[734] In hemp and *Luffa*, GA_3 is most active.[670,738]

In several species response to GA is strongly stage dependent.[35,616,732,733,748,753] For example, the response is maximum in the SDP *Xanthium* with application at the time of the single inductive long night and is no longer detectable if plants are treated 8 days later.[35] In normal plants of maize, pistil development occurs in the tassel when GA_3 is sprayed on leaves 3 to 10 days before the start of microspore meiosis (Figure 14).[748] Complete male sterility without pistil development is observed when application is made 1 to 3 days before the onset of meiosis (barren tassels are then produced); later treatments do not alter sex expression of the tassel. The effect of GA_3 on the stamenless mutant of tomato investigated by Sawhney and Greyson is also strongly stage dependent.[760] Applications before stamen primordia are 0.1 mm induce normal stamen production. With a progressive increase in stamen primordium size, more and more of mutant stamen features are produced following the GA_3 treatment. Flower buds with stamens longer than 0.8 mm fail to respond to GA_3, producing stamens indistinguishable from those of untreated mutants. Excised stamen primordia of a male sterile tobacco hybrid grown on media supplemented with GA_3 do not show a reversion to normal male organs.[747] At excision the stamen primordia were 0.4-mm long and perhaps they were already too old to respond to GA_3.

2. *Response to GA Inhibitors*

As anticipated the effect of growth retardants on sex expression in many species is opposite

Table 9
EFFECTS OF GROWTH RETARDANT APPLICATION ON SEX EXPRESSION IN VARIOUS INTACT PLANTS

Effect	Species	Sex type	Compound applied	Ref.
Promotion of femaleness	*Cucumis sativus*	Monoecious	AMAB, BCB, CCC, AMO, B995	116, 761—763
	Cucumis melo	Monoecious	CCC	763
		Andromonoecious	B995	380, 736, 764
	Luffa acutangula	Monoecious	CCC	619, 763
	Luffa cylindrica	Monoecious	CCC	763
	Momordica charantia	Monoecious	CCC	763
	Citrullus vulgaris	Monoecious	CCC	669, 763
	Trichosantes anguina	Monoecious	CCC	
	Citrullus lanatus	Monoecious	CCC	620
	Cannabis sativa	Dioecious	CCC	765
	Carica papaya	Dioecious	CCC	740
Promotion of maleness	*Zea mays*	Monoecious	B995, CCC, Ph	677
	Citrullus lanatus	Monoecious	B995	620

to that of GAs (Table 9). In cucurbits they promote femaleness. This tendency is generally expressed by a reduction in the number of male flowers and an increase in the number of female or hermaphroditic flowers which are formed at lower nodes. Increased female expression in response to growth retardants also occurs in various dioecious and hermaphroditic species, including hemp and spinach,[765] *Carica papaya*,[740] and *Lemna*.[631] An opposite effect to that of GAs is found not only in plants in which GAs increase maleness but also in systems where they promote femaleness as in maize and in isolated apical buds of spinach.[632,677]

Antagonism between GAs and growth retardants, such as AMO-1618 and CCC that block one or more steps in the biosynthesis of GAs, has been observed in several species when both compounds are applied in combination.[116,632,765] This was also seen in systems where the retardant applied alone is without effect upon sex expression.[628,630] In contrast, daminozide does not suppress the GA effect. The number of perfect flowers on andromonoecious muskmelon plants treated with both GA_{4+7} and daminozide remains the same as in plants treated with GA_{4+7} alone.[736] In a monoecious cucumber the male promoting effect of GA is even enhanced when combined with daminozide.[766] Daminozide action on sex expression is, however, complicated by alterations of other systems than those involving GAs.[709] This is, of course, a general objection that one has with growth regulator research.

F. Breaking Flower Bud Dormancy

GAs can be used instead of cold temperature or water stress to break flower bud dormancy of various woody, bulbous, and alpine plants including peach,[767] coffee,[520,524,768] *Hydrangea*,[769] *Azalea*,[254,770,771] tulip,[484,485] *Soldanella minima*,[173] *Saxifraga crassifolia* and *S. exarata*,[169] etc. As demonstrated by Nell and Larson, the conventional 6-week cool treatment used for 2 azalea cvs. can be replaced by spraying the plants weekly over a period of 6 weeks with 1000 mg/ℓ GA_3 and 1000 mg/ℓ GA_{4+7} (Table 10).[254] In some species better results are obtained when GA is used to replace only part of the cold requirement.[484,485,769] In 'Apeldorn' tulip, e.g., GA_3 totally substitutes for the first half of the usual 12-week treatment at 5°C, but only for part of the second 6 weeks.[484]

As expected, stage of bud development at the time of GA treatment can be critical. A uniform response is obtained in azalea mainly when applications are made after differentiation of anthers.[771] In tulip a preplanting application is more effective than one postplanting.[484]

Table 10
THE INFLUENCE OF FOLIAR APPLICATION OF GA_3 AND GA_{4+7} ON BREAKING FLOWER BUD DORMANCY IN TWO DIFFERENT AZALEA (*RHODODENDRON*) CVS., 'REDWING' AND 'DOGWOOD'

	Days from start of the treatment to anthesis		Flowering uniformity[a]	
Treatment	'Redwing'	'Dogwood'	'Redwing'	'Dogwood'
Control — no dormancy breaking treatment	—[b]	—	—	—
9°C for 6 weeks (conventional cooler treatment)	71	68	7.5	7.3
GA_3 for 3 weeks followed by GA_{4+7} for 3 weeks	57	49	6.9	5.8
GA_{4+7} for 3 weeks followed by GA_3 for 3 weeks	51	42	6.8	6.8

Note: Weekly foliar applications of GAs (1 g/ℓ) are applied to runoff with Tween-20 at 0.1% as a surfactant.

[a] Flowering uniformity = 1 to 10, poor to excellent.
[b] No flowering plants.

From Nell, T. A. and Larson, R., *J. Hortic. Sci.*, 49, 323, 1974. With permission.

Response to GA in *Soldanella minima, Saxifraga exarata* and coffee is good when flower buds are dormant at the premeiotic stage.[169,173,520]

In coffee it appears that the flower buds themselves are the receptor sites for the GA-induced response. Dormant buds painted with a GA solution (or treated with a GA-containing lanolin paste) resume growth while control buds in the axil of leaves opposite to the treated buds remain dormant.[520,524,768] Injecting the GA solution close to the developing flower inside the tulip bulb is more effective than smearing the basal plate of the bulb with a GA-containing lanolin paste and a soil drench with GA solutions is without effect[772,773] In other species, such as *Saxifraga, Soldanella*, and azalea, GA sprays to the entire plant are effective in breaking dormancy;[169,173,254,770] even though only a small part of the solution strikes the flower buds, we suppose that in these cases, too, the receptor sites are in the buds.

Relatively high GA concentrations and repetitive applications improve the response in coffee.[520,524] High concentrations are also required for an optimal response in azalea.[717,770] In tulip higher GA doses are required when the duration of the accompanying cold period is reduced.[484] In azalea and tulip GA_{4+7} is more effective than GA_3.[254,484]

When applied in combination with a cold treatment the timing of GA application influences the overall effect on breaking flower bud dormancy. In numerous deciduous woody species, treatment with GA_3 in autumn will prolong dormancy. In *Prunus avium* most flower buds remain completely dormant (many are dead), others have produced distorted or much reduced flower clusters.[774] Delayed spring flowering of peach, after a GA_3 treatment in autumn, seems to result from retarded bud differentiation and delayed flower development after initiation.[775] It is possible that the stage reached by treated flowers at the beginning of winter is inadequate for response to cold temperatures.

V. CYTOKININS

A. Rate of Development and Abortion

Generally cytokinins have no marked effect on time to anthesis, but they can increase both the size and the weight of the reproductive structures.[61,539,776] Their most spectacular action is to prevent, at least partially, flower or inflorescence abortion induced in a number

Table 11
EFFECT OF GAs AND CYTOKININS ON THE PERCENTAGE OF FLOWER BUD BLASTING INDUCED IN 'APELDOORN' AND 'EMMY PEECK' TULIPS (*TULIPA GESNERIANA*) BY AN ETHYLENE TREATMENT (0.5 $\mu\ell/\ell$ FOR 3 DAYS)

Treatment	Flower bud blasting (%) 'Apeldoorn'	Flower bud blasting (%) 'Emmy Peeck'
Control (water injected)	97	91
GA_3	6	7
GA_{4+7}	11	2
Kinetin	19	19
BA	0	0

Note: The growth regulators are injected (0.5 mℓ) at the concentration of 100 mg/mℓ into the flower bud at the beginning of the forcing period, prior to the ethylene exposure.

From de Munk, W. J. and Gijsenberg, J., *Sci. Hortic.*, 7, 255, 1977. With permission.

of species by adverse environmental conditions such as unfavorable daylength, low irradiance, inappropriate temperature, etc.

The inhibitory effect of unfavorable photoperiod was found to be overcome by cytokinins in various SDP, including chrysanthemum,[438,778] *Bougainvillea* 'San Diego Red',[541] *Phaseolus* P47,[777] *Caryopteris* × *clandonensis,*[400] and *Lemna perpusilla* P146.[779] In chrysanthemum BA injections in the leaf or applications to the terminal bud substitute for a limited number of SD in plants receiving insufficient induction for capitulum development,[778] while in *Bougainvillea* applications of PBA to the terminal bud of plants grown under LD strongly reduce the number of nodes to the first developing inflorescence and increase the number of such inflorescences and their rate of development.[539,541]

Abortion caused by low irradiance conditions in iris and tomato is counteracted by cytokinins.[436,542] In the latter, cytokinins not only increase the rate of inflorescence growth, but also the number of individual flower buds visible to the naked eye.[542] A similar effect of cytokinin treatment on flower number was found in *Bryophyllum crenatum*,[560] *Lilium speciosum*,[723] *Dendrobium* Louisae,[780] and in *Streptocarpus rexii* where the inflorescence, normally uniflowered, produces additional flowers after a BA treatment.[725] BA alone is, however, unable to stimulate full development until anthesis when applied just after macroscopic appearance of the inflorescence in tomato.[542] GA is required after cytokinin action (Figure 10) and some kind of joint activity seems apparent in chrysanthemum, too.[438,781] In tulip injection of BA or kinetin into the flower bud prevents ethylene-induced blasting and allows full development until flower opening[250,638] (Table 11).

Zieslin and Halevy observed that kinetin applied to the apex of shoots of 'Baccara' roses grown under 50% shade does not reduce blindness but stimulates axillary bud break.[641] When individual branches are sprayed with BA there is also an increase in the number of lateral buds breaking which results in an increase in both the number of developing flowers per branch and the proportion of blind shoots.[641] In this case cytokinin is not viewed as promoting abortion but the increased proportion of blind shoots is probably the result of increased competition between the numerous laterals that break after the application of the cytokinin. Perhaps a related phenomenon is seen in *Bougainvillea* where lateral branch competition strongly promotes abortion and delays inflorescence development (Chapter 6, Figure 1).[539] Kunneman showed that cytokinin injected in the stem of rose cuttings, near the single lateral bud remaining, prevents blindness induced by a low temperature treatment,[782] demonstrating clearly that, where internal competition is reduced, cytokinins stimulate flower development.

Cytokinins have a strong promotive effect on the overall growth of excised inflorescence or flower buds cultured in vitro.[209,215,563,566] In 'W 38' tobacco kinetin appears to act in a rather unspecific way; growth of all classes of organ primordia is stimulated resulting in well-developed flowers (Chapter 6, Figure 4).[215] Development of corn tassels, cultured when the most mature anthers are at an early initial stage, is strongly promoted by kinetin. Number of normal spikelets is increased and microsporogenesis proceeds in some of them.[566] In some species, gynoecium development is particularly stimulated (see Section V.C). Srinivasan and Mullins investigated extensively the effect of cytokinins on inflorescence formation and development in grapevine. They reported that inflorescences and flowers can be made to grow on young tendrils by treatment with cytokinins (grapevine tendrils may be interpreted as weakly differentiated inflorescences; see Volume II, Chapter 3, Section III.C.4).[783,784] Inflorescence development in dormant cuttings of grapevine is also stimulated by cytokinin treatment.[531]

As expected, the activity of cytokinins usually depends on the method of application, concentration, and structure. In *Iris* BA is most active in preventing flower bud blasting.[436] Mullins showed that when applied in solution culture BA at 10 to 20 mg/ℓ, but not PBA, promotes inflorescence development in grapevine cuttings; at higher concentrations, it is toxic.[531] When applied to the surface of grapevine inflorescences PBA is more effective than BA; kinetin and phenylurea are ineffective. Differences in ease of uptake and translocation probably account for some of these effects. In rose cuttings only the injection of BA in the stem prevents flower atrophy at low temperature. Dipping the roots in the BA solution or spraying the whole cutting is ineffective. Applying a drop of the cytokinin solution directly on the axillary bud causes the development of a bullhead flower.[782] Site of application is also of decisive importance in tomato where only treatments on the inflorescence are active in preventing abortion under low irradiance; spraying the whole plant is ineffective.[710]

In most studies repeated applications of cytokinins are made suggesting that a continuous supply of this growth regulator is an important aspect of its action. This is the case for grapevine,[784] chrysanthemum,[778] *Bougainvillea*,[539,541] *Phaseolus*,[777] rose,[641] tomato,[542] etc.

B. Abnormalities

Cytokinins affect the number of floral parts in some species. For example, tobacco flower buds, excised from the plant at the sepal-primordium stage and cultured on a medium supplemented with kinetin (10 mg/ℓ), produce a large but variable number of sepal-petal hybrid structures, not organized into proper whorls.[785] In 'White Sim' carnation, flower buds treated with PBA, when the calyx is still protected by the bracts or is just visible, show a marked increase in the number of primary petals and secondary growth centers.[61,776] Increased number of all floral parts is found in *Streptocarpus rexii* following a BA treatment; carpel number is especially affected.[725]

Changes in the growth pattern of the inflorescence of *Carex* are induced by BA; determinate branching is replaced by indeterminate branching.[529] BA also induces branched inflorescences in tomato.[433] The additional axes may be either reproductive or vegetative. Leaf-like organs also develop in some treated trusses. In *Sinapis* BA stimulates inflorescence reversion in plants grown under SD and low irradiance.[786] Cytokinin treatment increases the number of vegetative nodes present in the raceme, but, curiously, the vegetative portion of the inflorescence is frequently interrupted by flowers. Whether this developmental mosaic results from the discontinuous cytokinin supply or from some other physiological mechanism is unknown.

Table 12
VARIOUS SPECIES IN WHICH CYTOKININ APPLICATION PROMOTES FEMALE SEX EXPRESSION

Species	Sex type	Experimental system	Ref.
Bryophyllum crenatum	Hermaphroditic	Intact plant	560
Vitis various sp.	Dioecious	Intact plant	787—790
Mercurialis annua	Dioecious	Intact plant	791, 792
		Node excised from male plant	628
Cucumis sativus	Monoecious	Excised apex	661
Luffa acutangula	Monoecious	Intact plant	619
Luffa cylindrica	Monoecious	Intact plant	793
Citrullus lanatus	Monoecious	Intact plant	620
Cannabis sativa	Dioecious	Intact plant or excised embryo	532, 623, 624, 794
Spinacia oleracea	Dioecious	Intact plant or excised apical bud	533, 632
Cleome iberidella	Andromonoecious	Excised flower bud	630
Cleome spinosa	Andromonoecious	Intact plant	633
Asparagus officinalis	Male line	Intact plant	744
Melandrium dioicum	Dioecious	Intact plant	795
Melandrium album	Dioecious	Intact plant	795

C. Sex Expression

Cytokinins generally increase femaleness (Table 12). This was first reported by Catarino in 1964 who found that kinetin treatment stimulates gynoecium and calyx development of *Bryophyllum crenatum* and inhibits androecium and corolla growth.[560] Sex modification by cytokinins was later obtained, almost simultaneously, with grapevine[787] and *Mercurialis annua*.[791]

Negi and Olmo reported that flower clusters of a male clone of *Vitis* dipped in a solution of PBA about 3 weeks before anthesis produce typical fertile hermaphroditic flowers. Similar results were obtained with various *Vitis* species.[788-790] Male plants of *Mercurialis* sprayed with kinetin also produce fertile female flowers. Durand showed that this transformation occurs without alteration of the structure of the male inflorescence indicating that the cytokinin affects the primary sex character only.[626,791] It is possible to discriminate between different male genotypes of *Mercurialis* by their sensitivity to various cytokinins.[792] Some are easily feminized by BA, kinetin, and even by substituted ureas which have a weak cytokinin activity, while others respond to BA only suggesting that they are more strongly male. In this last case BA does cause development of a typically female inflorescence structure.

Cytokinin activity in *Vitis* is dependent on time of application. According to Doazan and Cuellar, sex transformation occurs only if the treatment begins before the start of microsporogenesis.[796] Negi and Olmo clearly show that sensitivity is highest at the time of megaspore mother cell differentiation.[788] Takahashi and associates also concluded that the stage of development of the flower at the time of the treatment is critical in *Luffa cylindrica*, as shown by the acropetal sequence in BA-treated male inflorescences of male, hermaphroditic and then female flowers.[793] In contrast cytokinin affects sex expression in hemp plants at very different stages of development. Young vegetative seedlings respond as well as plants which have already initiated flowers and in which sex characteristics are already apparent.[623,624] The persistence of the applied cytokinin or an active derivative in tissues can possibly account for its marked effect after the treatment of the young seedling and also for the result of Bose and Nitsch who found that plants of *Luffa*, from BA-treated seeds, exhibit a shift towards femaleness.[619]

Several studies stressed that the effect of the cytokinin is restricted almost exclusively to the treated inflorescence.[560,787,793,796] In hemp and spinach BA may substitute for roots in promoting femaleness.[532,533] This fits a generally accepted idea that the root system is the major source of cytokinins in most higher plants.

Published studies on cytokinins and sex expression in cucurbits, the models for work with all other growth regulators, are rare. Is it because cytokinins have little effect on cucurbits? Fragmentary evidence shows that in cucumber, *Luffa*, and *Citrullus lanatus* cytokinins promote femaleness.[619,620,661,793]

As shown in Table 3, the BA effect upon sex expression in excised nodes of *Mercurialis annua* is antagonized by IAA.[628] Antagonisms between cytokinins and GAs or ABA are also reported.[628,765,791]

D. Breaking Flower Bud Dormancy

Cytokinins can completely or partially substitute for the natural stimulus which releases dormancy of vegetative buds in several temperate zone species. Little information concerning their effect on reproductive buds is available. They are effective in breaking dormancy of flower buds of some deciduous fruit trees,[797,798] and of reproductive buds of *Salix babylonica* cultured in vitro.[799] In *Soldanella minima* zeatin has been found to substitute for the cold requirement of flower buds.[800] In azalea cytokinins increase the effectiveness of GA in overcoming dormancy;[801] however, this effect is not always very obvious.[254] Further studies are required on the role of cytokinins in flower bud dormancy break.

VI. ABSCISIC ACID

A. Rate of Development and Abortion

Work with ABA, the most widespread naturally occurring inhibitor, is very limited. Mainly inhibition of development has been reported so far, as in tulip where injection of flower buds with ABA induces blasting.[637] In *Phaseolus* P47, daily application of the inhibitor to differentiating flower buds of plants in inductive SD also results in inhibition and ultimately abscission of many of the buds at a later stage of development (Figure 15).[777,802] Exogenous application of xanthoxin, a naturally occurring neutral growth inhibitor related to ABA, is without effect.[777] In 'Baccara' rose apical applications of ABA do not increase blindness either under 50% shade or under full light in the summer.[641]

Application of ABA in lanolin paste to flower buds of coffee plants prevents breaking of bud dormancy which normally occurs with heavy watering following a drought period.[768]

B. Sex Expression

ABA effects on sex expression are summarized in Table 13. Its role does not seem to be as simple as for some of the other growth regulators. In gynoecious lines of cucumber ABA increases femaleness by lowering the node number to the first female flower and by decreasing the male to female flower ratio,[633,688,803] while in monoecious plants it has the opposite effect[663] or is inactive.[688,803] Andromonoecious and hermaphroditic genotypes also react oppositely to ABA; their response is moreover influenced by photoperiod (Table 13).[663] These results suggest an indirect effect of ABA. Since the young flower buds of cucumber, in which sex expression is not yet determined, are morphologically the same in different sexual types,[807] one would expect a direct effect to be similar in both monoecious and gynoecious plants. Monoecious squash, unlike monoecious cucumber, shifts towards femaleness in response to ABA, i.e., the male to female flower ratio is decreased.[804]

In intact hemp plants ABA has no effect upon sex expression when applied after the start of flower initiation,[623,806] but application through the root system of young seedlings decreases the percent of male plants and increases the proportion of individuals with perfect flowers.[624] Root treatment at the seedling stage also enhances femaleness in spinach.[533]

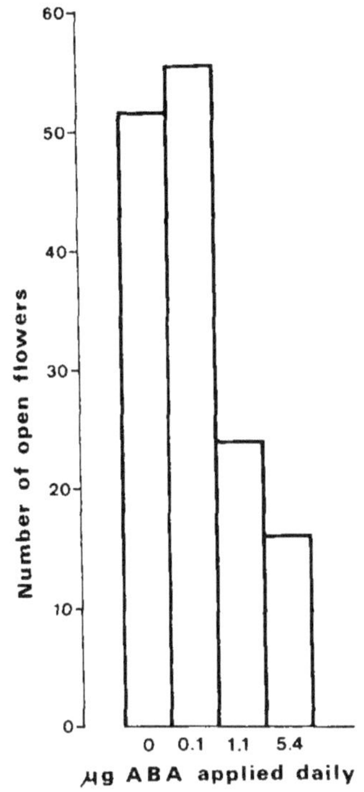

FIGURE 15. Effect of ABA on the production of open flowers in *Phaseolus vulgaris* P47 plants grown under inductive SD. ABA is applied in a droplet to the differentiating flower buds, at daily intervals for 7 days. (From Morgan, D. G., *Aspects and Prospects of Plant Growth Regulators*, Jeffcoat, B., Ed., BPGRG Monogr. No. 6, 1981, 151. With permission.)

Table 13
EFFECT OF ABA APPLICATION ON SEX EXPRESSION IN VARIOUS PLANTS

Effect	Species	Sex type	Experimental system	Ref.
Promotion of femaleness	*Cucumis sativus*	Gynoecious	Intact plant	663, 688, 803
		Andromonoecious	Intact plant in LD	663
	Cucurbita pepo	Monoecious	Intact plant	804
	Cannabis sativa	Dioecious	Intact plant (root treatment of seedling)	624
	Spinacia oleracea	Dioecious	Intact plant	533
Promotion of maleness	*Cucumis sativus*	Monoecious	Intact plant	663
		Gynoecious heterozygous	Intact plant	663
		Hermaphroditic	Intact plant in SD	663
	Cannabis sativa	Dioecious	Excised apical bud	805
	Spinacia oleracea	Dioecious	Excised apical bud	632
Absence of effect	*Cucumis sativus*	Monoecious	Intact plant	688, 803
		Andromonoecious	Intact plant in SD	663
		Hermaphroditic	Intact plant in LD	663
	Cannabis sativa	Dioecious	Intact plant	623, 806

Results from in vitro work do not clarify the problem. Isolated apical buds of spinach develop mainly into male plants when the culture medium is supplemented with ABA.[632] In hemp ABA induces flowering of excised apical buds in LD and male flowers only are produced, probably in response to daylength.[805]

Generally, ABA is found to counteract partially or totally the effect of GAs,[623,663,803,806] but it fails to decrease the male-promoting effect of GA_{4+7} in a monoecious cucumber.[663] In hemp ABA decreases the female-promoting effect of IAA and BA.[623,765] In cucumber, depending on the treated genotype, it either inhibits or increases the effect of CEPA. It is noteworthy that in both cases ABA induces a shift towards femaleness.[663]

VII. MISCELLANEOUS CHEMICALS

A variety of unrelated chemicals affect sex expression and flower bud dormancy in several species. We will not attempt a synthesis here but merely list some responses.

Substances such as 2,3,5-triiodobenzoic acid (TIBA) and morphactins (derivatives of fluorene-9-carboxylic acid), which in some systems inhibit polar auxin transport, generally simulate rather than oppose auxin action on sex expression. When sprayed on tomato plants about 1 week before the first flowers open, TIBA induces male sterility for about 8 weeks without impairing female fertility.[808] In several monoecious and dioecious species, including watermelon,[808] cucumber,[615] and *Carica papaya*,[672] TIBA also increases femaleness. In *Mercurialis annua* it induces the formation of male flowers in vitro on nodes excised from female plants thereby simulating auxin action.[628]

Morphactin has a feminizing effect in many species. It inhibits stamen growth and induces male sterility in hermaphroditic species, such as *Nicotiana paniculata*,[809] *Abelmoschus esculentus*,[810] and tomato.[811] In male hemp plants morphactin induces the transformation of stamens into carpellate structures or totally prevents stamen development, stimulating the formation of an ovary with a terminal bifid stigma.[812] In cucumber and muskmelon, morphactin reduces the ratio of male to female (or hermaphroditic) flowers.[813,814]

In *Carica papaya* and *Luffa acutangula* either increased femaleness or increased maleness occurs following a morphactin treatment.[619,672,815,816] In *Carica* the concentration is critical. Low concentrations markedly increase the proportion of female plants, while a high concentration has the opposite effect.[672] The reason for the conflicting results obtained with *Luffa* is unclear.

Maleic hydrazide (MH) induces male sterility in maize when applied before tassel emergence. The number of male flowers in the treated plants is approximately the same as in the controls, but no anthers are formed.[817] Fertility of the female flowers is not affected. Male sterility is also temporarily induced by MH in tomato.[808] In squash complete suppression of male flowers follows treatment with foliar sprays of MH.[615] According to Heslop-Harrison these effects of MH can hardly be regarded as specific influences on sex expression.[111] However, of interest is the greater sensitivity of male flowers to MH, which fits previous reports revealing greater sensitivity of microsporogenesis to external factors (Chapters 4 and 5).

In the dioecious plant *Melandrium rubrum* female sex hormones of animals (estrone, estradiol) favor gynoecium development, while the male sex hormone, testosterone, increases maleness.[818] Estrogens also enhance femaleness in *Ecballium elaterium*, while androgens increase the ratio of male to female flowers.[819] In cucumber both 17 β-estradiol and testosterone increase female tendency by lowering the node number to the first female flower.[820] Ecdysone, an insect hormone, is without effect on sex expression in squash. In contrast, altosid, a synthetic juvenile hormone for insects, appears to have a weak male-promoting action.[737]

In the LDP *Calendula officinalis* vitamin E advances anthesis of plants grown in SD, but this study does not make clear which stage of flowering is concerned.[821] Vitamin E was also found by Sironval to prevent flower abortion which normally occurs in young runner plants detached from flowering strawberry mother plants *(Fragaria vesca semperflorens).*[822]

Numerous unrelated chemicals (more than 60) listed by Doorenbos and Samish have some dormancy breaking action in woody species.[164,165] Various phenolic compounds and mineral salts, especially nitrate and sulfate, are active. Thiourea also has a pronounced effect but particularly on leaf-bud opening.[797] The question as to whether there is a common dormancy mechanism affected by such different chemicals is still unsolved.

CONCLUSIONS

Although application of growth substances affects all aspects of floral development some important specific responses should be emphasized. GAs and cytokinins are notable in accelerating development to anthesis, and/or in breaking flower bud dormancy, and/or in increasing ultimate size of the flower or inflorescence. They prevent abortion and blasting in many cases and this seems related in some way to their overall growth-promoting action. It appears that the growth rate of the flower relative to that of the competing vegetative tissues must be above some minimum for flower development to proceed to anthesis. In several systems, cytokinins and GAs appear to act in sequence. Usually cytokinins are required first and GA action is subsequent. ABA and particularly ethylene promote abortion and blasting at different developmental stages but most importantly at microsporogenesis. Auxin effects are strongly dose dependent. At high concentrations they usually inhibit reproductive development but are slightly promotive at low concentrations.

GAs generally promote reproductive axis elongation; auxins also promote elongation, but almost exclusively when they replace excised flowers and not when applied to the intact plant.

Ethylene is most frequently the feminizing and GA the masculinizing factor. These sex-related responses are most spectacular in cucurbits but similar trends have been seen in several other species. Auxins and cytokinins are also feminizing agents. The effect of auxins is suspected to be mediated by ethylene. ABA does not seem to play a critical role in sex determination, since it gives rather inconsistent responses.

It would appear that all flowers have the potential for female and male organ development, but in some flowers one kind of primordium is promoted (or inhibited) early at the expense (or to the advantage) of development of the other. Naturally occurring monoecy and dioecy may be the result of such growth regulator-related control of primordial development. Certainly there is a large degree of plasticity in sex primordial development.

Many reproductive abnormalities observed subsequent to growth substance applications confirm and extend our ideas concerning plasticity of many postfloral initiation stages of development. The ultimate fate of floral organ primordia, not just the sex organs alone, is not determined at the same time that their position is fixed. Also, some of the apparently unoccupied sites on the inflorescence or flower meristem can be forced to develop supplementary reproductive or vegetative primordia. Obviously normal development results from limitation and selection among a number of potential paths.

Different growth regulators often induce similar responses in the same or in different species. Hence, all generalizations that have been made concerning specific effects of the growth regulators have exceptions. Unquestionably, growth regulators play some essential role in normal reproductive development. Their ability to cause abnormal development should be expected based upon the inexact methods of application (and consequent impacts upon dosage, and timing and site of action). Most importantly, responses to exogenous growth regulators must reflect differences in sensitivity of tissues

as a function of their developmental stage and other environmental and endogenous parameters. For the present we simply do not know how to take all of these factors into consideration, not even in the most complex experimental designs.

Chapter 8

LEVELS, DISTRIBUTION, AND METABOLISM OF ENDOGENOUS SUBSTANCES

TABLE OF CONTENTS

I. GENERAL CONSIDERATIONS

The same general comments made for the equivalent chapter concerning floral initiation (Volume II, Chapter 7, Section I) apply here. Let us recall briefly that there is no obligatory relationship between the effect of externally applied substances and their endogenous levels in plants, as there may be no correlation between the endogenous level of one substance and the developmental phenomenon studied. Analysis of endogenous growth regulators requires extraction procedures that destroy the in vivo compartmentation of the compounds which may grossly affect our estimates of a biologically active fraction. Transport of a substance from active to inactive compartments may be one of the most important physiological processes affecting its role in development.

Extractions also provide a static picture of substances which are in a dynamic state; at any time their concentration is the outcome of their biosynthesis, accumulation, and metabolism. Hence variations of growth regulator level in extracts may be very misleading.

Also, there is an ever increasing literature showing that tissue sensitivity to a growth regulator may be a more physiologically significant variable than growth regulator concentration.[823] We are not aware of studies on flower development that contend with this problem.

There are often considerable losses of the active fractions resulting from extensive purifications required for the elimination of interfering substances. If extraction procedures do not permit isolation of individual compounds, then, because detection methods are often sensitive to the presence of other compounds, our conclusions concerning levels present in a fraction are affected by the cochromatographying unknowns.

Finally, in some studies, tissue sampling procedures are poorly defined. We find that the time of sampling is often not very close to or is not specified with respect to the flower development stage in question. Very often large tissue samples, including tissues far removed from the site of differentiation, are analyzed and hence it is difficult to say whether the levels measured can refer to the process in question.

II. AUXINS

A. Occurrence in Flowers and Inflorescences

Indirect evidence for production of auxin-like substances in flowers is that near normal stem elongation can be restored following flower removal by auxin application to the cut surface (Chapter 7, Section II.B).

Katunskij showed that maximum growth rate of the stalk of *Papaver* is associated with a high level of diffusible auxin.[824] Borriss and Bussman reported that, at the erect stage of the flower of *Calendrinia grandiflora* at anthesis, when growth rate of the stem is reduced, there is a minimum in diffusible auxin.[825] Measurements by Kaldewey of the level of diffusible auxin from the flower of *Fritillaria meleagris* indicate that auxin production increases before anthesis, falls a bit thereafter, and then increases again when fruit set has occurred.[603] These changes in the level of diffusible auxin are paralleled by modifications in the growth rate of the stalk. Similarly, the release of large amounts of IAA-like substances by narcissus flower buds coincides with the maximum elongation of the pedicel.[826]

Careful excision experiments demonstrated that ovules are the source of auxin moving into the stalk of *Fritillaria*. The first auxin peak coincides with intense cell division in the ovules and formation of the embryo sac.[603] In many other species, too, the gynoecium is the major site of auxin production during flower development.[595] Auxin production by anthers has also been reported especially during microsporogenesis.[595] For Lang, the anthers would be the principal auxin source during the early stages of flower bud development, whereas the ovary begins to produce larger amounts of auxin at later developmental stages.[595]

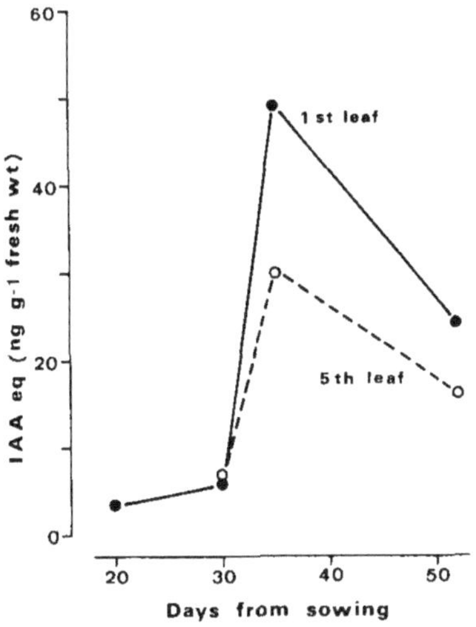

FIGURE 1. Changes in total auxin-like substances in the first and fifth leaves of *Sinapis alba* plants grown under inductive LD. (From Bronchart, R., *Bull. Soc. R. Sci. Liège*, 26, 241, 157. With permission.)

B. Changes in Levels in Relation to Normal Development or Abortion

In *Xanthium* Cooke observed that an increase in extractable auxin occurs at the time reproductive structures develop rapidly (Volume II, Chapter 7, Figure 1), [827] and a similar situation has been described in soybean and *Plumbago*.[827,828] Although in some studies it is difficult to relate changes in auxin with stages of flower development, it seems clear that in a number of species increased auxin levels occur generally after flower initiation and are associated with flower and/or inflorescence development.[829-833] This is the case for *Sinapis* plants grown continuously in inductive LD where a sharp increase in auxin content of leaves was found between the 30th and 35th day from sowing, i.e., just before anthesis of the first flower (Figure 1).[830] In *Chrysanthemum* auxin levels in inflorescence buds reach a maximum early in the development of the capitulum when its relative growth rate is highest.[136] Auxin content in the flowering shoot of 'Baccara' roses increases during early reproductive development through petal initiation and then remains at this level, [834] or decreases.[835] Zieslin and Halevy and Moe showed that blindness, induced by decreasing temperature or light intensity, is associated with a reduced auxin level in rose shoots (Figure 2).[834,835] Auxin levels are similarly decreased after a low cut on stems which promotes blindness on axillary branches.[835]

C. Sex Expression

The correlation of female sex expression with endogenous auxin levels is not as clear-cut as expected from the well-known feminizing action of auxin applications in many species.

Work with various cucumber sex types has yielded contradictory results. Galun, comparing extractable growth substances from stem apices, young and mature leaves, did not find significant differences in auxin-like substances between monoecious and gynoecious lines, only somewhat higher levels of an inhibitor in leaves of monoecious plants.[544] Ito and Saito reported that auxin content in the stem tip of cvs. which have a marked female tendency is reduced in comparison to levels in cvs. with a male tendency.[618] They also found that conditions which favor femaleness often are accompanied by decreased auxin in the stem tip. If female flower production is associated with lower rather than higher auxin levels, we

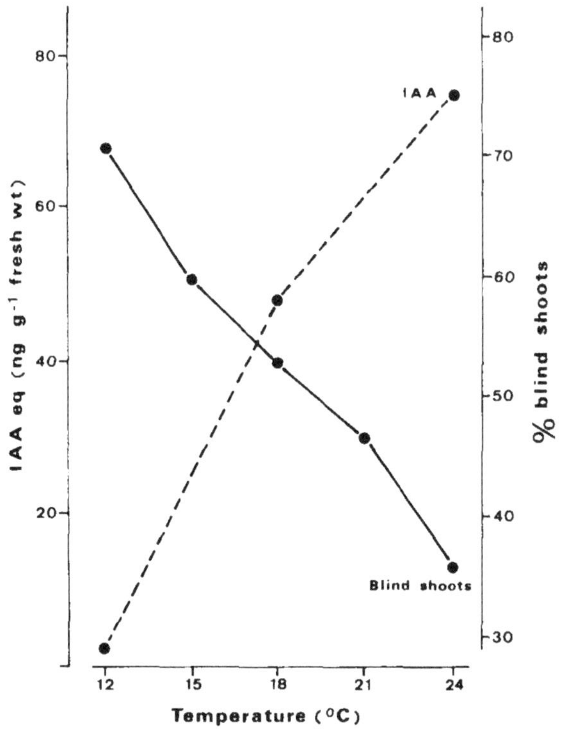

FIGURE 2. The relationship between the endogenous auxin-like substance level of the acidic fraction of ether extracts from 12 to 15-cm long rose (*Rosa hybrida* cv. 'Baccara') shoots and the formation of blind shoots. The different blindness levels are obtained by growing plants at different temperatures. (From Moe, R., *Physiol. Plant.*, 24, 374, 1971. With permission.)

would require a new interpretation for the response of cucumbers to applied auxins. On the other hand, Ito and Saito found increased auxin content in shoot tips of cucumbers as a function of plant age; femaleness also increases with plant age.[618] Results of more recent studies by Rudich and coworkers show that gynoecious cucumbers grown in LD have greater endogenous auxin activity in their apical tissues than monoecious lines.[682] Moreover, relatively higher auxin levels were found in the apices of monoecious plants grown in SD, which promote femaleness, than in LD conditions. Hence, there is an apparent contradiction between the work of Ito and Saito and others with regard to correlations between femaleness and auxin levels in cucumber.

In the dioecious species hemp, femaleness is associated with high auxin level.[836,837] Female plants may contain 30 times more auxin than male plants.[836] Developing female inflorescences of *Salix caprea*,[614] hemp,[837] and maize[840] also contain more auxin than the male inflorescences. In contrast, auxin level is higher in the shoot of male plants of spinach than in the female individuals.[838]

Several factors may affect auxin content in plants and thus could account for some of the above discrepancies. The effect of photoperiod, namely that higher auxin levels usually occur under LD, has been discussed in Volume II, Chapter 7. Developmental stage also affects auxin levels.[837] Galun and co-workers showed that in both hermaphroditic and andromonoecious cucumbers the upper part of the plant contains higher concentrations of auxin than the lower part (Table 1).[839] When homologous sections of the plants — i.e., sections of equal leaf size or containing equally developed floral buds — are compared, the auxin content of hermaphroditic individuals is much higher than that of andromonoecious plants

Table 1
AUXIN-LIKE SUBSTANCE CONTENT IN THREE SECTIONS OF THE MAIN SHOOT OF SUMMER-GROWN HERMAPHRODITIC AND ANDROMONOECIOUS PLANTS OF *CUCUMIS SATIVUS*

	IAA eq (log μg/100 g lyophilized material)	
Plant section	**Hermaphroditic**	**Andromonoecious**
Upper	1.83	0.94
Intermediate	1.56	0.85
Lower	0.90	0.85

From Galun, E., Izhar, S., and Atsmon, D., *Plant Physiol.*, 40, 321, 1965. With permission.

(Table 1). Thus it seems important, in these analyses, to distinguish among tissues in relation to their origin within the plant and/or age. If this precaution is taken, positive correlations between auxin level and femaleness are obtained.

There is no detectable increase in auxin level in cucumber, muskmelon, and squash following a treatment with CEPA which increases femaleness.[682,841] On the contrary, auxin levels in cucumber and muskmelon decrease for a few days and return to normal within 10 days of treatment. These results suggest that one should attach minor importance to increased endogenous auxin in female flower formation. Taken together with the finding that auxin applications increase ethylene production in cucumber plants, [681] they indicate as well that the female-inducing response to exogenous auxin is indirect and probably mediated through some action of ethylene.

The essential point is that sex expression, as with so many morphogenetic phenomena, is clearly not determined by the concentration of a single growth regulator, but involves complex interactions between several growth substances and other factors, such as growth regulator metabolism, as well. Thus variations in auxin levels relative to other endogenous regulators will certainly be more useful for development of hypotheses than absolute values per unit fresh or dry weight.

III. ETHYLENE

A. Changes in Evolution in Relation to Normal Development or Abortion

Exogenous applications of ethylene (or ethylene-releasing compounds) cause flower abortion in different species suggesting that this substance participates in the endogenous control of flower development. Ethylene evolution by *Lilium* 'Enchantment' and 'Baccara' rose plants grown under conditions which either favor floral development or cause abortion corroborate this assumption. van Meeteren and de Proft showed that flower bud abscission induced in *Lilium* by a dark period is associated with a marked increase in ethylene evolution by these buds when they are between 20- and 30-mm long (Figure 3).[651] STS or AVG applied during the dark prevents abscission and reduces ethylene evolution.[639,651] However, about 50% of flower buds blast in the AVG-treated plants, indicating that besides ethylene, other factors are involved in the control of abscission and blasting of lily flower buds.[639]

Ethylene production by the two upper shoots of 'Baccara' rose plants, grown under full light during summer in Israel, is relatively high.[842] With 50% shade, ethylene production decreases sharply in both shoots and remains low in the upper shoots (which all develop a

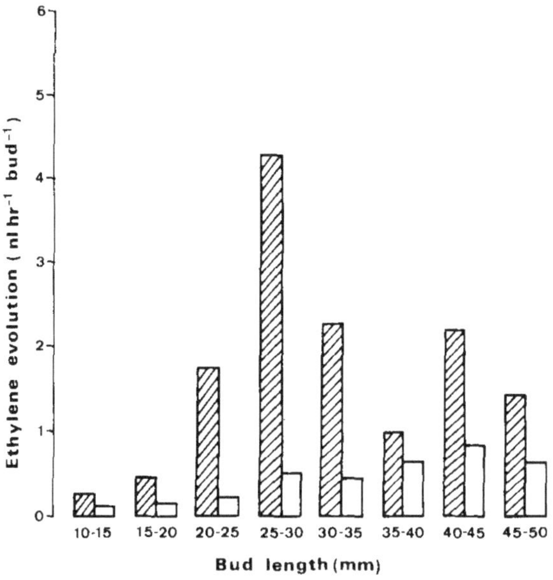

FIGURE 3. Effect of a dark treatment on ethylene evolution by *Lilium* × 'Enchantment' flower buds of varying lengths. Open columns: plants exposed to a daily light period of 18 hr. Hatched columns: plants exposed to darkness for 3 days. (From van Meeteren, U. and de Proft, M., *Physiol. Plant.*, 56, 236, 1982. With permission.)

terminal flower to anthesis). In contrast, ethylene production by the second shoots, which all become blind, increases by the second day following shading and by 4 days it is several times as great as in the upper shoot. We know now that flower atrophy in shaded rose shoots is probably controlled by some interaction between ethylene and reduced photosynthate transport (Chapter 9) rather than by ethylene level per se. Ethylene release by the abortive second shoots is about the same as that of the nonabortive shoots exposed to full light so that the role of assimilates may include that of determining bud response to ethylene.

A relationship between the cold requirement for flowering of tulip bulbs and increased ethylene production is suggested by the recent finding of Prince and associates that ethylene release increases markedly in bulbs after 12 weeks of 5°C storage.[843] This duration of precooling has been established as an optimum for obtaining normal flowering.[844] Other studies by Moe and co-workers indicate that ethylene evolution in tulip is triggered by moving bulbs from low to high temperatures and, hence, could not be used to interpret chilling effects per se;[845] but the new studies of Prince and co-workers showing increased ethylene evolution even in bulbs held at low temperatures encourage further consideration of its physiological relationship in the determination of rest completion.

B. Sex Expression

Sex expression in cucumber is correlated with endogenous ethylene production. Gynoecious types produce more ethylene than monoecious lines which produce more than androecious lines.[377,680,846] Ethylene evolution by gynomonoecious seedlings grown in vitro is intermediate between that from gynoecious and monoecious plants.[847] Thus, ethylene production follows closely the degree of femaleness (Figure 4). Also, ethylene evolved from plants grown in SD conditions, which promote femaleness, is greater than that from plants in LD, and female floral buds (up to 3 mm) produce more ethylene than male buds.[377]

Ethylene evolution by the apical bud changes in the course of plant development (Figure 4).[377] Soon after germination, it is about the same for all sex types; in the gynoecious

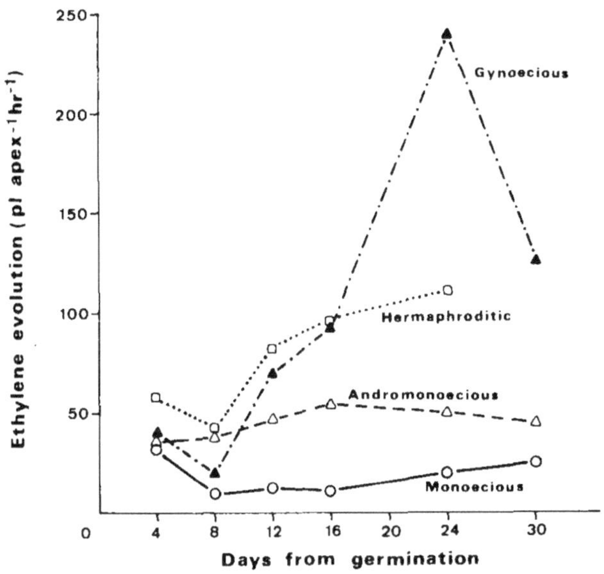

FIGURE 4. Ethylene evolution of apices of four different sex phenotypes of *Cucumis sativus*. (From Rudich, J., Baker, L. R., Scott, J. W., and Sell, H. M., *J. Am. Soc. Hortic. Sci.*, 101, 48, 1976. With permission.)

phenotype it increases markedly 8 days after germination and reaches a peak on the 24th day when the first female flower buds now visible are developing rapidly. This peak is absent in the hermaphroditic sex type. Ethylene evolution by the andromonoecious plants is constant and low compared to the gynoecious and hermaphroditic phenotypes, but is greater than for the monoecious plants.

The effects of GA_{4+7}, $AgNO_3$, and AVG upon sex expression and ethylene production in gynoecious cucumbers have been studied by Atsmon and Tabbak.[685] Despite their common effect upon sex expression (i.e., the induction of male flower formation in strictly female plants), GA_{4+7}, $AgNO_3$, and AVG affect ethylene evolution differently. GA is without effect, $AgNO_3$ promotes and AVG drastically reduces ethylene production indicating that there are different mechanisms for the male-enhancing effects of these compounds. Inhibition of ethylene biosynthesis by AVG (leading to decreased evolution) and inhibition of its action and/or metabolism by $AgNO_3$ (leading to increased evolution) both reduce the effective endogenous ethylene activity, but GA probably does not affect sex expression through ethylene.

In muskmelon the correlation between the degree of femaleness and ethylene production is not as good as in cucumber although a gynoecious phenotype has been shown to produce more ethylene than a monoecious phenotype.[680]

High levels of ethylene are consistently emitted by developing female flowers of *Cycnoches warscewiczii* and *Catasetum expansum;* conversely low levels are evolved by inflorescences of these species that bore male flowers.[848] Moving plants from shade to sunlight increases ethylene evolution and enhances production of female flowers,[446] suggesting that flower sex in these orchids is regulated by ethylene. However, applied ethylene and CEPA fail to stimulate female flowering indicating that the situation is not that simple. Use of ethylene inhibitors could probably help in determining the role of ethylene in sex expression in these plants but we emphasize again that sex expression is the result of interactions between several growth regulators.

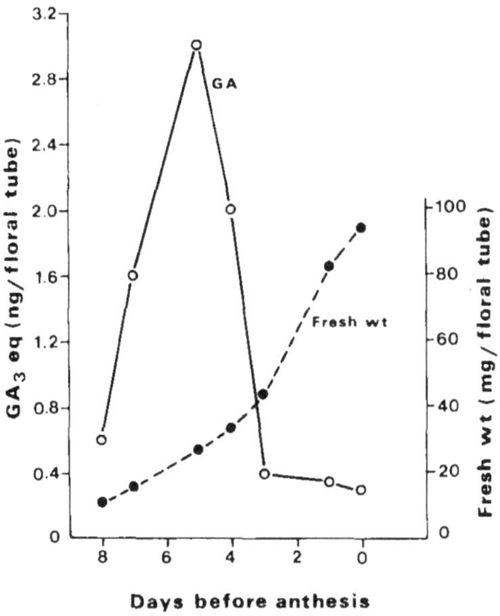

FIGURE 5. Changes in GA-like substances and fresh weight of the floral tube of *Mirabilis jalapa* during flower development. (From Murakami, Y., *Plant Cell Physiol.*, 16, 337, 1975. With permission.)

IV. GIBBERELLINS

A. Occurence in Flowers and Inflorescences

Relatively high GA levels have been found in flowers and inflorescences of several species as a function of stage of development. In *Mirabilis* the amount of GA in the floral tube increases rapidly after the calyx becomes visible above the bracts 8 days before anthesis, reaches a maximum 3 days later, and then declines markedly (Figure 5). The stamens, especially the anthers, have the largest quantity of GA and mainly account for the changes in concentration which occur during flower development.[716] High GA levels were also found in petals and stamens of some other species.[714,849-851] Since, when male sex expression is inhibited, GA levels are frequently reduced not only in the flowers but also in vegetative tissues (See Section IV.D). GAs are believed to play some critical role in stamen development.

GA distribution between the various floral organs fluctuates in the course of reproductive development. In *Vicia* during the early stages of flower growth, free GAs are found mainly in the androecium and the perianth.[851] Immediately before flower opening the highest GA levels occur in the gynoecium. In *Pharbitis* and rose, too, the GA concentrations in the gynoecium are high at anthesis.[850,852]

GA_3 is the major GA in *Avena* inflorescences,[853] GA_{20} in *Vicia* flowers,[851] and GA_9 has been found in the *Chrysanthemum* capitulum.[854] GA_{19} occurs in rice ear,[855] and in *Pharbitis* the calyx contains much larger amounts of GA_{19} than GA_3 which is at relatively high levels in both petals and stamens during periods of rapid growth and then declines shortly before growth ceases.[850]

B. Changes in Levels in Relation to Normal Development or Abortion

GA level in the leaves of 'Baccara' roses changes in the course of flower development. Up to carpel initiation both acidic and neutral aqueous GAs increase. One day before anthesis acidic GAs remain unchanged while neutral GAs decrease.[834] Zieslin and Halevy showed that GA declines under conditions, such as low temperature and low light intensity, that

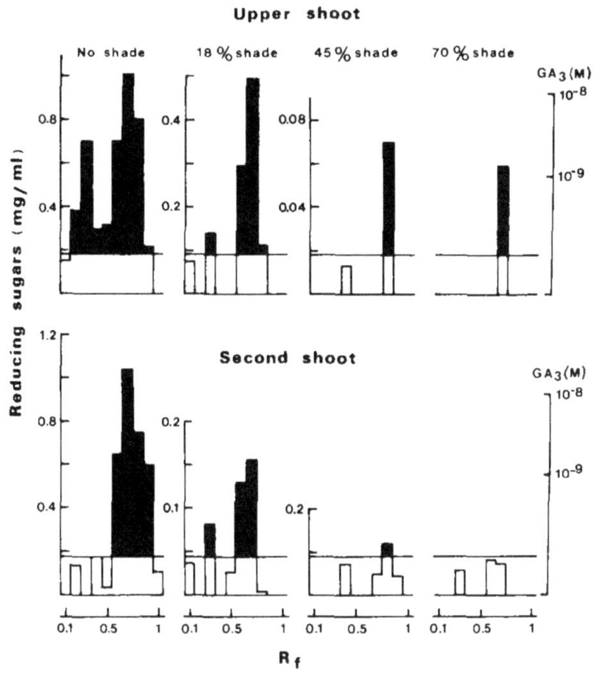

FIGURE 6. GA-like activity of extracts of leaves of the upper and second shoots of *Rosa hybrida* cv. 'Baccara' plants grown in greenhouse under various light intensities. The leaves are harvested when flowers on control shoots reach a diameter of 4 mm. Activity was tested in the barley endosperm bioassay. Black areas are above 2 × S.E. Maximum light intensity of nonshaded shoots was 340 W/m². (From Zieslin, N. and Halevy, A. H., *Physiol. Plant.*, 37, 331, 1976. With permission.)

promote flower atrophy.[842] At a relatively high temperature (24°C during a 12-hr photoperiod and 18°C during night) GA content is high in the two upper shoots which develop after pruning. A decrease of 5°C in night temperature results in a marked reduction in GA in the lower shoot only. Blindness, 3 weeks after start of the treatment, is low in the upper shoot at both temperatures. In the second shoot, blindness is much greater at 13 than at 18°C. GA levels in the flower buds were not measured but in a later study concerned with bullhead flower formation, Zieslin and coworkers found reduced GAs in flowers in low temperatures[852] (Section IV.C). Shading, which markedly reduces flower development, also reduces GA activity in the leaves.[842] The reduction is more pronounced in the lower than upper shoot and increases with shading. With 70% shading, GA activity completely disappears in the lower shoot (Figure 6). Similar results are obtained when only one shoot of the plant is shaded. In *Gladiolus*, GA content is always higher in the leaves of flowering plants than in those of nonflowering individuals.[703]

In summary, GA appears to be involved in the control of flower development in rose; a decrease in GA can be correlated with flower bud abortion and exogenous GA can prevent flower failure induced by unfavorable environmental conditions. In contrast, early inflorescence development in tomato is correlated with reduced levels of endogenous GAs. Abdul and Harris have observed that young leaves excised from plants grown at low temperatures, which increase the number of flowers in the first inflorescence, contain much lower levels of diffusible GAs during the early stages of truss development than leaves of plants grown at normal temperatures.[543] Yields of diffusible GA-like substances from the shoot tips are also reduced by treating plants with CCC, a treatment that increases the number of flowers

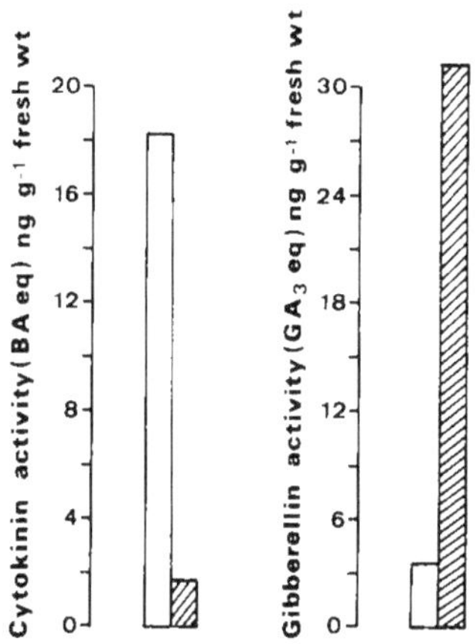

FIGURE 7. GA- and cytokinin-like activities in the first inflorescence of tomato (*Lycopersicon esculentum* cv. 'King plus') plants grown either under favorable light conditions which promote normal development to anthesis (open columns) or adverse light conditions which induce early inflorescence abortion (hatched columns). At harvest, 6 days after macroscopic appearance in favorable light conditions and between the 6th and the 10th day in adverse light conditions, the mean fresh weight of one inflorescence was respectively, 87.2 and 5.4 mg. (From Kinet, J. M. and Léonard, M., *Acta Hortic.*, 134, 117, 1983. With permission.)

and reduces the incidence of flower abortion in the first inflorescence.[707] Daminozide which does not affect flower number or abortion is without effect on yields of diffusible GAs from the shoot tips. Léonard and Kinet reported, moreover, that early inflorescence abortion occurring in tomato plants grown under a constant low light regime is correlated with higher GA content than is found in inflorescences developing in favorable light conditions (Figure 7).[856] The major peak of activity cochromatographs most closely with GA_3. Since GA_{4+7} is more effective than GA_3 in promoting late stages of inflorescence development in tomato,[202] studies of GA interconversions in relation to inflorescence growth would be valuable as well as analyses of changes in endogenous GAs during the late stages of floral development.

C. Petal Number

In carnation flowers the formation of secondary growing centers is promoted by GA_3 application to shoot tips during early stages of development (Chapter 7, Section IV.B). A period of low temperature which also stimulates secondary growing centers causes marked increases in levels of diffusible GA-like substances from shoot tips.[472] The work of Raman and Greyson reveals that the "double-flowered" and "single-flowered" genotypes of *Nigella* possess different levels and compartmentalization profiles of the methanol-extractable, acidic-ethyl acetate soluble (AES) and highly water-soluble (HWS) GAs.[857] "Single" tissues at all stages have less total GA than the "double" tissues. At flowering, the latter contain 10 to 20 times the levels of the former. Seedlings of both forms possess AES and HWS GAs at about equal levels; at flowering virtually all GAs of the "double" are located in the AES fraction, while in the "single" flower bud the GAs are essentially in the HWS fraction. Hence, the two tissues must differ with respect to GA metabolism. Also, kinetin has different effects upon the levels of GA-like substances in the two genotypes suggesting that it interacts with GA metabolism.[858]

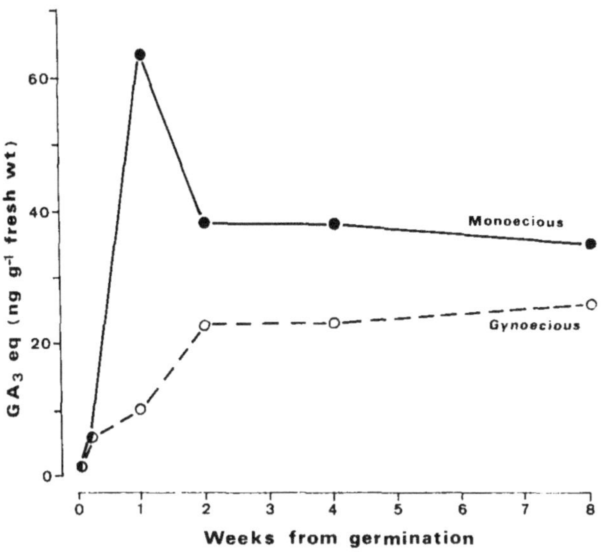

FIGURE 8. Changes in GA-like activity in extracts of monoecious and gynoecious *Cucumis sativus* at several growth stages. (From Hemphill, D. D., Jr., Baker, L. R., and Sell, H. M., *Euphytica*, 21, 285, 1972. With permission.)

As noted earlier, bullhead rose flowers induced by a low temperature treatment have decreased levels of GAs.[852] A higher inhibitor level (as revealed by the dwarf rice second leaf-sheath elongation bioassay) is found in extracts of flowers from low temperature as compared to high temperature conditions. In attempts to reconcile the apparently opposite GA relationships in rose on the one hand and carnation and *Nigella* on the other, we must consider that GAs are unlikely to be the only endogenous growth regulator controlling formation of additional petals since in carnation, increases in the incidence of secondary centers have also been observed in response to auxin and cytokinin application.[61,776] Cytokinin treatment of rose plants grown at a high temperature also promotes appearance of characteristic symptoms of malformed flowers.[852]

D. Sex Expression

Saito and Ito found that LD and high night temperature, which favor male sex expression in cucumber, increase the level of extractable GA-like substances in the stem tip.[859] A few years later Atsmon and associates were the first to relate cucumber sex type with GA content and production.[860] They found higher levels of GA-like substances in diffusates from seedlings and in exudates from the root system of 6-week old plants in a monoecious line than from a gynoecious type. A number of more recent studies with cucumber corroborate these initial findings (Figure 8).[682,861-863] They further show that GA levels are affected by plant age and treatments which influence sex expression. In seedlings and shoot tips of andromonoecious and monoecious lines GA activity reaches a peak 1 to 2 weeks after germination and declines thereafter when tendency to female flower formation increases.[862,863] Chilling gynoecious seeds results in increased maleness and increased GA activity,[862] while treating monoecious plants with CEPA results in decreased maleness, lower GA levels, and increased GA-inhibitor content.[682] CEPA treatment of *Cucurbita pepo* plants cause similar changes in sex expression and endogenous GA level.[841]

Hayashi and associates have identified GA_1 as the active compound of the monoecious cucumber.[861] When seedlings are treated with $[^3H]GA_1$ more radioactivity is found in the diffusate from monoecious seedlings than from gynoecious ones.[860] The same is true of

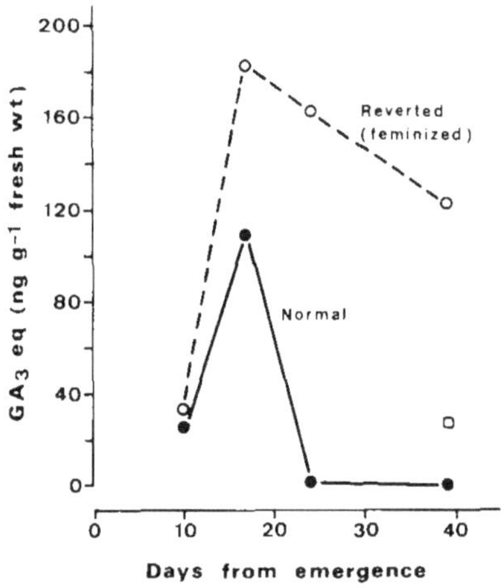

FIGURE 9. Changes in total GA-like activity in extracts of normal and sex-reverted apical inflorescences of *Zea mays* (inbred 66A4-2). Sex reversal was induced by low light intensity. Tassel initiation occurred 17 days after emergence and anthesis on day 39. The open square shows the level in lateral female inflorescences of normal plants on day 39. (Adapted from Rood, S. B., Pharis, R. P., and Major, D. J., *Plant Physiol.*, 66, 793, 1980.)

biological activity in root diffusates from older plants which have been treated with GA_{4+7}.[860] A considerably greater amount of applied GA is recovered from monoecious plants as compared to gynoecious ones, suggesting increased catabolism of GAs in the female lines.

Higher levels of endogenous GAs in either vegetative parts or reproductive structures have also been associated with maleness in hemp,[837] spinach,[838] and *Populus nigra*.[864] Tomato plants of the stamenless-2 mutant (sl_2/sl_2), which bear abnormal stamens, have lower levels of extractable GA-like substances than the wild type in vegetative as well as floral parts.[865]

In some species, however, higher GA levels are associated with femaleness.[246,862,866] For example, seeds and 4-week old plants of andromonoecious and monoecious muskmelon are GA deficient relative to the more female gynoecious and hermaphroditic lines.[862] This result fits well the observation that exogenous GA fails to induce male flowers on gynoecious muskmelon (Chapter 7, Section IV.E). In corn GA-like activity in the apices rises to a maximum at tassel initiation (day 17) and then falls rapidly (Figure 9).[246] At anthesis (day 39), only a trace of GA-like activity is observed in the male apical inflorescence whereas it remains high (particularly in the nonpolar GA region), in the lateral female inflorescence. In terminal apices undergoing sex reversal in low light conditions, large increases in GA-like activity are seen (Figure 9). When microspore meiosis begins in the normal tissue (day 24), GA-like activity of reverting apices is almost 100 times that of normal apices. There are also published results at variance showing that tassel differentiation is correlated with high GA-like substance activity, and female inflorescence formation with decreased GA levels.[840] However, the findings that GA_3 promotes femaleness in corn (Chapter 7, Section IV.E) and that maleness predominates in dwarf genotypes, [867] most of which appear to be deficient in endogenous GAs,[868] along with the results of Rood and associates (Figure 9), clearly favor the idea that high GA levels, particularly nonpolar GAs, favor femaleness in corn.

All these results, taken in conjunction with the well-known effect of exogenous GAs upon sex expression of a variety of plants, strongly support the concept that GAs are an integral

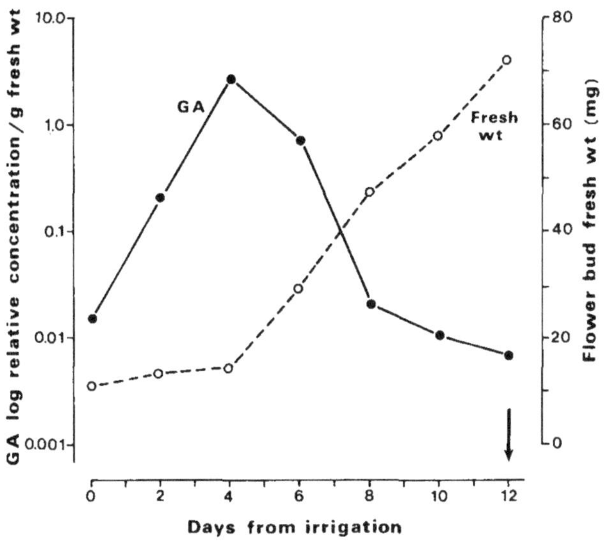

FIGURE 10. Changes in GA content and fresh weight of *Coffea arabica* flower buds collected just before overhead irrigation and at 2-day intervals afterwards until anthesis (arrow). (From Browning, G., *J. Hortic. Sci.*, 48, 29, 1973. With permission.)

part of the regulatory system for sex expression. Why high GA concentrations favor maleness in some plants and female sex expression in others remains unsolved. Further studies are required to determine whether or not these inconsistencies really reflect differences between species which react to GA level per se. No conclusions are possible here since studies of GA biosynthesis, metabolism and compartmentalization, and changing sensitivity of tissues to GA in relation to sex expression are still lacking. Furthermore, the relations between GA and other growth regulators require further elaboration since all classes of growth regulators seem to be involved in the control of sex expression.

E. Breaking Flower Bud Dormancy

In coffee, Browning has shown that flower buds collected 2 to 4 days after overhead irrigation, which releases dormancy, have higher levels of extractable GA than dormant buds (Figure 10).[869] GA levels rise before the rapid increase in fresh weight of the buds which occurs 4 to 5 days after dormancy release. Once rapid growth gets under way GA levels decline as a result of increased bud fresh weight. The data are clearly consistent with the idea that GA is the stimulus for dormancy release in coffee flower buds. Attempts to prolong dormancy by spraying buds with growth retardants have failed, which argues against *de novo* GA biosynthesis at the time release occurs. GA may be converted or released in the buds from a bound or other inactive form or perhaps the growth retardants simply do not penetrate or are inactive in the GA biosynthetic pathway. In 'Gloria' *Azalea* , GA_3 level is high just before anthesis, whereas GA_{4+7} rises during the chilling treatment needed to break dormancy, and peaks at the time plants are moved to higher temperature for forcing; this increase does not occur in unchilled plants.

GA levels in tulip bulbs, as a function of low temperature storage, have been reported in several studies.[871-875] Inconsistent results have been obtained; they will not be reported here in detail because a diversity of techniques for extraction, purification and detection of GAs, and different cvs. have been used as well as different temperatures and durations of the low temperature treatment. Together these differences create a mystifying labyrinth of possibilities. GAs have been found in acidic and neutral forms and the relative amounts of each vary, but that is all we can say that makes sense.

During forcing at 18°C, free GAs increase in whole tulip bulbs while bound GAs decrease rapidly after a slight initial rise.[876] A similar change in the ratio between free and bound GAs occurs in the shoot tissue.[874] At 13°C, there is a marked decline in free GAs and a twofold increase in bound GAs, but when bulbs are transferred to 18°C, free GAs increase to their initial level and bound GAs decrease to a low level. Apparently tulip bulbs possess a temperature-sensitive interconversion mechanism for regulating free GAs using bound GAs as the reserve.

In 'Wedgwood' *Iris* the level of two different substances with GA characteristics increases during the early differentiation of the flower at 13°C.[877] Changes in the incorporation of [^{14}C] mevalonate into terpenoid compounds in the shoot and scale of *Iris* parallel changes in endogenous GA content in these tissues.[878] Although this study does not say what changes in GA synthesis may be expected as a result of low temperature treatment, it is the kind of research that must be extended if we are to understand more about the control of GA levels and their significance.

V. CYTOKININS

A. Changes in Levels in Relation to Normal Development or Abortion

Owing to the marked effect of cytokinin applications in preventing flower abortion in different systems (Chapter 7, Section V.A), there is little question that they can play a significant role in flower development. Cytokinins have been identified in flowers and inflorescences of some species; however, changes in their endogenous level during reproductive development do not permit simple interpretation. A further handicap is that such data are found in relatively few publications.

Zieslin and Halevy observed that total cytokinin activity is higher in leaf extracts from flowering than from non flowering shoots of 'Baccara' rose when flower buds are first apparent to the naked eye (when discrimination between flowering and blind shoots is possible).[834] With 'Golden Times' rose, van Staden and associates have found higher cytokinin activity in shoot tips which are darkened for 8 days than in control shoots kept under full light (Figure 11).[879] All control shoots flower, while only 60% of the darkened shoots produce flowers. Since the determinations are made at an early developmental stage which does not allow discrimination between flowering and non flowering shoots, it is impossible to know whether the increased cytokinin activity under darkness is directly related to the enhanced blindness. This increase in cytokinin level is partly reversed by exposing the etiolated shoots to light for 2 days suggesting that cytokinin accumulation in darkness could result from reduced cytokinin consumption (Figure 11).[879] When comparisons are made between illuminated shoots only, high flower production is correlated with high cytokinin activity in the shoot tips. Upper shoots produce more flowers and have higher cytokinin levels than lower shoots (Figure 11). Similarly, lower shoots on branches with a darkened upper shoot produce more flowers and have a slightly higher cytokinin level than when the upper shoot is illuminated (Figure 11). Hence, it seems that normal flower development in rose is associated with high cytokinin levels as in tomato where aborting inflorescences are severely cytokinin deficient.[856] (Figure 7). Davey and van Staden reported that cytokinin concentration in root exudate declines during inflorescence development in tomato.[880] Since, at the same time, the volume of root exudate per unit time increases, we have calculated that cytokinin flux (flow rate × concentration), after a marked decline during the early stages of inflorescence development, increases from about 2 weeks before anthesis. In contrast, cytokinin flux in root exudate of *Helianthus annuus* markedly increases during the early stages of inflorescence development, peaks long before anthesis and then strongly declines.[881] This decrease probably results from the cessation of root growth which occurs concurrently with inflorescence formation. Flux of cytokinin in root exudate of lupin de-

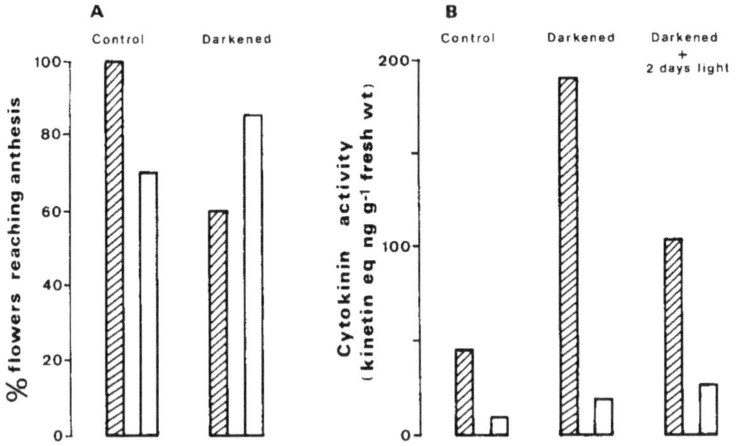

FIGURE 11. Effect of light on (A) flowering response and (B) cytokinin-like activity in extracts of shoot tips of *Rosa hybrida* cv. 'Golden Times'. Well-developed shoots on 3-year-old plants were decapitated above the upper 5-leaflet leaf. Ten days after bud break, the subtending leaves of the two uppermost young shoots were removed. An 8-day dark treatment was imposed to the upper shoot only. Hatched columns: upper shoot. Open columns: shoot second from the top. (From van Staden, J., Zieslin, N., Spiegelstein, H., and Halevy, A. H., *Ann. Bot. (London)*, 47, 155, 1981. With permission.)

creases during inflorescence development,[882] whereas it markedly increases in bleeding xylem sap of *Perilla frutescens* which develops flowers in SD.[883] Cytokinin concentration in xylem sap of *Phaseolus* P_{47} is also higher in SD conditions which stimulate flower bud development to anthesis, than in LD which cause flower bud abscission;[802] unfortunately, we do not know the volumes of exudate collected per unit time in these two photoperiodic treatments. SD, which promote inflorescence development in *Bougainvillea* 'San Diego Red', cause a decrease in cytokinin level in all plant tissues.[884] This decrease is least pronounced in the shoot tips, establishing an increased ratio of shoot tip/leaf cytokinin activity. This change in cytokinin balance, as a result of SD treatment, suggests that the tip could become a relatively better sink, a phenomenon which could be of great importance in the control of reproductive development (Chapter 9). A GA_3 treatment which prevents flower development also reverses most of the SD influence on reduced cytokinin levels.[884]

Relatively high concentrations of nucleotide cytokinins have been found in the inflorescence stalk of *Yucca flaccida* before the opening of the flower buds.[885] High cytokinin levels also occur in leaves of this species during the period of rapid inflorescence elongation.[288] The evidence accumulated by Vonk suggests that cytokinin nucleotides occurring in phloem exudate from the inflorescence stalk originate from the leaves. Thus, removal of leaves decreases cytokinin nucleotides while removal of the underground part (roots plus a part of the rhizome) results in a transient decline followed by an increase up to the initial level in the exudate.[886] Moreover, when the tip of a mature leaf of *Yucca* is removed and the cut surface treated with [^{14}C] zeatin, radioactive zeatin ribotide is found in the exudate collected from a decapitated inflorescence stalk.[887]

Detached *Yucca* stalks fed with [^{14}C] zeatin convert the cytokinin into nucleotides and flower parts hydrolyze the nucleotides to ribosides, [888,889] indicating that cytokinins are metabolized in reproductive parts. Further studies are required to determine the role of cytokinin metabolism in *Yucca* inflorescence development and the true role for the nucleotides and ribosides. Similar questions have been raised in research with *Bougainvillea* reproductive axes where BA is rapidly converted to mono-, di-, and trinucleotides, such that no free BA is detectable 6 hr after application.[420,890]

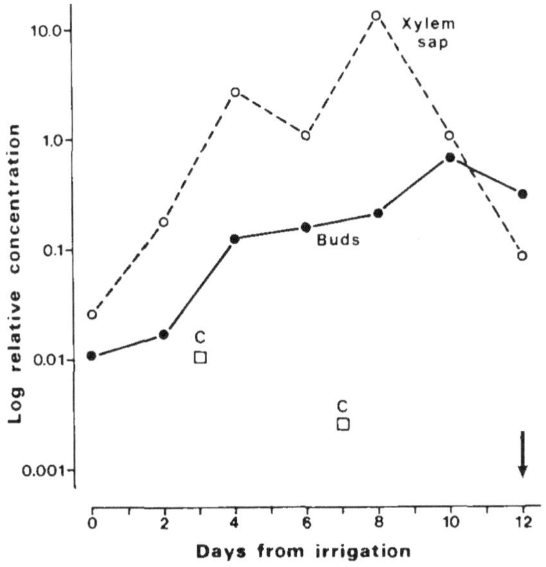

FIGURE 12. Cytokinin-like activity in xylem sap and in extracts of flower buds of *Coffea arabica,* collected just before overhead irrigation and at 2-day intervals afterwards until anthesis (arrow). Open circles for concentrations on a stem fresh weight basis in xylem sap. Closed circles for concentrations on a fresh weight basis in flower buds. The open squares (C) show the cytokinin concentration in xylem sap of control samples collected from adjacent unirrigated trees. (From Browning, G., *J. Hortic. Sci.*, 48, 297, 1973. With permission.)

B. Sex Expression

Endogenous cytokinin levels, in relation to sex expression,have been examined in very few species. Cytokinin activity is high in open female flowers of cucumber and low in male flowers.[891] In hemp cytokinin content increases in young mature leaves during the transition to flowering in both male and female plants.[892] This increase is transient in male but not in female plants where inflorescence development is correlated with a persistently high cytokinin level. Kahlem and associates have found that shoot tips of female plants of the dioecious species, *Mercurialis,* yield higher cytokinin activity than tips of male plants.[893] Using the soybean callus bioassay Louis and Durand have compared cytokinin activity in shoot tips of female plants with that in fertile, semi-sterile, and sterile male plants.[894] They found different cytokinins in different lines and reported that isopentenyl adenosine (IPA) is undetectable in fertile male extracts and increases with the degree of male sterility; high levels of IPA are observed in female plants. These results were not corroborated by Dauphin and associates who showed by GC-MS that IPA levels are always slightly greater in male than in female *Mercurialis* shoot tips. Isopentenyladenine nucleotides and zeatin riboside are also present in both sexes.[895] The most interesting finding of this more recent work is that free zeatin is detected only in females, while the corresponding nucleotides are present in males only. This result suggests that different sexes in *Mercurialis* are associated with different capacities for cytokinin metabolism.

C. Breaking Flower Bud Dormancy

Browning found that when coffee flower bud dormancy is broken by irrigation or rainfall, the level of cytokinin in the xylem sap rapidly increases (Figure 12).[896] Cytokinin activity also increases in the buds but more slowly at first. The xylem exudate and buds yield

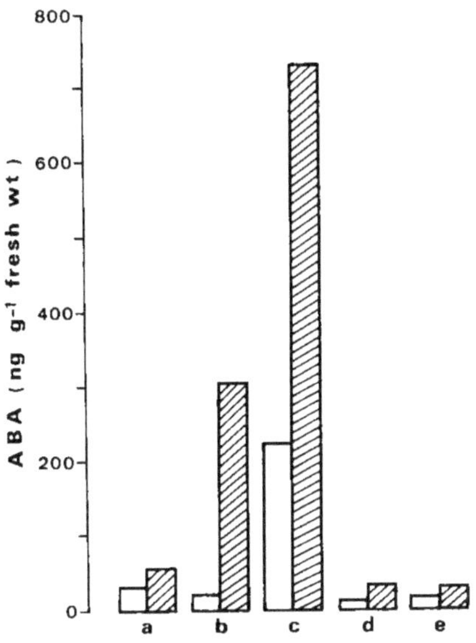

FIGURE 13. Effect of daylength on ABA concentration in the leaves, buds, and stem of *Phaseolus vulgaris* P47 plants. Open columns: favorable 8-hr LD. Hatched columns: unfavorable 18-hr SD. (a) Combined unfolded trifoliate laminas. (b) Unfolding trifoliate laminas. (c) Terminal flower buds. (d) Combined petioles. (e) Stem. (From Morgan, D. G., *Aspects and Prospects of Plant Growth Regulators*, Jeffcoat, B., Ed., BPGRG Monogr. No. 6, 1981, 151. With permission.)

qualitatively different cytokinins (i.e., different Rf in the same chromatographic system) before dormancy release but not after. Thus, if cytokinin transport from the xylem into the buds is required for bud break, we can understand why water stress usually must be relieved before buds can emerge from dormancy. However, the buds themselves must produce some cytokinin since there is an approximately 8-fold increase in cytokinin 2 to 3 days before rapid absorption of water by the bud (on the 4th to 5th day).

VI. ABSCISIC ACID

A. Changes in Levels in Relation to Normal Development or Abortion

ABA has been detected in flower buds of some species.[777,851,869,897] In *Vicia* it is localized mainly in the gynoecium with highest levels at anthesis.[851] ABA concentration also increases in coffee flower buds a few days before anthesis.[869]

Bentley and associates and Morgan have found that leaves and buds of P47 *Phaseolus* plants grown under LD (favoring abortion) contain more ABA than those in SD (promoting flower development) (Figure 13).[777,802] The highest concentration of free ABA is in the aborting buds; these buds also contain considerable amounts of bound ABA. In rose, inhibitor level is higher in the leaves of blind shoots than in those of flowering shoots, particularly after all floral organs are initiated.[834] At least two inhibitors are present. One, identified as ABA, and the other in the neutral fraction may be a derivative.

Increased free ABA is found after a period of water deficit which reduces the growth of both the terminal male and the axillary female inflorescence primordia of maize.[898] Whether the inhibition of growth is due to the elevation of the endogenous ABA content rather than decreased water potential in the meristems is not answered. Identical effects of water deficit, and of applied ABA with no water stress, on growth of the axillary female inflorescences suggest that ABA is the responsible factor. Once the water stress has been relieved growth of the terminal male inflorescence remains inhibited while that of the axillary female inflorescence is promoted. The latter may result from water stress-induced damage to the terminal tassel with consequent reduced apical dominance.

B. Sex Expression

According to Galoch, leaves and inflorescences of female hemp plants contain more ABA-like substances than those of male plants.[837] Similarly, ABA content in leaves of a gynoecious cucumber is twice that of a monoecious line which would suggest a role for ABA in the differentiation of female organs.[682] Friedlander and associates showed, however, that the situation is not that simple.[863] Although SD, which promote femaleness, cause higher ABA content than LD in shoot tips of both monoecious and gynoecious cucumbers, the tips of the monoecious line always have higher ABA content than the gynoecious type regardless of photoperiod. Also, ABA content decreases as cucumber plants, monoecious or gynoecious, age with a concomitant increased female tendency.

Thus, ABA does not appear to be a major factor in sex expression and flower development in cucumber. Different endogenous ABA levels and opposite responses to exogenous ABA of various cucumber sex types may be accounted for by a dose-response curve with an optimum for female flower development,[863] but other growth regulators, especially GAs and ethylene seem to be much more active.

C. Breaking Flower Bud Dormancy

The role of ABA in bud dormancy is still unclear. It accumulates in peach flower buds in the fall until about the time of leaf abscission.[897] ABA levels fluctuate in dormant buds, being relatively high near the end of the rest period, and then decreasing with sufficient chilling to break dormancy.[174,897] Thus, the correlation with the onset and termination of rest is quite good. Furthermore, in seven clones of peach, ABA concentration is directly correlated with the chilling requirement to break dormancy of the flower buds. In December when dormancy is deepest the correlation coefficient is $r = 0.89$.[899] In 'Gloria' *Azalea*, ABA drops to undetectable levels during the chilling treatment at 9°C, but it increases in unchilled plants kept at 19°C.[870]

Several studies suggest, however, that inhibitor levels are not a primary factor in the control of flower bud dormancy. High temperature and high light intensity, which significantly delay floral bud growth in peach, have no significant effect upon ABA concentration in the floral buds,[487] while fall applications of GA_3, which delay flowering the following spring (Chapter 7, Section IV.F), decrease ABA concentration in the flower buds.[899] Mechanical defoliation of sour cherry (*Prunus cerasus)* trees prior the onset of leaf abscission prevents autumnal ABA accumulation (which normally occurs in flower buds of undefoliated trees), but does not stimulate dormancy breaking.[900] Finally, ABA disappearance from dormant buds of *Prunus* during the winter months is equally unlikely to be correlated with the satisfaction of a chilling requirement since it occurs in plants kept at high temperature (22°C) as well as in plants at 4°C.[900] A change in the balance of hormones seems essential since growth promoters, such as GAs and cytokinins, appear to increase as a function of environmental conditions promoting dormancy release.

VII. STEROIDS

Various sterols have been identified in both male and female inflorescences of maize.[901] 24-Methylene-cholesterol, detected in the male inflorescence only, may be associated with pollen. Free sterols of male and female flowers of cucumber are similar in composition.[902] In squash Kopcewicz and Chromiński found that CEPA promotion of femaleness is correlated with higher estrogen content in aerial plant parts.[903] Owing to the limited available information, no role for these compounds in sex expression can be assigned.

CONCLUSIONS

The presence of high amounts of endogenous growth regulators in reproductive structures and changes in their levels during development (combined with the known responses to exogenous treatments) suggest that these chemicals play some essential role in the control of flower and inflorescence differentiation and growth. However, interpreting these findings remains difficult, even impossible, owing to the severe limitations of the techniques used for their extraction and detection and our poor understanding of the mechanism of growth regulator action. There are relatively few studies on metabolism although available evidence indicates that altered metabolism of growth regulators, such as GAs and cytokinins, may be an essential feature in the control of development of reproductive structures.

Variations in the level of all growth regulators characterize each developmental step. Auxin content increases during active growth of the reproductive structures, probably as a result of synthesis in sexual organs, especially the ovary. At this time the elongation of reproductive axes is quite rapid.

Abortion processes are associated with high ethylene evolution, high ABA level and low GA and/or cytokinin levels. A balance of growth regulators seems also to be involved in the control of flower bud dormancy. Growth promoters such as GAs and cytokinins appear to increase as a function of environmental factors promoting dormancy release; ABA level does not always correlate very well with rest.

With regard to sex expression auxin and cytokinin levels as well as ethylene evolution are often higher in systems with increased femaleness; GA level is usually higher when male tendency is promoted. However, none of these growth regulators appear to be morphogens, in the sense that one could say that GA is the male sex "hormone" or that ethylene is the female sex "hormone". As tempting as those statements would be, we believe that growth regulator control of sex expression and flower development in general, is no less complex than that of floral evocation itself.

Growth regulator levels, expressed per unit weight of tissue extracted, appear to be of little value; relative amounts of a number of endogenous regulators may be more useful for the interpretation of the phenomena in question. But, as indicated in the introductory section, it is no longer a question of measuring growth regulator levels alone (no matter how well done): we must have some idea of how tissue sensitivity to the growth regulators is changing.

Our concern for growth regulators seems to be premature but the comprehensive review presented accounts for research in progress. Hopefully some of this research will help, at an unspecified future time, in explaining how flower development proceeds and is controlled.

Chapter 9

PHOTOSYNTHESIS, ASSIMILATE SUPPLY, AND UTILIZATION

TABLE OF CONTENTS

I. INTRODUCTION

In most crop plants patterns of assimilate movement as a function of reproductive development are studied essentially after anthesis. At that time the developing fruit, which we have declared beyond the boundary of this book, becomes the major sink.

Carbohydrates and other assimilates limit fruit growth. This is supported by the inverse relationship between fruit size and number, and by increased fruit yields following girdling, which results in increased carbohydrate levels above the girdle in woody species.[904,905] In many cereals the inflorescences themselves contain very active photosynthetic tissues, providing assimilates principally to developing grains.[906] Impaired photosynthesis and assimilate movement during the postanthesis stages will result in improper fruit and seed development and premature abortion in many species. Thus the ultimate goal of reproduction, viable seeds, is often thwarted even at these late stages by nutritional deficiencies.[907,908]

Do photosynthetic assimilates (and reduced pyridine nucleotides and ATP) limit merely the growth of differentiated inflorescence parts and fruit or is their supply part of the signal(s) for differentiation? Can we identify hormonal or environmental parameters for flower development that operate at the level of assimilate supply and partitioning and can we separate effects upon flower (inflorescence) development from those upon vegetative growth? We will approach these questions by building a case for the importance of photosynthesis for flower development without dealing with many other environmental parameters at the same time. This is often quite unrealistic for thermo- and photo-periodic-sensitive species because of well-known light flux × temperature × daylength interactions (Chapters 4 and 5). These interactions, in fact, are often at the heart of plant adaptation to relatively restricted climatic zones (Chapter 11).

There are good reasons for believing that photosynthesis and photosynthetic assimilates, per se, limit flower development in a very special way. We have already presented evidence showing the relatively high light flux requirement for flower development compared to that for vegetative growth (Chapter 4). Now we shall deal with additional, closely related evidence.

II. LEAF AREA

Since Winkler's early study with grapevine it is usually inferred that minimal leaf area during the early stages of inflorescence development is essential for floret initiation.[909] Continuous defoliation in grapevine inhibits late stages of inflorescence development, too, so that anthesis is prevented and buds abort prematurely.[910] Leaf initiation continues unabated through the period of inflorescence bud failure indicating that the terminal vegetative bud is a better competitor for stored reserves than the inflorescence bud.

Ramina and co-workers showed that inflorescence development in *Bougainvillea* is a direct function of leaf area, a relationship expected if photosynthetic assimilates limit development.[539] They found that inflorescence number and earliness to flower are linear functions of leaf area in the 10- to 40-cm^2 range (Figure 1). Dry weight and soluble solid percentage in the shoots bearing inflorescence axes are linear functions of leaf area, too, indicating clearly that synthesis of major metabolites for these shoots is also a function of leaf area.

van Meeteren suggested that flower bud abscission in 'Enchantment' lily is not strongly correlated with leaf area.[911] In conditions of relatively low irradiance (and continuous illumination), leaf area can be reduced by 50% with little effect on flower bud abscission. However, further reduction in leaf area, or putting plants in darkness for about 9 days, greatly increases abscission. Immersion of bulbs in a STS solution before planting reduces net photosynthesis and increases dark respiration of the emerged leaves, yet essentially eliminates abscission. Thus, he concluded that there are other more important factors limiting flower development, some of which probably involve inhibitory actions of ethylene. Leaf

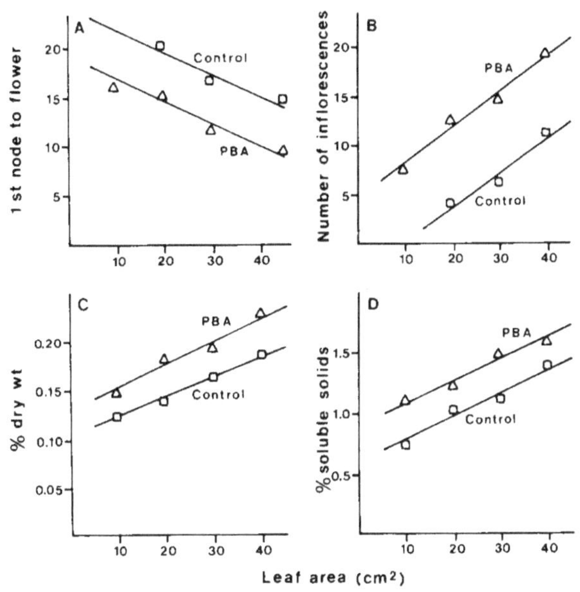

FIGURE 1. Regression lines for relationships between leaf area and inflorescence development in *Bougainvillea* 'San Diego Red'. Results are expressed as first node to flower (A) or number of inflorescences developing (B); dry weight gain (C), and soluble solids (D) in the shoot apical region. PBA at 500 mg/ℓ was applied every 4 days to shoot tips in 10-μℓ droplets. (From Ramina, A., Hackett, W. P., and Sachs, R. M., *Plant Physiol.*, 64, 810, 1979. With permission.)

removal experiments by Mastalerz with 'Croft' lilies, combined with results from other experiments affecting assimilation and assimilate transport, indicate that current photosynthesis plays a vital role in flower bud development; failure in the assimilatory system, that reduces carbohydrate availability to the flower, increases bud abortion.[425]

III. ALTERING PHOTOSYNTHETIC RATES WITH CO_2 AND DCMU

A. Increased CO_2

Influences of CO_2 concentration on inflorescence development are expected if indeed light flux relationships are mediated by photosynthetic processes. The results of a few studies indicate that flower and inflorescence development is enhanced quantitatively and qualitatively by CO_2 levels above normal ambient levels. Overall increases in dry weight gain are often so large when CO_2 levels are raised that it may be difficult to detect specific influences upon reproductive structures. Doubling the CO_2 level in field-grown wheat between the time of inflorescence initiation and anthesis increases the number of fertile spikes as well as the number of fertile florets per spike.[912] In field-grown barley preanthesis CO_2 enrichment greatly increases harvestable grain number suggesting that, here too, fertile floret number per inflorescence is increased.[913] In this instance the elevated CO_2 treatment was not timed as closely with inflorescence initiation as is desired for specifying an effect upon inflorescence development.

Hughes and Cockshull found that in *Chrysanthemum,* inflorescence development is accelerated with increasing CO_2 treatments (Table 1), particularly under low light conditions.[914] Inflorescence weight, number of inflorescences, and ray floret number per inflorescence are also increased in response to increased CO_2. As expected from the early increase in inflo-

Table 1
EFFECT OF CO_2 LEVEL ON EARLY CAPITULUM DEVELOPMENT IN 'BRIGHT GOLDEN ANNE' *CHRYSANTHEMUM MORIFOLIUM* GROWN IN 8 HR-SD

	CO_2 concentration ($\mu\ell/\ell$)				
	Experiment 1		Experiment 2		
Light integral (J/cm/day)	325	600	325	600	900
250	8.7	8.9	9.0	10.0	9.7
125	7.5	8.0	6.6	8.3	—
63	3.3	3.8	2.4	2.6	4.0

Note: Results of 2 different experiments are expressed as mean score at 76 days.

From Hughes, A. P. and Cockshull, K. E., *Ann. Bot. (London)*, 35, 899, 1971. With permission.

rescence development, a higher proportion of the plant dry weight is found in the inflorescences in the high CO_2 treatments. These results show clearly the importance of photosynthetic input at early stages of inflorescence morphogenesis when floret number (and inflorescence number) is determined.

CO_2 enrichment of greenhouse-grown roses has given some increase in number and quality of harvested flowers.[915] Since the period of enrichment is for 3 months or longer, it is impossible to specify an influence on inflorescence development alone. The studies of Kinet and Calvert and Slack with tomato show clearly that CO_2 enrichment promotes inflorescence development.[430,916] In Kinet's work, enhanced CO_2 levels partly negate the inhibitory effect of a 10-day low light treatment during the period most critical for inflorescence development to anthesis (Chapter 4, Section III.B).[430] Since 1% CO_2 was used (30-fold greater than normal ambient levels), interactions with ethylene-controlled systems could be expected. A role of ethylene in the induction of tomato inflorescence abortion was already suspected (Chapter 7, Section III.A). Other physiological systems are affected as well when CO_2 concentration is at or above 0.1%. Data on plant dry weight and/or photosynthetic rates, e.g., as for *Chrysanthemum* and roses,[914,917,918] are necessary if we are to say that high CO_2 benefits on inflorescence development are mediated through photosynthesis.

B. Reduced CO_2

In 'Wedgwood' *Iris*, Mae and Vonk found that reducing CO_2 levels to 0.003%, at a time when flower abortion is expected, does not increase bud abortion.[436] This is surprising since relatively high irradiance is required for good flower development to anthesis. Photosynthesis is promoted at high irradiance since, when CO_2 levels are normal, there is increased dry weight gain in plants compared to those in low irradiance. Obviously, at the low CO_2 levels imposed, photosynthesis is severely restricted; hence, Mae and Vonk have concluded that the major beneficial effect of high irradiance on flower development is on some other physiological system which favors assimilate redistribution to the developing inflorescences (reviewed below).

C. DCMU

Sawhney and Cumming found that DCMU [3(3,4-dichlorophenyl)-1,1-dimethylurea] persists in the SDP *Chenopodium rubrum* for several days after application,[919] and questioned whether earlier results, showing DCMU-induced inhibition of flower initiation,[920] were due to pre- or postinitiation effects. They applied ascorbic acid or dichlorophenol-indophenol, two compounds that overcome DCMU-induced inhibition of photosynthesis, at the end of the SD-inductive treatment and found that flowering is restored to normal. Thus, they concluded that DCMU-induced inhibition of flowering is probably the result of inadequate photosynthesis for flower development during the period following flower initiation. In a critical experiment, when DCMU was applied immediately following the inductive treatment, they could not demonstrate DCMU-induced inhibition of flowering without at the same time inhibiting vegetative growth.

IV. SINK ACTIVATION: INHIBITION OF COMPETING SINKS

Developing inflorescences are not necessarily stronger sinks than newly initiated leaves, which are net importers of assimilates. Indeed, at early stages, inflorescences are clearly weaker sinks in *Lolium*,[921] *Bougainvillea*,[541] grapevine,[922] and probably tomato.[923] Sink strength of many flowers increases very sharply upon development of the ovary and even then may not be strong enough to "outcompete" developing leaves of vegetative shoots. We review below several studies showing that environmental and chemical parameters that promote inflorescence development appear to do so by activation of the reproductive structures or by inhibition of competing sinks.

A. Chilling

Studies with bulbous plants provide some of the most convincing data showing that conditions promoting flower development do so by favoring assimilate transfer from the leaves (or storage organ) to the flower bud, i.e., they appear to promote sink strength of the flower. In chilled bulbs of 'Wedgwood' *Iris*, Rodrigues Pereira found that starch and soluble sugar levels start to increase in the apical region before flower initiation and continue to rise dramatically afterwards during development (Volume II, Chapter 8, Figure 5).[924] This carbohydrate accumulation, not found in bulbs remaining vegetative, is accompanied by a decrease in the dry weight, soluble sugars, and starch contents of the scales suggesting that these organs are major sources of carbohydrates for the apex before and after flower initiation.

Ho and Rees' thorough studies of the carbon economy of tulips leave little doubt that the developing flower and scape derive most of their substrates for growth from current photosynthesis.[925-927] This is particularly true during the late stages of flower development, before anthesis, when storage bulb reserves have fallen to a very low level. By this time about 70% of the exported reserves have been translocated to developing leaves with the roots importing a large portion of the remaining fraction; hence the developing flower must depend heavily on current photosynthesis. Nevertheless, during the early stages of flower bud development, before new leaves have emerged and when there are no current assimilates, mobilization of stored reserves appears to be critically important to the tulip flower. Moe and Wickstrøm and Aung and co-workers have shown that the precooling treatment of stored tulip bulbs, well known to promote high quality flowers, causes carbohydrate interconversions (polyfructosans to oligosaccharides and starch).[84,491,928] Ho and Rees' data show that small amounts of reserve materials from the scales have moved to the developing flower by the time rapid growth is to begin and, although there are no unequivocal results showing that the soluble oligosaccharides are among the transported materials that may be essential for early flower development, the circumstantial evidence favors this interpretation.[925-927]

B. Irradiance of Reproductive Structures

Several papers pertaining to high irradiance-induced activation of reproductive structures

Table 2
DISTRIBUTION OF LABELED ASSIMILATES IN 'WEDGWOOD' *IRIS* PLANTS AS AFFECTED BY A DARK TREATMENT FROM THE 14TH TO THE 7TH DAY BEFORE ANTHESIS

	Dry weight (mg)		% of ^{14}C translocated out of source leaf in 24 hr	
Plant part	Control	Darkened plant	Control	Darkened plant
Leaves	2143	2163	3.7	18.7
Stem				
Internodes 1.2	94	136	2.8	14.8
Internodes 3.4	238	256	1.5	5.5
Internodes 5.6	305	271	2.2	2.3
Internode 7	165	77	15.5	0.1
Flower	517	108	58.2	0.1
Mother bulb	453	388	1.7	5.5
Daughter bulb	65	41	2.7	14.0
Bulb disk	232	217	5.0	19.5
Root	1060	695	6.7	19.5
Total	5272	4352	100	100

Note: $^{14}CO_2$ pulse was applied to the 2nd leaf 5 hr after the end of the dark treatment.

From Mae, T. and Vonk, C. R., *Acta Bot. Neerl.*, 23, 321, 1974. With permission.

were reviewed in Chapter 4. Work with 'Wedgwood' *Iris* demonstrates clearly that high irradiance increases flower bud dry weight much more so than whole plant dry weight.[436] Also, a dark period a week prior to anthesis almost totally prevents dry weight gain by the flower bud, but has very little effect on whole plant dry weight. The results with $^{14}CO_2$ supplied to control and dark-treated plants confirm that the flower buds of the latter have almost zero sink strength (Table 2). The dark treatment in some manner inhibits growth of the developing bud more than that of other surrounding tissues; note that bud dry weight in the dark-treated plants is about 20% that of the controls. There are other changes, too, showing that allocation of assimilates to other sink tissues is reduced by the dark treatment. That is, the phenomenon of light activation of sinks is not restricted to flower buds alone. Also, the light activation phenomenon does not seem to involve photosynthesis directly. Mae and Vonk have shown that high light requirements for flower bud development can be fulfilled at very low CO_2 levels, where overall plant dry weight is reduced but flower bud weight is equal to that of the controls.[436] With reduced irradiance and normal CO_2 flower bud weights are severely reduced.

Mor and Halevy found that darkening young rose shoot tips sharply reduces import of [^{14}C] assimilates from a nearby source leaf (Table 3).[434] There is no significant difference in $^{14}CO_2$ fixed and in the proportion of assimilated carbon exported from the labeled leaf in plants with normally lighted or darkened shoot tips, but the distribution of label within the plants is quite different. With darkened shoot tips all parts of the plant show increased ^{14}C with a sharp decrease in the developing shoot tip. This study is not complete with respect to assimilate utilization, but it does show clearly that both the methanol-soluble and insoluble

Table 3
DISTRIBUTION OF LABELED ASSIMILATES IN THE ROSE (*ROSA HYBRIDA* cv. 'MARIMBA') PLANTS AS AFFECTED BY DARKENING UPPER YOUNG SHOOT FOR 7 DAYS

	% of ^{14}C translocated out of source leaf in 16 hr		RSA	
Plant part	Control	Darkened shoot	Control	Darkened shoot
Young shoot (total)	81.8	38.1	3.6	3.0
Parts of young shoot				
Tip	5.1	0.6	22.8	1.7
Leaf (3rd from bottom)	10.8	10.0	1.9	6.7
Other parts (stem and other leaves)	65.9	27.5	3.7	2.1
Other plant parts (excluding roots)	13.4	46.0		
Roots	4.8	15.9		

Note: The length of the shoot at start of darkening is 2.8 cm. $^{14}CO_2$ pulse is applied to a mature leaf one node under the young shoot (the leaf subtending the young shoot is removed). Relative specific activity (RSA) in a given plant part is computed by dividing specific activity (cmp/g dry weight) of this organ by specific activity of the whole plant.

From Mor, Y. and Halevy, A. H., *Plant Physiol.*, 66, 990, 1980. With permission.

^{14}C fractions change nearly equally as a result of treatment. When the data are transformed into relative specific activity (RSA), the effect of darkening appears even more pronounced, i.e., ^{14}C in the darkened shoot tip is only about 7% that of the control (Table 3). ATP levels in the control and darkened shoot tips are approximately the same. With sink enhancement one would expect to find increased ATP turnover, and, with frequent sampling, at least a transient drop in ATP level in the shoot tip.

C. Daylength

Late-flowering cvs. of *Chrysanthemum,* that initiate terminal inflorescences in LD but more rapidly in SD, have an absolute requirement for SD for development to anthesis,[482] except when lateral buds below the inflorescence are removed.[153] This suggests that SD enhance sink strength of the capitulum relative to that of the lateral buds.

Shillo and Halevy found that LD-induced improvement of inflorescence quality in *Gladiolus* is accompanied by relatively greater dry weight accumulation in the inflorescence stalk.[237] In LD the dry weight ratio of inflorescence stalk to corm increases quite rapidly when compared to that of the plants in SD. This is particularly apparent from spike emergence through anthesis (Figure 2). Paradoxically inflorescence development is delayed up to 2 weeks in LD compared to SD. Thus, although there is increased sink strength in the inflorescence axis, the overall rate of development, as measured by anthesis, is not increased. LD appear to prolong the "filling" period for each floret and the stalk (permitting more florets to develop to anthesis), most likely through increasing sink strength of the entire stalk relative to that of the corm. An equal possibility, of course, is that SD promote sink strength in the corm relative to that of the stalk. This is a question of not knowing the

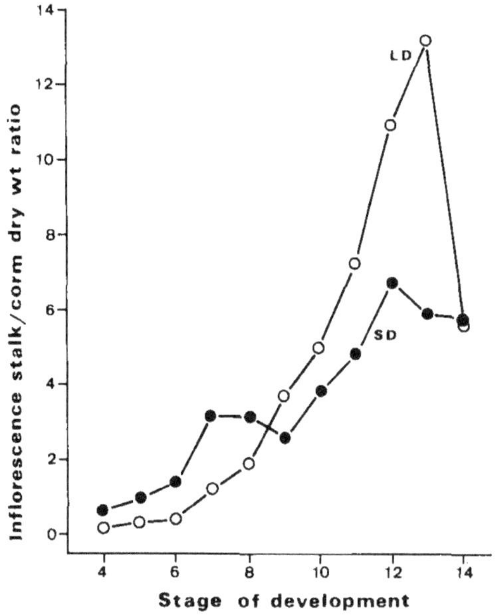

FIGURE 2. The effect of daylength on distribution of dry matter between inflorescence stalk and corm at various stages of development of *Gladiolus grandiflorus* cv. 'Spic and Span'. Results are expressed as the ratio of dry weight of the stalk to the corm at each developmental stage. Stages 1 to 8: 1 to 8 expanded foliage leaves; Stage 9: spike emergence; Stage 12: 1 or 2 florets open; Stage 14: all florets open. SD = natural short day conditions of the winter in Israel. LD = natural short days accompanied by dark periods interrupted in the middle by 4 hr of 100 to 200 lx incandescent light. (From Shillo, R. and Halevy, A. H., *Sci. Hortic.*, 15, 187, 1981. With permission.)

mechanism of sink enhancement and inhibition. LD will also promote leaf growth in *Gladiolus* when given at an early stage of plant development, thereby increasing the total source tissue and increasing assimilate supply to both corms and inflorescences.

Studies on transport of [^{14}C]-labeled assimilates confirm that both the corm and developing inflorescence compete for current photosynthates.[929,930] Both daylength and cv. appear to play large roles in determining which organ has the greater relative sink strength. In the cv. 'Spic and Span', grown outdoors in Israel during the wintertime, the corm is relatively an equal competitor with the inflorescence, whereas in 'Eurovision', in 12-hr daylengths and 22 to 27°C constant temperature regimes, the inflorescence, from the time when it is still hidden within leaves up to anthesis, has a 4- to 6-fold greater relative sink strength than the corm. It would be of interest to know whether in 'Eurovision', held in shorter daylengths, the corm would become a relatively stronger sink.

SD promotion of inflorescence development in *Bougainvillea* is accompanied by a fairly rapid increase in [^{14}C]-assimilate accumulation in the reproductive axes, but this phenomenon is observed most clearly when the young leaves are removed and the shoot tips are treated with a cytokinin (see below).[541]

D. Growth Regulators

Hale and Weaver demonstrated that a developing grapevine inflorescence is a relatively weak sink and does not compete well with young leaves for [^{14}C] assimilates exported from

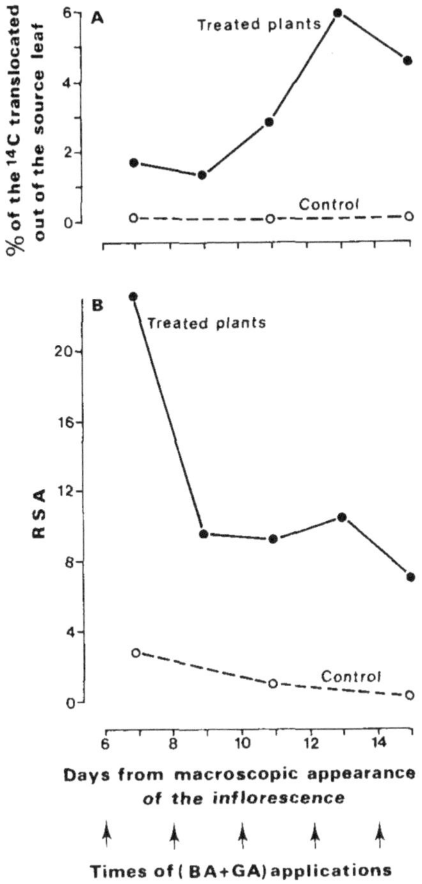

FIGURE 3. [^{14}C] accumulation in the inflorescence of *Lycopersicon esculentum* 'King plus' as a function of a BA + GA treatment in low light conditions. The inflorescences of treated plants develop until anthesis; those of control plants abort. Results are expressed: (A) as a percentage of the [^{14}C] translocated out of the fed leaf and (B) as RSA calculated as in Table 4 of this chapter. BA + GA treatment: 5 applications of BA 10 mg/ℓ and GA_{4+7} 25 mg/ℓ on the inflorescence from the 6th day after its macroscopic appearance. (From Léonard, M., Kinet, J. M., Bodson, M., and Bernier, G., *Physiol. Plant.*, 57, 85, 1983. With permission.)

a mature leaf.[922] Inflorescence clusters become relatively strong sinks only after floret anthesis when the ovaries are stimulated to rapid growth by pollination.[922,931] Srinivasan and Mullins found that cytokinins promote differentiation of anlage primordia to inflorescences rather than tendrils (Chapter 7, Section V.A),[238,784] presumably through well-known cytokinin-induced sink enhancement in grapevine.[932] Thus, throughout most of inflorescence initiation and development in grapevine, one may assume that photosynthetic assimilates can be limiting factors for differentiation. It is during the preanthesis stages of inflorescence development that GA and cytokinin treatments to the inflorescences enhance yield in the grapevine, the result expected if sink enhancement is critical at this time.

In *Chrysanthemum* plants held in LD a combined BA and GA treatment of the shoot apex advances inflorescence initiation and floret development (Chapter 7, Section V.A), and has been found by Tjosvold to increase [^{14}C]-assimilate movement from a $^{14}CO_2$-fed leaf (all other leaves removed).[933] If the plants are not girdled below the source leaf, no increase in [^{14}C] is seen in the apical tissues of the hormone-treated plants until the third week of treatment when inflorescence development is at the floret initiation stage. Even though the leaf remaining is the only major photosynthetic source, apparently reserves in shoot tissues below are major suppliers to the apical bud, diluting substantially [^{14}C] contributions from the fed leaf. This problem of alternate sources of substrates is potentially troublesome in all species in which a single leaf is labeled. The nonlabeled pools of substrates and sources of supply must be identified and quantified before firm conclusions on sink enhancement can be taken from research with labeled compounds. There are difficulties in demonstrating unequivocally that growth regulators, SD, and lateral bud removal increase sink strength of the young capitulum, but this seems a better hypothesis in accounting for results with *Chrysanthemum* than one proposing specific morphogenetic action.

In constant low light conditions, BA + GA treatments cause a five- to tenfold increase in sink strength of the young inflorescence cluster in tomato, as measured by RSA (Figure 3).[203] Note that the induced increase is greatest 1 day after the first of the five applications required for full development of the inflorescence to anthesis. Léonard and co-workers have shown that [^{14}C]-assimilate movement to the developing inflorescence of tomato plants is a function of stage of inflorescence development, irradiance, and growth regulator applications.[203,934] Under high light conditions there is a marked decline in relative sink strength in the inflorescence between the 6th and 12th day after its macroscopic appearance, which corresponds with the period when low light treatment promotes maximum abortion.[934] It appears, too, that the stem below the inflorescence becomes a progressively stronger sink as the inflorescence approaches anthesis, suggesting that there are more complex physiological changes occurring in other portions of the tomato plant.

E. Competing Sinks — Nutrient Diversion

Inflorescence development as a function of nutrient diversion to competing sinks has been studied extensively in *Bougainvillea* by indirect and direct methods. Ramina and coworkers found that inflorescence development is an inverse function of number of branches developing (Chapter 6, Figure 1);[539] altering branch number is used as a means of diverting or partitioning assimilates away from reproductive axes. Percent dry weight and percent soluble solids in flowering shoots are also inverse functions of axillary branch number, an expected result if increased branch number does indeed reduce assimilate supply to each branch. Also, apical applications of a cytokinin promote inflorescence development, percent dry weight and soluble solids in treated branches, supporting the idea that sink strength as well as assimilate supply limit inflorescence development. Studies with [^{14}C]-fed plants indicate that assimilate accumulation in the terminal bud and inflorescence primordia is rapidly enhanced by environmental (SD) and chemical parameters that promote inflorescence development (Figure 4). Young leaf removal in *Bougainvillea* promotes inflorescence development and [^{14}C]-assimilate accumulation in shoot tips and inflorescence primordia independently of cytokinin treatment. Since there are greater than additive responses to the combined defoliation and cytokinin treatments, it is unlikely that both are affecting the same systems.

Young leaves are net importers of assimilates and presumably their removal saves assimilates for inflorescence development by removing strong competing sinks. The data with [^{14}C] assimilates support this view (Figure 4). However, young leaf removal reduces growth regulator supply to apical tissues and this may also contribute to increased inflorescence development. Stems of plants from which young leaves are removed elongate more slowly, as if sources of shoot growth promoters were removed. Indeed bioassays of diffusates from young, expanding leaves of other species indicate that they are centers of auxin and GA synthesis.[935] GA_3, which is a powerful inhibitor of inflorescence development in *Bougainvillea,*[231,705] does not inhibit transport of assimilates into flowering shoots nor prevent ac-

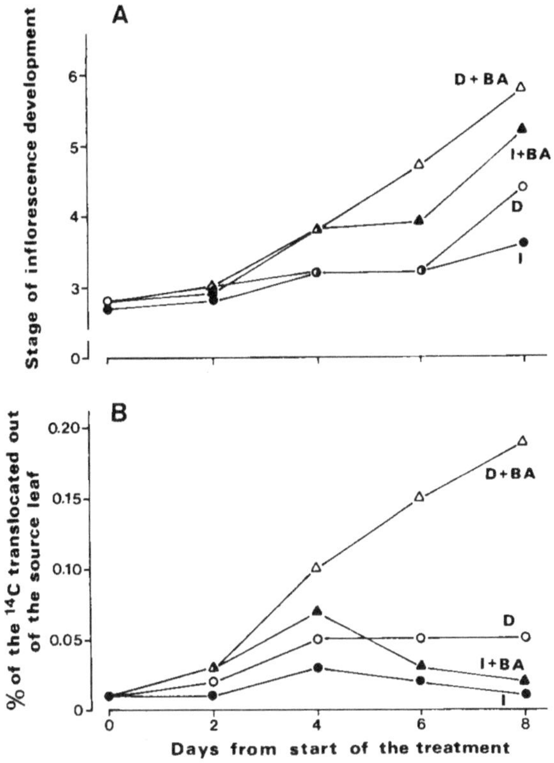

FIGURE 4. Development (A) and [^{14}C] accumulation (B) in the inflorescence of *Bougainvillea* 'San Diego Red' grown in SD as a function of defoliation and BA treatment. The stage system for morphological development in A is the same as in Chapter 2, Figure 1. In all plants — both the so-called intact plants (I) and the defoliated plants (D) — the leaves below the 10th, counting basipetally from the first unfolded leaf, are removed. The node of the first unfolded leaf is termed n. Nodes below n are designated n − 1, n − 2, etc. In defoliated plants, leaves at nodes n through n − 5 are removed at the time of treatment and new leaves are continuously removed as they unfold during the course of the experiment. BA at 500 mg/ℓ is applied every 4 days to shoot tips in 10-μℓ droplets.

cumulation of soluble solids.[539] It does, however, increase the rate of growth of young leaf primordia, inflorescence bracts and the peduncle at the same time that it prevents development of the reproductive meristem beyond bracteole initiation.[420] Thus, GA_3 stimulates sinks competing with the developing inflorescence primordia, and, if young leaves in *Bougainvillea* are sources of GAs that act in the same way as exogenous GA_3 applications, a nutrient diversion hypothesis readily accounts for the promotion of inflorescence development by young leaf removal (see Volume II, Chapter 8, Section VIII for a more complete treatment of the argument).

Studies by Harris and Jeffcoat show that import of [^{14}C]-labeled assimilates (current photosynthates) by carnation flowers begins quite early during their development and is determined to some extent by growth and metabolism of the stem tissue immediately below the flower.[936] Their [^{14}C] data do not permit evaluation of relative sink strength, but the results show without question that stem tissue is the major competitor of the flower for photosynthetic assimilates (at least for the source leaf selected).

Young leaf removal also increases the number of flowers in the first inflorescence of

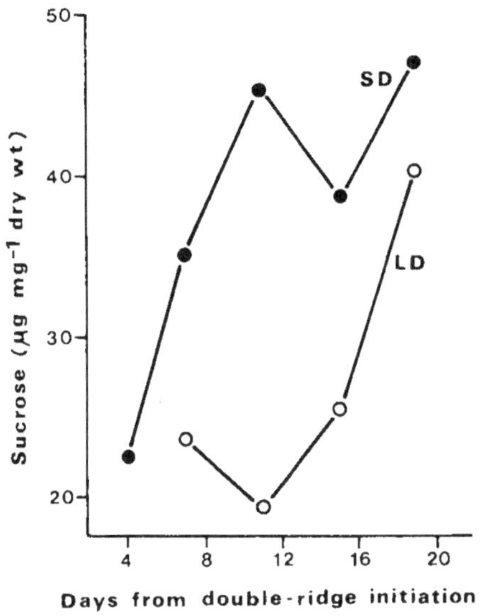

FIGURE 5. Sucrose concentration in the apex of *Triticum aestivum* cv. 'Warimba' as affected by daylength. LD = plants grown in constant 16-hr days at an irradiance of 225 μE/m²/sec¹; SD = plants transferred to 8-hr days at an irradiance of 475 μE/m²/sec¹, 4 days after double-ridge initiation. Stamen initials are present 10 days after double-ridge initiation in LD and 5 days later in SD. (From Mohapatra, P. K., Aspinall, D., and Jenner, C. F., *Ann. Bot. (London)*, 49, 619, 1982. With permission.)

tomato (Chapter 6, Section II.B).[542,596] This result, as in *Bougainvillea,* is explained in part by removal of competing sinks and, hence, improved nutritional advantage to the inflorescence. Another correlative growth phenomenon in tomato, possibly closely related to nutrient diversion, is the reduction of inflorescence abortion in decapitated plants held in low light.[546] Since the developing inflorescences occupy lateral positions on the stem, it is likely that they would benefit by removal or reduction of apical dominance. This beneficial effect is partly eliminated by adding IAA to the cut surface of the decapitated plants. BA + GA treatments that promote inflorescence development also promote growth of vegetative lateral buds on intact plants, indicating that similar controls may be exerted on both kinds of buds.[546]

V. CARBOHYDRATE LEVELS

A. Carbohydrates in Inflorescences in Relation to Supply and Demand

Mohapatra and co-workers found that sucrose concentration in the apices (including inflorescence primordia) of spring wheat is generally higher in SD than LD during inflorescence development, even though the light integral is the same in both conditions (Figure 5).[937] Sucrose concentration falls in LD apices until a terminal spikelet is initiated, and then increases until it is not significantly different from that in SD apices. The authors reasoned that fluctuations in sugar levels: (1) indicate less than adequate control from reserves and (2) are responses to changes in supply and demand. They assumed that supply is the same in the two treatments because they have established equal light integrals in LD and SD, but demand differs greatly in the LD and SD plants since leaf (shoot) and apical growth rates

Table 4
CALCULATED SUCROSE CONSUMPTION IN THE GROWTH OF SHOOT AND APEX OF *TRITICUM AESTIVUM* CV. 'WARIMBA' DURING THE PERIOD FROM 4 TO 10 DAYS AFTER DOUBLE-RIDGE INITIATION FOR PLANTS GROWING IN LD AND SD AT THE SAME IRRADIANCE (375 μE/m²/sec)

Daylength (hr)	Estimated total cost (mg sucrose)		Apex estimated mean daily cost (μg sucrose)	Free sucrose in apex (μg)	
	Shoot	Apex		6 days from double-ridge initiation	10 days from double-ridge initiation
16	331.2	0.1204	20	0.27	2.81
8	214.8	0.0068	1.1	0.22	0.51

From Mohapatra, P. K., Aspinall, D., and Jenner, C. F., *Ann. Bot. (London)*, 49, 619, 1982. With permission.

are considerably higher in LD than SD. They calculated that apices in LD have a nearly 20-fold greater demand for sucrose than apices in SD (Table 4). The daily sucrose requirement for growth exceeds by 7- to 70-fold the amount found in the LD apices and by 2- to 5-fold that found in the SD apices (this calculation excludes respiratory requirements). Hence, the apex is certainly dependent on a continued supply of C skeletons from photosynthesizing tissue; the LD inflorescence apex would appear to be more limited by supply than the SD apex particularly up to stamen initiation.[937]

In 'Wedgwood' *Iris* flower buds Mae and Vonk measured carbohydrate levels in relation to abortion. An important point is that the measurements combine polysaccharides and soluble sugars (Table 5).[436] The three treatments resulting in ascending degrees of abortion are high light (56 W/m²), low light (11 W/m² starting at 20 days before anthesis), and darkness (from the 14th to 7th day before anthesis and then 56 W/m²). The data for abortion come from several experiments that may not have been run at the same time as those in which carbohydrate levels were determined. Note that total carbohydrates are lower in conditions favoring abortion, yet no exact correlation is apparent. In fact, the lower levels found in the weak light compared to dark treatment at 2 days before anthesis is not reflected in lower flower bud anthesis; fewer flowers abort in low light than in dark conditions. We suppose then that the data for carbohydrate levels represent varying rates of (1) sugar and starch consumption by the flower, (2) mobilization of reserves from the stem, and (3) contributions from current photosynthesis. The physiologically significant carbohydrate pool for development is not apparent in these data. If there is such a pool, it can be identified by separation of starch from sugars and by including [^{14}C] data reflecting contributions from current photosynthesis.

In tomato inflorescences developing in low irradiance conditions combined with a BA + GA treatment, soluble sugars and starch increase for only a few days (Figure 6).[202] Later, when the growth regulator treatment is halted and the inflorescences develop rapidly sugar levels eventually fall below that of other treatments (even those that induce abortion) indicating that, at this time, sugar demand exceeds supply. Starch levels continue to climb in inflorescences receiving constant high irradiance. In this case apparently sucrose (and other sugar) supply exceeds inflorescence demand and finds its way into the starch reserves.

Table 5
EFFECT OF ILLUMINATION TREATMENT ON ABORTION AND CARBOHYDRATE CONTENT OF 'WEDGWOOD' *IRIS* FLOWER BUD

		Carbohydrate content (mg glucose eq/g dry wt)	
Treatment	Abortion (%)	9 days before anthesis	2 days before anthesis
Control (56 W/m^2 throughout)	0	265	297
Weak light (11 W/m^2 starting at 20 days before anthesis)	42	169	73
Darkness from 14th to 7th day before anthesis	79—100	237	90

From Mae, T. and Vonk, C. R., *Acta Bot. Neerl.*, 23, 321, 1974. With permission.

Sucrose, glucose, and fructose concentrations in *Bougainvillea* inflorescences are not markedly higher just before or during stages of rapid development towards anthesis (high light flux; BA application) than when inflorescences are on a path towards abortion.[420] Also, in *Chrysanthemum* apical buds, sucrose, glucose, and fructose levels are not increased by BA + GA treatments that greatly accelerate apical growth rate (leaf initiation) and advance floret initiation and development in noninductive conditions.[438]

The general significance of these findings is that sugars, amino acids, and other vital nutritional requirements for growth may decrease, rather than increase, when demand (inflorescence growth) exceeds nutrient supply. At best, in rapidly growing tissues, if there is an adequate reserve (e.g., starch, storage proteins) concentrations of some essential materials may remain constant or decrease for short periods (until reserves are mobilized). In slow-growing tissues the translocated metabolites may increase for short periods if there is a sudden increase in supply, exceeding growth demands of the tissues. Studies of metabolite turnover are more likely to reveal the true nature of assimilate utilization (and sink activation).

B. Enzyme Activities and Their Relationship to Substrate Levels

Russell and Morris found that, in tomato inflorescences, reducing sugars and invertase activity increase in parallel fashion with treatments preventing abortion.[938] They used low and high planting densities to effect changes in light levels within the canopy, finding very high abortion in the high density group.

There is a noteworthy discrepancy between Kinet and co-workers (Figure 6) and Russell and Morris' findings for reducing sugars that may be accounted for by differences in inflorescence growth in the two studies.[202,938] The former found no reduction in soluble sugar levels in inflorescences that have stopped growth in low light conditions (zero demand for assimilate), whereas the latter observed decreased reducing sugar concentrations with high density plantings. In this case, the inflorescences continue to grow at a slow rate; then we may have an example of lowered demand still exceeding assimilate supply, with consequent reduction in sugar levels.

Russell and Morris suggest that regulation of invertase activity, particularly in the developing stamen, may be controlling sucrose (and other assimilate) transport to the inflorescence.[938] They find no evidence to support the view that invertase activity is substrate limited and, hence, do not believe that increased sucrose movement into the inflorescence is the

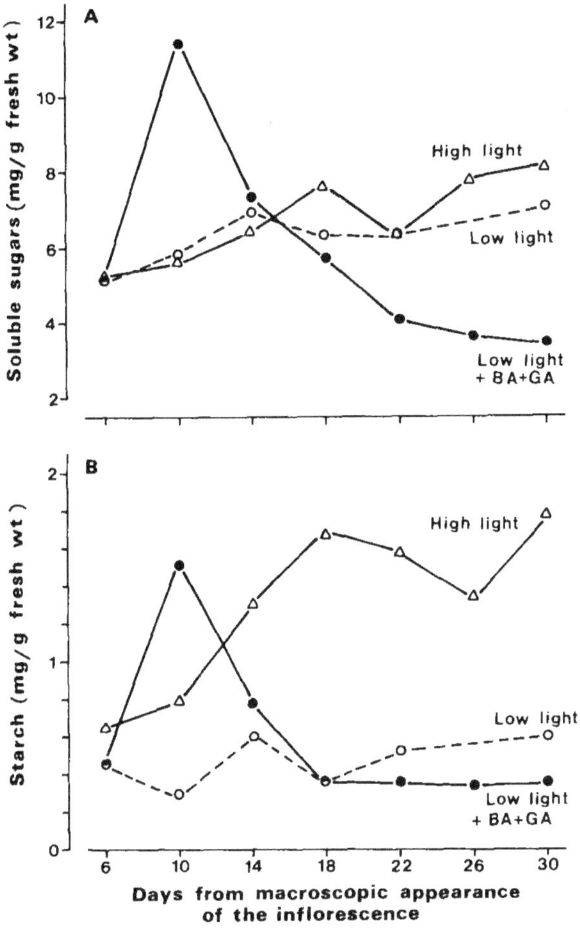

FIGURE 6. Effect of light conditions and a mixture of BA + GA on (A) soluble sugar and (B) starch contents of inflorescences of 'King plus' *Lycopersicon esculentum*. High light: 12-hr day, 20 W/m^2; low light: 8-hr day, 12 W/m^2; BA + GA treatment as in Figure 3 in this chapter. The inflorescences develop until anthesis in high light and in low light + BA + GA, and abort in low light alone. (From Kinet, J. M., Hurdebise, D., Parmentier, A., and Stainier, R., *J. Am. Soc. Hortic. Sci.*, 103, 724, 1978. With permission.)

cause of increased invertase activity. However, they did not do the kinds of experiments required to test for substrate-induced enzyme activity. Their data show only parallel variation in invertase activity and sucrose levels. Other observations suggest that substrate supply may indeed limit male organ development as well as enzyme levels. For example, Howlett reported that sporogenous tissue fails to reach meiosis in tomato suffering severe carbohydrate deficiencies.[257] Many studies show that environmental conditions resulting in reduced carbohydrate levels also result in repression of male organ development in hermaphroditic, monoecious, and dioecious species, with little effect upon female organs (Chapter 3).

Fawzi and El-Fouly have shown for carnation that amylase and invertase activities increase in mature shoot and apical bud tissues as a function of stage of flower development.[939] As expected, starch content decreases and soluble sugars increase concurrently. Hence, increased sink strength of the developing carnation flower is coordinated with increased mobilization of stored reserves as well as supply of current photosynthates. This is in keeping with above observations for mobilization of polyfructosans in tulip bulbs. However, the important fact here is that enzymatic activities affecting carbohydrate allocation to starch, sucrose, and reducing sugar pools seem to be closely coordinated with flower development itself.

CONCLUSIONS

Photosynthetic assimilates appear to have more than a supportive role in reproductive development. Increased source strength, sink activation, and reduced sink competition are useful terms and concepts when accounting for environmentally and chemically induced promotion or inhibition of flower (inflorescence) development. In cases where assimilate supply itself appears to play a dominant role in controlling development we do not know whether the primary concern is with substrates for biosyntheses and energy transformations in the developing flower (inflorescence), or whether the assimilate stream flow, in which the flux parameters are dependent upon photosynthesis, is merely passively carrying sink activators from the leaves. What comes first: assimilates or sink activators? Suppose it is the assimilates. There should be an early increase in, let us say, sucrose and that should be followed by meristem activation. Suppose it is the sink activators that come first. Then we should find early decreased sugar (or starch) levels with concomitant increased turnover. Much of the research required to provide good data on these questions remains to be done. We lack studies on flower development comparable to those on evocation and flower initiation where more precisely timed phenomena have been recorded (e.g., the early, preinitiation increase in sugar and starch levels in *Sinapis*). Some major questions for future research concern distinguishing between sink activation and increased assimilate supply in relation to various stages of flower development. Sink activity will also affect photosynthesis. A recent review by Herold on regulation of photosynthesis by sink activity emphasizes the importance and complexity of feedback inhibition.[940] Thus, should we invoke sucrose level as a trigger for development, there is surely an adjunct requirement for research on the complex metabolic controls of this compound.

Chapter 10

CONCLUDING ANALYSIS OF THE FLOWERING PROBLEM

TABLE OF CONTENTS

> Un fait morphologique peut ouvrir un champ immense, car la forme est l'expression matérialisée de la vie, et toutes les lois biologiques se traduisent en elle.
>
> L. Plantefol*

> The genesis of form is one of the foremost problems, but its solution is evasive. It is necessary to study development at all levels to elucidate the factors underlying morphogenesis.
>
> S. N. Postlethwait and O. E. Nelson**

* From Plantefol, L., in *La Theórie des Helices Foliaires Multiples*, Masson et Cie, Paris, 1948, 144. With permission.

** From Postlethwait, S. N. and Nelson, O. E., *Am. J. Bot.*, 51, 238, 1964. With permission.

I. INTRODUCTION

Morphologists have described in great detail the genesis and final form of the various reproductive structures found in higher plants (Chapter 1). Physiologists have concentrated their investigations on the environmental (Chapters 3 to 5) and chemical (Chapters 7 to 9) parameters affecting flower and inflorescence development such that we often have now a rigid control of this process. Geneticists have provided us with an abundance of genotypes with varying size, form, and function. The combined efforts of all these scientists have made, as we shall see in the final chapter, many noteworthy contributions to agriculture and horticulture whether through development of better adapted cvs. or through improved cultural practices. This produces a feeling of solid achievements, with more to come, but for the fundamentalist the question also arises of how far have we come to grip with the mechanisms of flowering in terms of developmental biology.

II. FLOWERING AS A MULTISTEP PROCESS

Even though superficially flowering seems to go automatically to completion once it is started, it is a multistep process that can be arrested at several stages of development before anthesis (see Chapters 4, 5, and 7).

The very first step towards reproduction in shoot apices is what has come to be known as "floral evocation". As explained in Volume II, Chapter 5, many events of this step — whether these events occur at the molecular, cellular or higher levels — seem fairly universal despite the diversity, among species, of responses to environmental and chemical factors and of morphology of reproductive structures. This universality enabled us to propose a model for evocation, briefly summarized below.

The model (Figure 1) proposes that evocation is not a single sequence of events, but consists of multiple parallel sequences. The sequences are independent, at least initially, from each other, since any one can be set in motion in the absence of the others *(partial evocation),* and they are sequentially initiated within the apex during the floral transition. At various substeps of the evocation process the different sequences interact (indicated by oblique arrows in Figure 1), and these interactions are believed to be essential for evocation to proceed and reach a "point of no return" where the meristem is irreversibly committed to initiate reproductive structures.

As far as the limited available evidence goes, this point seems to be reached in general some time before the morphogenetic processes of initiation can be first detected (Volume II, Chapter 5). The nature of the critical event(s) associated with reaching this point is not known, but it is noteworthy that the point of no return follows by a few hours the appearance of new antigenic proteins in the meristem of *Sinapis.*[941]

Although the morphogenetic changes are largely inevitable and flowers are inexorably produced once evocation is completed, the reproductive condition of meristems is not necessarily stable indefinitely. Occurrence of mixed reproductive-vegetative structures in many species indicates that the apex, under certain circumstances, may regress partly or completely to the vegetative condition. In most cases only meristems in early reproductive stages are capable of reverting, whereas those having functioned reproductively for some time are unable to do so (Volume II, Chapter 5, Section VII). Thus, after the initial commitment to flower, reached at the completion of evocation, there remains some flexibility in the ensuing realization processes. *Determination* in the sense usually accepted by developmental biologists — meaning that the developmental potentiality of a group of cells has become irrevocably restricted to a particular pathway — is clearly not reached at the point of no return, but only at a later stage. Determination is, as evocation, a cryptic state whose essential underlying events are totally unknown. Recall also that determination may occur in plants

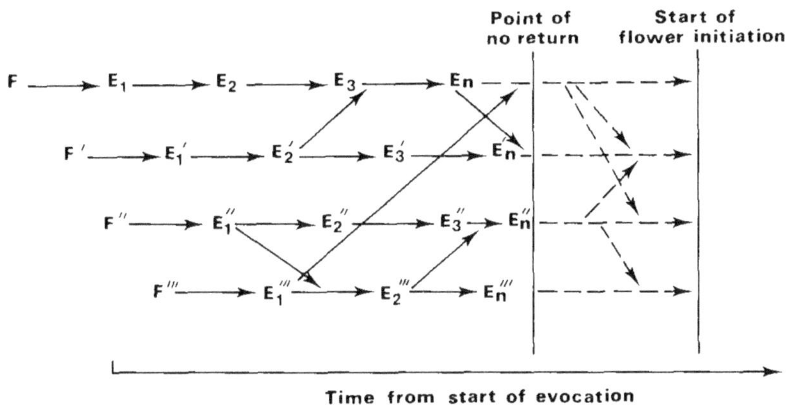

FIGURE 1. A model of floral evocation and its control. Successive events in each sequence are numbered E_1, E_2, ... E_n. Initial events of the different sequences are represented by E_1, E'_1, E''_1, E'''_1, and the factor(s) controlling these sequences are F, F', F'', F''', respectively. The point of no return represents the end of evocation; morphogenetic processes, i.e., flower initiation, then occur after a lag period. (From Bernier, G., Kinet, J. M., and Sachs, R. M., *The Physiology of Flowering, Vol. II. Transition to Reproductive Growth*, CRC Press, Boca Raton, Fla., 1981.)

at tissue, organ, or meristem levels, albeit a much greater plasticity *(totipotency)* can be retained at the cell level.[942]

Subsequent to the initial evocation/determination step in the meristem there is great variety in the number and nature of steps constituting the flowering process depending on the kind of reproductive structures to be produced. There are evidently more steps and they are much more interlocked within each other in cases where a complex inflorescence is formed than in those in which the meristem is entirely transformed into a single flower. Let us deal first with this latter, relatively simple, case.

Despite the limited number of plants investigated, surgical work involving bisection of flower buds and floral organs (Chapter 6, Section III.B.1) is clear in its implication that the flower meristem passes through a sequence of distinct physiological phases that somehow govern the nature of successively initiated floral organs. Thus in tobacco, e.g., Hicks and Sussex distinguished a sepal-forming phase from a petal-stamen- and a carpel-forming phase.[586] Interestingly, these phases seem generally to precede the actual emergence of the corresponding primordia. We do not know the determinants of these phases, but some preliminary observations point to a possible critical role of changes in gene expression in localized cell populations within the meristem. It is well known that each class of mature floral organs has its specific protein complement, and Kahlem has shown that proteins specific to the mature male sporogenous tissue and tapetum appear very early during stamen morphogenesis, possibly even before the altered cell division pattern establishing the emergence of stamen primordia.[196]

The similarities between each of these physiological phases of flower meristems and the floral evocation step of shoot meristems is obvious, so that it is appropriate here to refer to them as "sepal evocation", "petal-stamen evocation", and "carpel evocation".

As for the initial step of floral evocation, completion of each of these successive phases of evocation of floral organs is normally followed by the formation of the corresponding class of organs, but the developmental path of these is again not rigidly fixed initially. This path remains optional for some time after initiation as amply demonstrated in many experiments in which the fate of these organs was altered by environmental, chemical or surgical treatments. At some stage determination occurs and the nature and functions of each class of floral organs are then irreversibly fixed and resistant to manipulations. Flower development

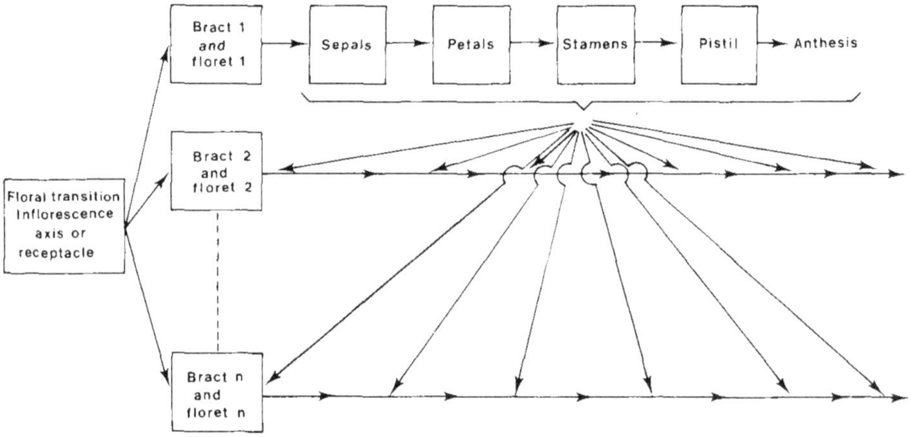

FIGURE 2. A possible organization of evocation/determination steps in the production of a hypothetical inflorescence. Each box represents one evocation/determination step and subsequent differentiation and growth of the tissues involved. The complete catenary sequence of steps of flower morphogenesis is only shown for floret 1 but the same sequence is assumed to occur in each other floret. Each individual step in the scheme might interact with all other steps but for the sake of clarity only a limited number of such interactions are shown by oblique arrows.

can thus be viewed as a sequence of several specific evocation/determination steps (Figure 2, floret 1).

When flowers are grouped in inflorescences a similar, although more complex, sequence of such steps governing the initiation and morphogenesis of each floret or branch may also be envisaged. These steps occur successively and/or simultaneously in different parts of the inflorescence (Figure 2).

III. RELEVANCE OF THE MODEL FOR FLORAL EVOCATION AND ITS CONTROL

Put that way, the problem of producing complete and functional reproductive structures can be considered as a complex and intricate amplification of the initial floral evocation/determination step. Compared to what we know about the events of this initial step (Volume II, Chapters 2 to 5), our knowledge about the subsequent steps is in general weak. We have found no reason, however, to reject application of the model of Figure 1 to each step of reproductive morphogenesis and development.

Endogenous regulation of all these steps might also be similar to that envisioned for floral evocation, i.e., each independent sequence is under the control of a particular factor or set of factors (Figure 1). These controlling factors, symbolized F, F′, F″, etc., are of a positive or negative nature, i.e., the sequences are set in motion either by the increase in concentration of a promoter above a threshold value or by the decrease in concentration of an inhibitor below a critical value. The most remarkable lesson we have learned in our detailed survey of physiological regulation of flower development is that it is basically identical to that of floral evocation, reviewed in Volumes I and II. Indeed, flower development depends upon the same complex correlative influences that are in play in the process of floral evocation and both are controlled by the same environmental parameters, same classes of growth regulators, and assimilates. Also, in flower development, as in flower initiation, the action of many of these factors is markedly time dependent, suggesting that in both cases there are "windows" during which the factors must be received if they are to be effective.

All of this strongly supports the idea that each step of the flowering process is under the control of several interacting promoters and inhibitors.

Of course, each of these steps has its own specificity which is presumably expressed initially in the particular complement of proteins that is synthesized. Evidence reviewed in Chapter 1 indicates that specificity can also be expressed by other characters, such as (1) the size of primordia at initiation, i.e., the number of cells involved in the organogenetic events, (2) the placement of primordia (phyllotaxis), and (3) the growth rate of nascent primordia (cell growth and division). No doubt many other molecular and cellular events, typical of each step, remain to be discovered. We believe however that the specificity of each step not only resides in the specificity of its component events, but mostly in the unique array of interactions in space and time among these events. Similarly, although all steps are subjected to the same general sort of multifactorial controlling system, there is presumably a specific balance of factors for each step. For example, a given growth regulator or a ratio between two or several regulators may promote one step, inhibit another and have no influence on still another one.

Finally, correlative influences acting within the developing inflorescences and flowers themselves (see Chapter 6) demonstrate that the arrays of evocation/determination steps do not unfold independently of one another but depend upon some internal coordination. Without this coordination no complete and functional flower or inflorescence is formed. The way coordination between these steps (represented by oblique arrows in Figure 2) is achieved is totally unknown. Speculations on this topic are discussed next.

IV. SYSTEM OF RELAYS OF GENE ACTIVATION: THE MODEL OF WARDLAW AND HESLOP-HARRISON

For Wardlaw and Heslop-Harrison the evocation/determination, in space and time, of successive whorls of lateral members in a hermaphroditic flower meristem could be viewed as a chain of relays of gene activation.[104,943] In this model, pictured in Figure 3, the agents controlling such a relay system are believed to be specific inducers with short intercellular ranges which set in train processes leading to the step-by-step release of genetic information in defined cell lineages within the meristem. Thus, the initial inducer, w, which in photoperiodically sensitive species may be identified with the leaf-generated floral stimulus (i), initiates an irrevocable pattern of changes ultimately affecting all meristematic cells. The first-formed laterals, e.g., the sepals, in which the gene complex A is activated, feed to the primordia at the succeeding level, i.e., the petals, the inducer x, activating the gene complex B, and so on. In this way the onset of each stage is conditional upon the completion of the previous one and leads in turn to the next. Also, the initial inducer w or an early product of gene complex A is hypothesized to repress the gene systems normally active in the vegetative leaf-forming meristem. To explain the fact that each lateral appendage is not usually determined along one characteristic developmental pathway at the time of emergence, Heslop-Harrison has developed the model by assuming that each evocation step leads not to the simultaneous expression of a block of genes, but to the initiation of an ordered pattern of release of genetic information, the whole apical control system then acting hierarchically.[104]

In the cases of incomplete and unisexual flowers the model must be simplified by suppression of one or several stages or must include additional control mechanisms inhibiting the development and differentiation of either stamens or pistil after their initiation.

The observation that the different kinds of floral organs appear in the flowers of many species in a fixed identical sequence and that this sequence cannot be reversed nor disrupted readily by surgical manipulations (Chapter 6) supports the idea that position in the sequence is of ultimate developmental significance and, thus, suggests the operation of relay systems in flower morphogenesis. However, the varied cases of floral organ metamorphosis (sepalody, petalody, staminody, pistillody) and of abnormal flowers with missing or extra organs caution against an interpretation of flower structure based on rigid 'position-dependent'

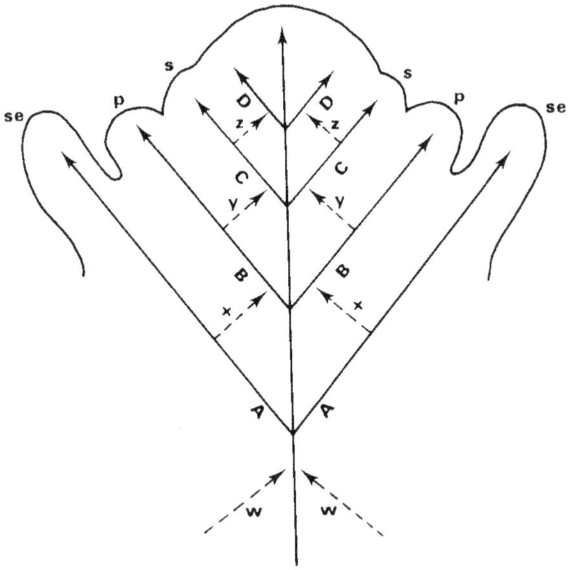

FIGURE 3. A model for a developing flower meristem based upon the idea of control through a relay system of gene activation. The floral stimulus(i), w, initiates the floral transition. Cell lineages are defined sequentially destined to generate the different classes of lateral members: sepals (se), petals (p), stamens (s), pistil. In the first lineage gene complex A is activated and produces the inducer ×, which diffuses to the next group of primordia, where it activates gene complex B, and so on. (From Heslop-Harrison, J., *Meristems and Differentiation*, Brookhaven Symp. Biol., No. 16, 1964, 109. With permission.)

evocation/determination events (see Chapter 1, Section III.D). For example, because extra appendages necessarily arise in unusual places in aberrant flowers, it is clear that the fate of a primordium is not strictly position dependent, contrary to what is generally the case in animal tissues.[942] Another objection to hierarchical control of a fixed epigenetic sequence comes from the examples of centrifugal and mixed centripetal/centrifugal initiation of floral organs cited in Section III.A of Chapter 1. Also, inducers with small intercellular ranges, such as those postulated in the model of Figure 3, have not been identified so far, and the mechanisms for expression of a number of genes presumably scattered throughout the genome remain obscure and entirely hypothetical.

The above model is not intended to explain long-distance interactions, such as those disclosed in Chapter 6 between the different flowers of an inflorescence. For these interactions, other explanations have to be devised.

Thus, while successful flowering requires precise interactions and close coordination between remote and neighbor parts of the reproductive structures, it must be recognized that the mechanisms of these interactions have generally eluded a detailed description. It is absolutely essential for real progress in this field to use improved microanalytical techniques,[943a] and remain open to entirely new concepts on integrated regulation.

V. GENESIS OF NEW FORMS

Flowering cannot be simply reduced to a series of evocation/determination steps, presumably involving for the most part events at the cellular and molecular level. Flowering is also, perhaps essentially, the production of three-dimensional structures whose shapes are totally different from those of vegetative organs. The fascination we and many others feel about the flowering process is certainly related to the incredible beauty and variety of these forms, either when mature or when developing and then revealed by SEM.

The forms of reproductive structures are basic to plant classification and are evidently ultimately directed by the genome. Thus, changes in form should initially derive from change in gene expression, i.e., changes in the protein complement of cells. But, from proteins to macroscopic forms, there is an enormous gap in our knowledge despite much information accumulated in recent years on the processes of macromolecular aggregation (self assembly) that may yield microtubules, flagellae, membranes, and organelles. For the moment we can only emphasize the overwhelming complexity of the problem of tracing macroscopic form inception at the molecular level. We recognize our ignorance in pinpointing the right combinations of biochemical reaction systems that give rise to shape and integrated physiological functions.

An appealing approach to the problem of morphogenesis, described in Chapter 1, Sections IV.C and V.B., stresses the importance of localized variations in growth rates and/or the preferred directions of growth. The growth rate of a group of cells or cell lineage may be controlled, at least partly, by the nature and/or amounts of growth regulators and assimilates present accounting for the importance of these compounds in flower and inflorescence formation. Growth responses to regulators might in turn be dictated more by cell sensitivity, as suggested by Trewavas,[823] rather than directly by the concentrations of regulators. We still have difficulty when trying to define the determinants of plant cell sensitivity to growth substances.[944,945] It is clear that "binding sites", essentially "binding proteins", for growth regulators (auxins, ethylene, cytokinins) are present in some plant tissues, but it is uncertain whether these are physiologically active receptors similar to those known for animal hormones. If plant cell sensitivity can ultimately be related to proteinaceous receptors, then modulation of these receptors would be considered a focal point for genetic control.

The participation of growth regulators and assimilates in the control of the direction, i.e., polarity, of growth is less straightforward. For Lintilhac and Green the factors involved here are the direction of cellulose microfibrils in the cell wall and of cytoplasmic cortical microtubules and their orientation in turn would be essentially dependent on the initial shape of the structure (meristem) considered, the pattern of mechanical stresses created inside this structure by the processes of cell growth and division and also the shape of individual cells (Chapter 1, Section V.B).[179,186,187] The factors in play are thus essentially geometric and biophysical, not biochemical, but since the rates of cell growth and division (partly responsible for the stress pattern) are known to be controlled biochemically there is an obligate interplay here of chemical, physical and geometric factors in the control of morphogenesis.

Incoming fluxes of growth regulators and/or assimilates might be different in different zones of flowering meristems or cells in the different zones might react differently to these fluxes due to their differential sensitivity. Either way differential effects upon the rates of cell growth and division are expected in the different zones or groups of cells and this will ultimately modify the overall stress pattern and finally the polarity of cell growth and division. The shape of these meristems and their appendages is then bound to be altered. In this way fluxes of growth regulators and/or assimilates may act indirectly as morphogenetic signals. This is an attractive idea for developmental physiologists but so far it is essentially speculative because of an almost total lack of relevant investigations.

Similar hypotheses — involving an array of physical, geometrical and chemical controlling factors — can be developed (and are possibly the only way) to explain the precise and ordered changes in phyllotaxis that characterize inflorescence and flower morphogenesis.[77,78,180]

VI. A LAST WORD ON FLORIGEN

In view of the remarkable similarities in the regulation of both initiation (evocation) and development of reproductive structures by environmental and chemical factors and also by correlative influences, it is really puzzling that physiologists long believed that initiation was controlled by a unique, universal, and specific hormone, florigen, while further steps of the flowering process were generally thought to be modulated by balances between multiple nonspecific compounds.

The experimental evidence accumulated during the last two decades, documented in Volume II of this series, makes the florigen concept outdated and untenable. We like to stress that there is no basic difference in regulation between the first and subsequent steps of the flowering process, all of which depend on a complex balance between a number of common growth regulators and other chemicals. One needs not alter the statement if new plant-growth regulators are discovered and found to cooperate with the already known regulators in the control of the various steps of flowering and of other developmental processes.

VII. A LOOK INTO THE FUTURE

A look at Figure 2 explains why most workers in the field have concentrated their efforts towards an understanding of the first step (floral transition) of the flowering process. Subsequent steps are so numerous and so interlocked with each other that disentangling them for analysis seems a hopeless goal.

As we have seen in Volumes I and II, the refined analytical approach to evocation was made possible by selecting model plant systems that could be highly synchronized by environmental signals for the vegetative to reproductive transition. Similarly, what we need to perform similar studies on some specific steps of flower (inflorescence) development is a carefully synchronized population of plants for the steps in question. This problem of synchronization of developmental stages still raises difficulties in many plant systems because of insufficient knowledge on the regulation of flowering in these cases. However, it has been addressed successfully in recent years in a few selected experimental systems, e.g., *Lilium*, tomato, *Silene*, *Bougainvillea*, carnation, etc., and in each case for one or a few selected developmental stages. Clearly, not all stages of the flowering process can be synchronized in a single plant system but, by combining the investigations on different stages in different systems, we may expect to be able one day to describe the similarities and the specificities of all these steps and to provide a complete and integrated picture of the whole process of flowering from its earliest manifestations up to anthesis.

Chapter 11

APPLIED ASPECTS OF FLORAL INITIATION AND DEVELOPMENT

TABLE OF CONTENTS

> Few scientists think of agriculture as the chief or the model science. Many, indeed, do not consider it a science at all. Yet it was the first science — the mother of sciences; it remains the science which makes human life possible, and it may well be that, before the century is over, the success or failure of Science as a whole will be judged by the success or failure of agriculture.
>
> A. Mayer and J. Mayer*

I. INTRODUCTION

Plant physiologists are employed in relatively great numbers in agricultural enterprises around the world. Their investigations of plant functions, often at the most fundamental level, have as their ultimate goal improving crop yield and/or quality. Some research has paid off handsomely in the areas of plant nutrition, irrigation, and cultural practices (adjusted planting density, advanced sowing date, fruit thinning, postharvest handling, etc.). There has definitely been a payoff from studies on flowering, too; most advances in this area as in so many others result from combined studies with geneticists.

In species where the reproductive structure (flower, fruit, seed) is the commercially valuable plant part, dates of flower initiation and anthesis and numbers of flowers (fruits) produced per unit time are important inputs for coordinating sowing dates and harvests with known climatic parameters and desired crop rotations. Selection and design of species for climates, a common practice in agricultural planning, requires ready access to such data. Since a crop at harvest is the result of many layers of system functions of both vegetative and reproductive stages of development, this chapter's aim is to analyze ideas and data using wherever possible an integrated (rather than analytical) approach.

II. MORPHOLOGY

Some years ago Popham wrote that "a knowledge of developmental plant anatomy will allow the geneticist and the physiologist to view the component parts of a complex problem, which will enable them to design more precise experiments and to draw more precise inferences from data thus obtained."[29] Popham goes on to note the importance of a time study of the stages of flower development to the improvement of horticultural practice. In the sections that follow, concerning juvenility, thermo- and photoperiod, and especially chemical control of flower initiation and development, we depend heavily on careful morphological studies of vegetative, transitional and reproductive meristems. It would not be possible to schedule treatments on a rational basis, with any precision, without having done concomitant morphological examination of apical meristematic tissues. Phenological tables for most of the major crops require that the dates of flower initiation and anthesis be specified. These dates, bracketing inflorescence development, are essential for predicting days to maturity (harvest) as a function of climate.

* From Mayer, A. and Mayer, J., *Daedalus*, 103, 83, 1974. With permission.

It is not surprising that microscopic examination of plants from grain fields and greenhouses is a common practice. Rawson and Kirby's studies on development of the wheat inflorescence (in reference to plant age, timing of vernalization and subsequent photoperiodic treatments) are prerequisite for analysing potential sink limitations to grain yield.[17-19,157] Their analyses for some cvs. make clear that spikelet and floret numbers per plant are determined directly by days between sowing and vernalization; also plant size (leaf number) is directly related to this interval. Other studies with wheat that have contributed greatly to our understanding of this species performance in different climatic conditions have been based on thorough examination of the inflorescence axis at the primordial stage of development. It is clear that economic yield for many grain crops is limited by the availability of sink tissue which is directly related to grain number.[913] Grain number and perhaps grain size can be augmented in breeding programs in which the morphological determinants of these parameters, floret number and ovary size, are well defined by cv. Hence, morphological studies will continue to be important adjuncts to crop improvement programs.

Bodson depended almost wholly upon morphological observations for unraveling the complex sequence of daylength and temperature treatments required for flower initiation, flower development, and flower bud break in two azalea cvs. of particular value to the Belgian horticultural industry.[239] What we would like to emphasize with the example for azalea is that even if the morphological studies have been done for several cvs., they must be repeated each time new cvs. are introduced. It may be easier to do the second and third time around, but for each cv. care must be taken to define the time to appearance of the critical developmental stages if commercial applications are the goal.

There is an aspect of gross, rather than microscopic, morphology that is important to agriculturists and is directly related to time to flowering. Simply put, there is an inverse correlation between flowering and plant stature. In annuals early cvs. are relatively short and late cvs. tall. Commercially grown cvs. are further divided into two classes: indeterminate, which tend not to develop terminal inflorescences and, hence, are generally taller at maturity, and determinate, which are shorter, more heavily branched and appear to have terminal flowers. In general, the determinate, early flowering types are well adapted to regions with long growing seasons where the indeterminate and/or late-flowering types grow too tall and fall over (lodge). These are important considerations for hand or mechanical harvesting of crops and they often determine the true, marketable yields obtained from the fields.

III. MODIFICATION OF FLOWERING TIME IN SOME MAJOR CROPS

Four environmental parameters appear to be of overriding practical importance in flower or fruit production; namely, the light integral (the integral of irradiance × time), photoperiod (daylength), temperature (whether treated in terms of cardinal temperatures or accumulated heat or chill units), and water stress. Our review here will be altered considerably with respect to ordering and weighting of topics, particularly with the introduction of phenoclimatology, a subject best illustrating the integrative approaches required for agricultural practice.

A. Phenoclimatology

Frequent reference has been made to "tailoring" crops to climate by selection of existing or creating specific genotypes. To do this well, the collaboration of many scientists is required and a special research area, phenoclimatology, which assesses plant performance as a function of specific climatic parameters, has developed. Phenoclimatology is ultimately strongly dependent upon field research but controlled environment studies have been extremely useful analyzing specific climatic parameters. Data on appearance of defined developmental stages

— this is phenology — usually as a function of days after sowing or emergence (for annual crops) or simple calendar dates (for perennial crops) are essential. Climate assessment includes some measure of accumulated thermal and irradiance and/or daylength factors during the calendar year or growing season. By combining phenological data with climatic data — this is phenoclimatology — one can predict: (1) when a certain stage, usually crop maturity, will occur, (2) whether thermal or other climatic parameters are adequate for crop production, and, ultimately, (3) whether a cv. is well adapted to a climatic region.

1. Developmental Stages (Phenology)

Phenological tables most often use the intervals from: (1) sowing to flower initiation, (2) flower initiation to anthesis, and (3) anthesis to crop maturity. The length of the juvenile phase is often critical in fitting crop to climate, particularly where the growing season is relatively short or where chilling or daylength sensitivity determines time to flowering (even later stages of development). In only a few instances can we say that the juvenile phase alone, as distinct from the induction, flower initiation, and development phases to anthesis and ripe fruit, has been affected by breeding programs.

Data for tomato and pea varieties, where daylength and temperature influences on node to first flower have been apparently essentially eliminated by selection in breeding programs, clearly demonstrate the genetic control of earliness-to-flower.[946] The value of early-flowering cvs. for field plantings of tomatoes resides not only in their ability to exploit a shorter growing season, but also in their smaller size which is convenient for machine harvest. Yields comparable to those obtained in larger cvs. can often be obtained by increasing planting density.

The vegetative growth phase in rice has been subdivided by Vergara and colleagues into: (1) the juvenile phase, from germination to induction, which by definition is daylength insensitive, and (2) the photoperiod-sensitive phase (from the beginning of induction up to panicle initiation).[947] In some cvs. the juvenile phase accounts for the major differences in number of days to anthesis. Where the growing season is quite short or multiple cropping is desired, a short juvenile phase (Norin 20) is preferred, whereas with a longer season, cvs. with a long juvenile phase (Tainan 3) and/or long critical night (BPI-76) are acceptable (Figure 1).

The days to flowering in rice and most crops change with the temperature parameter which, when vernalization is not involved, merely compresses or lengthens all phases of development; no drastic reduction or increase in minimum leaf number to flowering is found.

A factor not overlooked by cereal crop specialists is that individual plants of cvs. with longer vegetative phases, whether due to length of the juvenile phase or delayed induction, are likely to be much larger at full bloom, with more leaves and tillers and a deeper root system; these may perform better at lower planting densities than the cvs. with a short vegetative phase and may be better adapted to dry land conditions where a relatively deep root system is essential.

Specific soybean cvs., basically all SDP, have been developed for wide latitudinal distribution and consequent large differences in length of the growing season (Figure 2).[948] There are nine (sometimes ten) soybean maturity groups based on the length of the interval from sowing to ripe seed; early- or late-maturing soybeans may differ not only with respect to the daylength requirement for flower initiation but also with respect to the requirements for seed development. Definition of developmental stages in soybean is essential for the analysis of cv. adaptation to region. Major and coworkers in evaluating thermal unit methods for predicting soybean harvest dates, found that cooler temperatures and longer daylengths at higher latitudes are additive in delaying anthesis; the effect of cool temperatures predominates in early spring plantings, whereas the long-day-induced delay in flower initiation and pod development is especially important in late-spring plantings.[949,950] They found, too, that

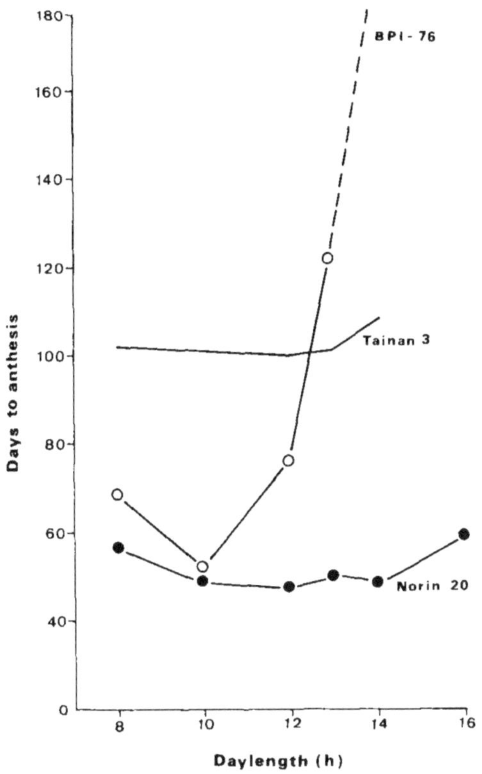

FIGURE 1. Days to anthesis as a function of daylength in three different *Oryza sativa* cvs. (From Vergara, B. S., Chang, T. T., and Lilis, R., *Int. Rice Res. Inst. Technol. Bull.*, 8, 1, 1969. With permission.)

from anthesis to seed ripening SD are necessary to offset any delaying influence of cool autumn temperatures. This suggests that SD promote seed (or pod) development; hence, one may want to give much more weight to the postanthesis stage than is usually the case in tailoring soybean cvs. to specific regions.

Sorghum cvs. are apparently all SDP in which earliness is determined by the critical night length for induction and not by length of the juvenile phase. This is clearly seen by the fact that, in 4 genotypes tested, flowering time is approximately the same at 10-hr daylengths, but differs widely at 14-hr daylengths (Table 1).[321] But the story of earliness in some field-grown *Sorghum* cvs. would appear to be considerably more complex. Caddell and Weibel found that the elapsed time to photoperiodic sensitivity is strongly associated with vigor, i.e., the rate of leaf initiation and generation of leaf surface area.[951] The three cvs. they examined responded to SD induction as a function of leaf area at the time of induction; this means that growth rate after sowing has a great impact on days to flowering and crop maturity.

Most maize cvs. are treated as DNP, although in some varieties SD advance anthesis (see Chapter 4, Figure 6).[349] Earliness is related almost entirely to a thermal (heat unit accumulation) index computed for each region, a topic which will be covered in more detail in a special section below. Selection of cvs., adapted according to the average growing degree day accumulation for the region, may include selection for longer or shorter juvenile phases. Troyer's analysis of early flowering in maize includes vigorous growth at low temperatures;[952] i.e., early cvs. appear to have the capacity to grow faster at low spring temperatures and achieve physiological maturity at an earlier date, perhaps similar to some *Sorghum* cvs. Rate

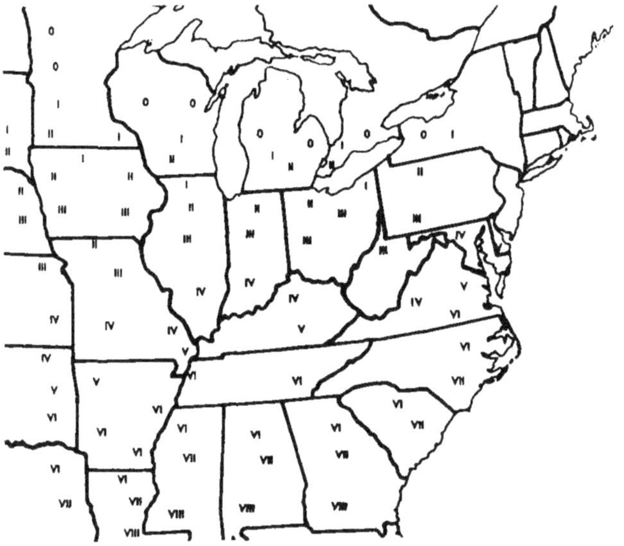

FIGURE 2. Map showing maturity rating of *Glycine max* cvs. recommended in different regions of U.S. Group 0 represents the earliest maturity group and Group VIII the latest maturity group. Areas of adaptation are only approximate since cvs. of two or more maturity groupings may be grown in the same area to provide for early and late planting. (Reproduced with permission from Poehlman, J. M., *Breeding Field Crops*, 2nd ed., The AVI Publishing Company, Inc., P.O. Box 831, Westport, Conn., 1979, 263, Fig. 14.7.)

Table 1
DAYS TO ANTHESIS AS A FUNCTION OF DAYLENGTH IN FOUR DIFFERENT *SORGHUM* GENOTYPES

	Genotype			
Daylength	**Early**	**Intermediate**	**Late**	**Ultralate**
10 hr	44	44	46	47
Natural days (14 hr)	47	67	83	102
20—24 hr	>100	>100	>100	>100

From, Vince-Prue, D., *Photoperiodism in Plants*, Copyright (c) 1975 McGraw-Hill Book Co. (UK) Ltd. Reproduced by permission.

of growth, including leaf initiation and expansion, rather than fewer leaves produced is the key physiological feature for corn geneticists interested in early flowering. Low temperature vigor is a heritable trait in corn, one that has been incorporated into cvs. grown at higher latitudes where heat unit accumulations are relatively low, and also in typical "corn belt" zones when early harvest is desirable. In fact early spring growth is desirable in all temperate zone crops to insure foliar canopy closure by June when the available light for photosynthesis is maximum.

In bulbous crops we are concerned with flower development in the mature bulb in storage as well as in plants growing under field conditions. In both cases climatic effects are judged

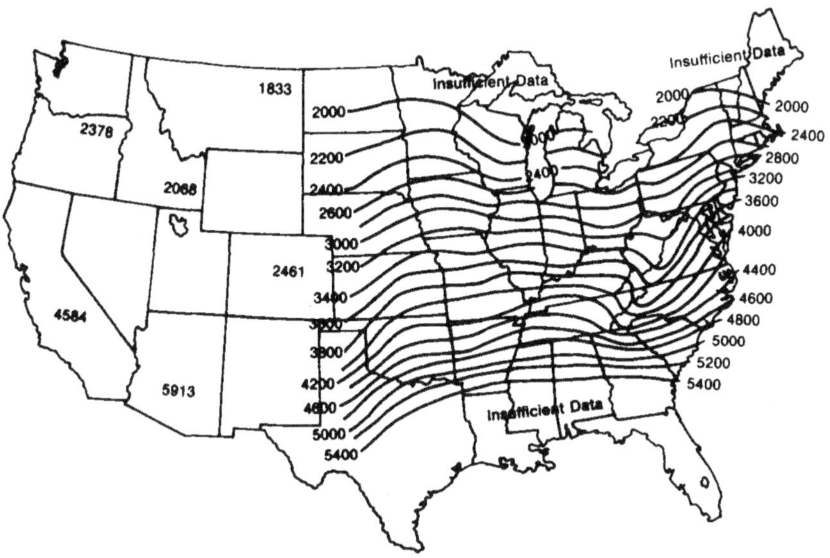

FIGURE 3. Mean number of growing degree days (GDD) in U.S. using a base temperature of 50°F (10°C). One GDD is defined as 1 day in which the average temperature is 1°F above the base temperature. If average temperature is 70°F for 30 days, the GDD accumulation is 600. (From Aldrich, S. R., Scott, W. O., and Leng, E. R., *Modern Corn Production*, 2nd ed., A & L Publications, Inc., Champaign, Ill., 1975. With permission.)

well past the juvenile phase. For many vegetatively propagated flower crops the juvenile phase is of little concern. The induction period, whether or not well defined by a specific photoperiodic or temperature requirement, is tied to some morphological stage, usually leaf number, that is used to predict the progress of the crop towards flowering or to time an inductive treatment. Because flower bud break is relatively easy to assess macroscopically, termination of rest in dormant fruit trees does not pose a morphological or developmental stage problem. Most of the practical work of a phenoclimatological nature done with temperate zone fruit trees is on climate assessment and modification, but, should chemical treatments be used to prolong or terminate rest, careful attention to the stage of flower development at the time of application may be required.

2. *Climate Assessment*

a. Thermal Unit Models

Phenological tables most often use accumulated heat or chill units, e.g., growing degree days (Figure 3), rather than calendar days for predicting crop performance.[953] This is true even in species with daylength sensitivity (although in these cases the fit is not as good) because temperature is the most significant parameter for all phases of plant growth and varies widely from day to day and year to year. A major problem with accumulated heat unit systems is how best to assess the thermal factor, particularly the choice of the base, physiological temperatures for growth and development (or termination of rest). The base temperature in theory is that below (or above, for chilling requirements) which no physiological and/or morphological development will occur. It must be determined empirically for each crop. Decisions must be made whether to (1) use the average daily mean temperature, (2) treat day and night temperatures separately, (3) cut thermal unit accumulation off at a certain temperature maximum, (4) subtract for temperatures too high or too low for good performance, and (5) use specific coefficients or base temperatures for different stages of development. For example, in wheat different base temperatures must be used for each phenological stage. For corn 10°C works well throughout the entire crop development cycle, 5.5°C is used for peas, 7°C for chilling for flower initiation in olives, etc.

The effort that goes into a heat unit system depends in large part on how important it is to the crop planner to know harvest date within a few days. A thorough study of phenological events in corn, computed by 6 different heat unit systems, indicates that a gain in accuracy of predicting tasseling and silking is accomplished by treating average day and night temperatures separately and by setting the maximum temperature for heat unit accumulation at 27°C.[954] In other crops, such as soybean, we expect that major advances in phenoclimatology will come from expansion of the critical stages of development, with concomitant improvement in evaluating the thermal unit factor.[949,950]

Temperature fluctuations during winter may have important practical drawbacks. Delayed and erratic flowering in various deciduous fruit trees frequently results from insufficient chilling while — where the winter is cold enough so that the result is early rest completion — flower development may be too far advanced. The possibility of frost damage in spring is increased since flower buds become more susceptible to cold as they develop.[955] They are most hardy when fully dormant, as shown for peach buds that may survive midwinter temperatures as low as −20°C, but, as warmer temperatures occur in the spring after rest completion, they begin development, lose cold hardiness, and can be killed at −2 to −4°C.[956,957]

Considerable research has been devoted to delaying flower bud development after completion of rest to avoid spring frost damage to emerging flower buds. A relatively simple method is to spray trees with water, wetting the buds, thereby preventing radiant heating of the resting buds on sunny days through evaporation of the water layer.[958,959] To use such techniques effectively, methods for evaluating climate with respect to rest and bud development must be available.

Numerous correlations between numbers of hours of chilling and termination of rest, the physiological state of greatest interest, has fostered the development of rest prediction models.[486,488,960] The key issue, as expected, is selection of the appropriate base temperature, at or below which the rest-terminating, physiological processes commence. The model elaborated in Utah by Richardson and associates for 'Redhaven' and 'Elberta' peaches,[486] and later extended to other species,[960] is based on 6°C as the optimum chilling temperature. One chill unit (CU) thus equals 1 hr exposure at 6°C. Equally difficult is estimation of the chilling factor in the climate from hour to hour or day to day. Since termination-of-rest processes have some optimum below and above which they are inhibited, different temperatures must be weighed differently, and temperature sequences must be considered in any model that pretends to accuracy and generalizability. The CU contribution becomes less than one as temperatures drop below or rise above the optimum value (Table 2). A negative contribution of the CU accumulation occurs at temperatures above 15.9°C (60°F), and 0 unit contribution occurs below 1.4°C (34°F). Using the model and measured temperature data in the field, the date of rest completion can be estimated. In the model CU are calculated daily and summed from beginning of CU accumulation until sufficient CU are accumulated to complete rest.

Bud development subsequent to rest completion has also been modeled by summing growing degree hours (GDH°C) from end of rest until sufficient GDH are accumulated to reach a specific stage of phenological development (e.g., full bloom).[960,961] The accumulation of GDH is based on a lower limit of 4.5°C and an upper limit of 25°C, i.e., 1 GDH is defined as 1 hr at a temperature 1°C above the base temperature of 4.5°C. GDHs are calculated by subtracting 4.5°C from each hourly temperature between 4.5 and 25°C. All temperatures above 25°C are assumed equal to 25°C; thus the greatest accumulation for any 1 hr is 20.5 GDHs.

CU necessary to complete rest and GDH required to reach full bloom are constants specific

Table 2
SPECIFIC TEMPERATURE VALUES AND THEIR EQUIVALENT CU CONTRIBUTIONS AS USED IN THE MODEL FOR ESTIMATING THE COMPLETION OF REST FOR 'REDHAVEN' AND 'ELBERTA' PEACH TREES *(PRUNUS PERSICA)*

Temperature		
°C	°F	CU contributed
<1.4	<34	0
1.5—2.4	35—36	0.5
2.5—9.1	37—48	1
9.2—12.4	49—54	0.5
12.5—15.9	55—60	0
16.0—18.0	61—65	−0.5
>18.0	>65	−1

From Richardson, E. A., Seeley, S. D., and Walker, D. R., *HortScience*, 9, 331, 1974. With permission.

Table 3
ESTIMATES OF CU AND GDH°C (ACCUMULATED FROM END OF REST TO ANTHESIS) REQUIREMENTS FOR VARIOUS FRUIT TREES

Cv.	CU	GDH°C
'Tilton' apricot	720	4111
'Italian' prune	788	6411
'Elberta' peach	800	5110
'Bing' cherry	880	6072
'Bartlett' pear	1210	5644
'Delicious' apple	1234	6933

Adapted from Ashcroft, G. L., Richardson, E. A., and Seeley, S. D., *Hort-Science*, 12, 347, 1977.

for each fruit cv. They can readily be determined provided end of rest dates, full bloom dates, and temperature data are available for several years. However, since end of rest data are not readily available, they have been estimated for various deciduous fruit trees by appropriate statistical methods using full bloom dates and temperature data for several years (Table 3).[960]

An exchange of three letters in a horticultural journal suggesting restrictions in the use of the Utah model attests to the practical importance and ultimate difficulty in climate assessment in deciduous fruit tree management.[962] This model has not accurately predicted the end of bud rest in some marine climatic regions,[962,963] perhaps because it does not consider other environmental factors that may interact with temperature (Chapter 5, Section

I.G.3). A model developed for low chilling 'Sungold' nectarine indicates a broader range of effective temperature and a higher optimum for rest completion than in Table 2.[488] Remarkably, however, the Utah model for 'Red Delicious' apples is quite accurate in predicting flower bud development after rest completion in 'Cox's Orange' pippin apples in southeast England,[964] a climatic zone differing drastically from the apple-growing regions of Utah.

Genetic tailoring of deciduous fruit crops to climatic zones is quite difficult and at best a lengthy undertaking; this sets it apart from much of the studies with annuals. Selection for dormancy characteristics has been done with fruit trees and will probably expand as fruit-growing regions are extended to lower latitudes, where reduced chilling requirements for terminating rest are an essential quality.

Another example of complexity and refinement of technique is found in the Dutch bulb industry where controlled-environment storage guarantees proper thermal regimes for flower development and subsequent bulb performance in gardens or commercial "forcing" greenhouses. The basic discoveries of Blaauw and colleagues, relating a sequence of flower bud developmental stages to specific temperature requirements,[4] laid the foundation for modern bulb storage facilities. It was an early success of phenoclimatology, one that contributes to the reputation of the bulb industry for high quality products.

b. Temperature-Daylength Models

Nuttonson was among the first to use a temperature-daylength system for predicting maturity dates in crops over a range of planting dates and locations (Table 4).[965] His results show the value of using the simple multiple of thermal units × daylength for both wheat, sown from Alaska to Mexico, and peas, sown from Alaska to Maryland. Note that the minimum days from sowing to ripe grain in 'Marquis' wheat is for lower heat unit accumulations (cooler climates) and longer days which support some of the notions of climate adaptations for wheat discussed below.

Aitken's work is one of the most thorough in this area.[966] She provided a much more complex picture of a region, using photothermographs (Figure 4), than would otherwise be possible by growing degree day × daylength sums. Combined with phenological data provided (Table 5), one can readily visualize applying a similar analysis to many crops in many regions.

3. Case Histories

a. Wheat

Wheat cvs. in general yield best when some of the reproductive stages occur in moderately cool spring and summer conditions. Warrington and co-workers found the most important temperature effects during inflorescence development, rather than before initiation or during grain growth.[463] Plants held at 15 to 20°C during the inflorescence development phase have 3-fold more florets (and grains) per inflorescence head than those held at 25°C. Shoot dry weight is affected relatively little by the temperature treatments. The conclusion is that the major temperature effect is upon spikelet or floret initiation, and leads to the recommendation that for high yields: (1) wheat should be planted in the early autumn to permit completion of a greater part of the vegetative phase before winter, and (2) inflorescence initiation should commence as soon as possible in the spring to permit complete development during the cooler months.

We turn now to some specific genetic controls of flowering in wheat and a prototype case for tailoring crops to specific climatic conditions (in this case Mediterranean climates). An important feature of the wheat-breeding programs analyzed was recognition of the complex, but independent, inheritance of vernalization and photoperiodic requirements for flowering.[967] A few major genes with multiple alleles permitted separation and recombination of

Table 4
PHENOLOGY OF 'MARQUIS' WHEAT *(TRITICUM AESTIVUM)* IN A NUMBER OF NORTH AMERICAN AREAS

					From emergence to anthesis				From emergence to ripe grain			
Location	Date of seeding	Date of emergence	Date of anthesis	Date ripe	Days	Day-degrees (°)[a]	Av length of day (hr)	Av length of day multiplied by day-degrees	Days	Day-degrees (°)[a]	Av length of day (hr)	Av length of day multiplied by day-degrees
Fairbanks, Alaska Lat. 64°	June 3	June 12	July 15	August 29	33	955	20.3	19,386	78	2054	19.0	39,026
Saskatoon, Sask Lat. 52°	May 3	May 15	July 4	August 2	50	1192	16.6	19,787	89	2375	16.4	38,950
Winnipeg, Manitoba Lat. 50°	May 13	May 28	July 13	August 14	46	1402	16.2	23,712	77	2448	16.0	39,168
Dickinson, North Dakota Lat. 48°	April 20	May 4	June 29	July 31	56	1395	15.5	21,622	89	2540	15.4	39,116
Lincoln, Nebraska Lat. 41°	April 1	April 14	June 13	July 9	60	1743	14.2	24,751	86	2816	14.5	40,832
Tlalnepantla, Mexico Lat. 19°	December 1	December 10	March 26	May 7	106	2121	11.6	24,604	148	3711	11.7	43,419

[a] Day-degrees are calculated above 32°F.

From Nuttonson, M. Y., *Vernalization and Photoperiodism*, Murneek, A. E. and Whyte, R. O., Eds., Chronica Botanica, Waltham, Mass., 1948, 129. With permission.

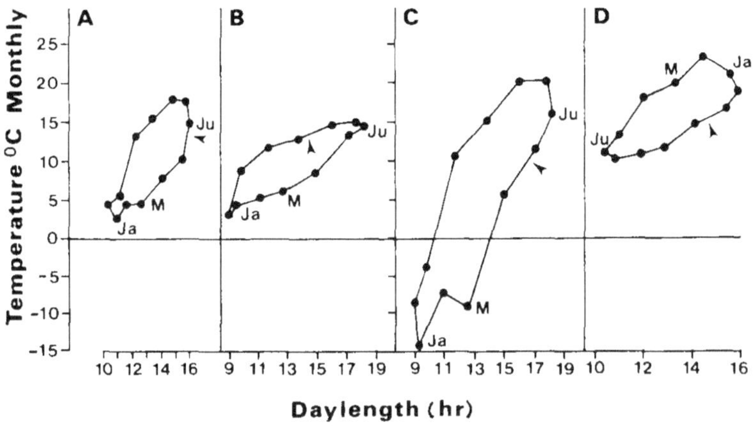

FIGURE 4. Mean daily values of temperature and daylength (including civil twilight) in successive months and in various locations. For Sacramento (A), Wales (B), and Canada (C), in 1963, and for Melbourne (D), in 1968. Ja = January; M = March; Ju = June. The arrow indicates the sowing date at each location. (From Aitken, Y., *Flowering Time, Climate and Genotype*, Melbourne University Press, Melbourne, 1974. With permission.)

the genetic determinants as in the studies by Klaimi and Qualset.[968,969] Equally important was recognition of the peculiarities of Mediterranean climates where much of the wheat of the world is grown. These areas have relatively mild winter temperatures, predominantly winter rains, and hot dry periods in the spring and early summers. In relation to wheat performance in other regions (see above) this suggested an early fall planting.[970] Varieties with the physiological attributes of spring wheat are well suited to these areas because little or no freezing resistance is needed, but the vernalization requirement of winter wheat cvs. would prevent premature flowering in the fall. Daylength-insensitive cvs. would be most productive, because they would flower earlier in the spring, exploiting stored soil moisture and cooler temperatures for grain production. Conversely, cvs. requiring LD for flowering would flower later in the spring and would be damaged by high temperatures and limited water supplies at critical stages of inflorescence development and postanthesis grain maturation. These facts were clearly recognized by Pugsley who suggested that some combination of vernalization-requiring winter wheats and day-neutral spring wheats should be attempted.[967] Thus, with genetic analyses on hand, genes for the vernalization requirement could be introduced into spring wheat cvs. with little or no LD requirement; spring wheat cvs. are among the most productive and desirable owing to research done to adapt the varieties to lower, almost tropical, latitudes. In Qualset's group the program was designed to select for early spring anthesis in a certain latitudinal range rather than for photoperiodic insensitivity.[971] Indeed, it may be highly desirable that some minimum daylength requirement be retained[972] to delay flowering time beyond late spring frosts which can be frequent in certain Mediterranean-type climatic zones. The results of this genetic tailoring have led to the introduction of new high-yielding wheat cvs. With no suggestion of improved photosynthetic apparatus, these cvs. "merely" represent a better fit of crop to climate.

b. Rice

All rice cvs. are somewhat earlier where temperatures are elevated throughout the growing season. However, cvs. with short critical night lengths for flower initiation are particularly useful in areas — such as central California and northern Arkansas — where low spring temperatures or early fall frosts, and resulting short growing seasons, may terminate a crop before grain is ripe. Controlled climate data are of great practical value in predicting how rice cvs. will perform in a region. For example, late flowering in the cvs. Caloro and Texas

Table 5
DEVELOPMENT RATE OF EARLY CVS. OF *PISUM SATIVUM*, *TRITICUM AESTIVUM* AND *TRIFOLIUM SUBTERRANEUM* AS A FUNCTION OF SITE OF CULTIVATION

Species and cv.	Site of cultivation with season of sowing and mean values of temperature (for 90 days) and daylength (for 30 days) after sowing			
	Melbourne winter (9°C—10.6 hr)	Canada spring (17°C—17.6 hr)	Wales autumn (11°C—12.8 hr)	Mather Sacramento, Calif. spring (17.1°C—15.8 hr)
	Days for Development Stage 1[a]			
Pisum 'Alaska'	13	10	18	8
Triticum 'Gabo'	71	28	45	24
Trifolium 'Dwalganup'	27	27	28	>34
	Days for Development Stage 2[a]			
Pisum 'Alaska'	67	25	35	30
Triticum 'Gabo'	56	19	—[b]	35
Trifolium 'Dwalganup'	90	25	180	30
	Days for Development Stage 3[a]			
Pisum 'Alaska'	66	33	—[b]	32
Triticum 'Gabo'	53	33	—[b]	41
Trifolium 'Dwalganup'	>53	>38	55	64+
	Number of Vegetative Nodes on Shoot			
Pisum 'Alaska'	9.1	8.5	8.4	8.7
Triticum 'Gabo'	9.0	7.8	8.0	7.2
Trifolium 'Dwalganup'	2.5[c]	4.5[c]	4.1[c]	11.2

[a] Stage 1: sowing to flower initiation; stage 2: flower initiation to first anthesis; stage 3: first anthesis to ripe seeds.
[b] Killed by frost.
[c] Value on lateral shoot.

From Aitken, Y., *Flowering Time, Climate and Genotype*, Melbourne University Press, Melbourne, 1974. With permission.

Patna 49 makes them questionable choices at central California latitudes where the longest day is 15 hr and where there is a 60-day period when daylengths exceed 14 hr.[973] Since about 40 days must be added postanthesis for grain filling and maturation, the elapsed time from sowing to maturity could be greater than 6 months for these cvs. in some regions of central California, where the sowing to harvest interval is not considered that long (i.e., the risk of early fall and late spring rains becomes too great). In general, maturity groups in most rice cvs. in the U.S., based on number of days from sowing to ripe grain, reflect the delay in anthesis due to long summer days which prevent initiation and slow inflorescence development. The length of the juvenile phase does not seem to be critical.

c. Chrysanthemum

Although *Chrysanthemum* cvs. used for cut flower and pot plant production have similar

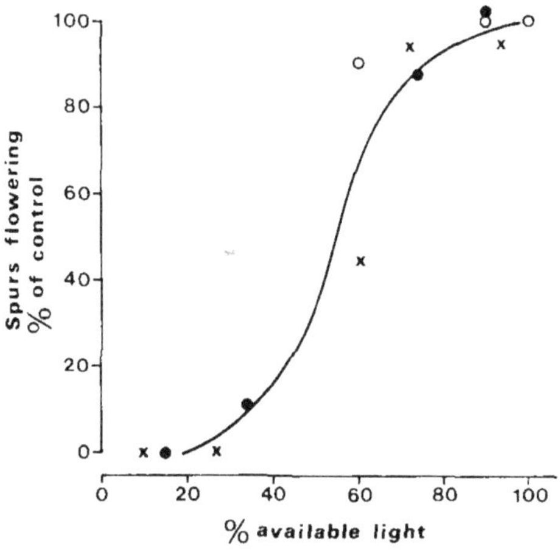

FIGURE 5. Effect of light on apple (*Malus*) spurs at different positions within the tree on their flowering the following spring. Spurs were deflowered in the spring of the year in which light values are measured. The curve is a combination of data from three different cvs.: 'McIntosh' (●), 'Greening' (○), and 'Macoun' (X). Control spurs were exposed to 100% available light. (From Cain, J. C., *J. Am. Soc. Hortic. Sci.*, 96, 664, 1971. With permission.)

daylength (SD) requirements for inflorescence initiation and development, they differ widely in their response to temperature. Three categories of cvs. have been described based upon adaptation to temperature:[404]

1. "Thermozero" types, relatively unaffected in the temperature range of 10 to 27°C, grow rapidly at 15°C. They are, thus well adapted for year-round production.
2. "Thermopositive" cvs. perform best at elevated temperature; inflorescence development beyond the capitulum stage is inhibited below 15.5°C. For year-round production they must be grown in winter-heated greenhouses.
3. "Thermonegative" cvs., in which inflorescence development is inhibited above 15.5°C, can be grown in summer greenhouses only when temperatures can be kept near 16°C.

B. Specific Environmental Factors

We began this section on crop adaptation with phenoclimatology because it presents best the integrated studies that have had such great success in agriculture. But there are many good examples in agriculture where practice has been improved merely by recognizing how flowering time is modified by single environmental factors.

1. Irradiance and Light Integral

In coniferous forests, cone production in Ponderosa, Sugar, and White pines is far greater on single, isolated trees than on those that are shaded by neighbors. The significance of this response to light is that 90% or more of the seeds produced from these species comes from the dominant trees which may or may not be the best when faced with strong competition in a commercial forest.[974]

In fruit orchards and vineyards shaded branches are nonbearing. Good light penetration to the leaves subtending the developing fruit is essential for fruit set and growth, but Cain's studies with apples show unequivocally that there is a high light requirement for flower initiation and development before anthesis and fruit set (Figure 5).[975] A study with container-

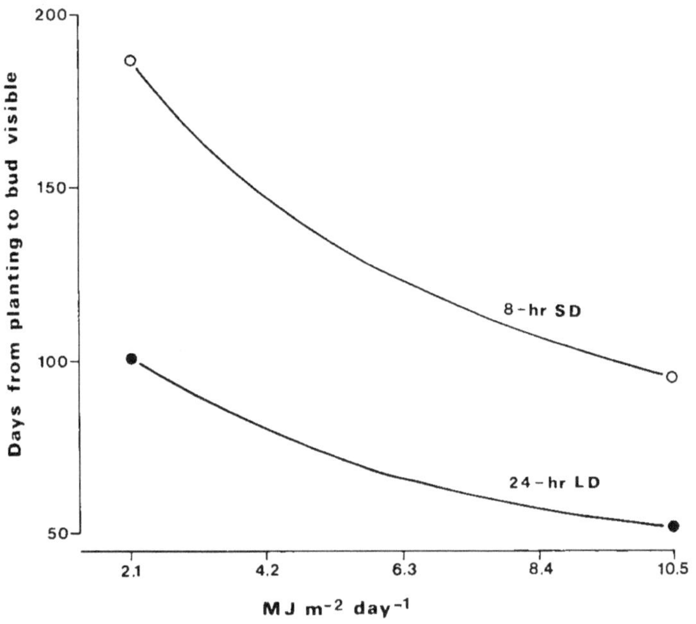

FIGURE 6. Fitted curves of days from planting to visible flower bud in carnation (*Dianthus caryophyllus* cv. 'White Sim') as a function of mean daily radiation at two photoperiods. (From Bunt, A. C. and Powell, M. C., *Acta Hortic.*, 141, 1983. With permission.)

grown apricot trees indicates that light reduction in summertime greenhouses, coupled with higher daytime temperatures, reduces flower bud initiation by more than 50% compared to that on similar-sized trees in outdoor nurseries.[976] A review by Buttrose cites a number of studies with species as diverse as blueberry, cacao, coffee, peach, pear, apple, apricot, and grapevine, in which in each instance the number of flower buds is reduced by shading.[977] It is reasonable to assume that pruning practices have been directed as much towards improving light penetration to the fruiting wood than towards any other consideration. The same is probably the case for trellising and orchard and vineyard supports. Jackson's review emphasizes overall yield-photosynthesis considerations rather than flower initiation per se.[978]

Data from year-round greenhouse production of *Gerbera, Chrysanthemum,* carnation, and rose provide ample evidence that low solar inputs reduce the number of commercial quality flowers produced per unit area as well as increasing the time to anthesis. Although in carnations short winter daylengths account for much of the delay in production, the lower numbers of flowers produced, as well as part of the delay in flowering, are a function of the reduced light integral as well (Figure 6).[979] In tomato and rose, premature flower abortion at low light levels sharply reduces fruit and flower production and this seems to be the case in several bulbous crops, too.

Since insufficient light integral is among the most important factors limiting wintertime greenhouse production of vegetables and flowers in northern latitudes, it is well to ask what can be done practically to alleviate the problem. Illuminating large growing areas with light of sufficient energy to augment wintertime lighting to that of spring or fall (let us exclude summer) is very expensive and, except for research purposes, impractical. Cathey and Campbell suggest that about 50 W/m^2 for 8 to 12 hr is required for good growth responses, including flowering, where the sole source of light will be from artificial sources.[980] A computation of electrical power costs to achieve 50 W/m^2 for 12 hr daily, for 4 months, assuming $0.075/kWhr, is about $56/m^2 for metal halide lamps which are among the most economical. Capital costs are about $218/m^2, which must be amortized and included in the

Table 6
THE INFLUENCE OF CANE LENGTH, AFTER PINCHING, ON FLOWER PRODUCTION OF THE ROSE *(ROSA HYBRIDA)* CV. 'BACCARA'

Cane length (no. of 5-leaflet leaves remaining)	Flowers per cane	Days to anthesis	Average flower stem length (cm)
2	0.7	47	81.5
4	1.3	40	79.2
6	1.6	36	77.8
8	2.2	31	82.1

Note: Cane length is expressed by the number of 5-leaflet leaves remaining. The shorter the cane, the lower the pinching and the lower the axillary buds which develop.

From Zieslin, N. and Byrne, T., *Sci. Hortic.*, 15, 179, 1981. With permission.

operating costs. Thus, large-scale commercial irradiance schemes, to augment wintertime greenhouse illumination only, to produce high quality crops may cost close to $800,000 per hectare annually.

To avoid these costs many commercial flower and vegetable production facilities have been moved to lower latitudes where the light integral is uniformly higher in winter. Although there are rumors of governmental subsidies, apparently increased transportation rates can be compensated on the northern European and North American markets. Also, wintertime transport of products from the southern to northern hemispheres, taking advantage of the complete inversion of seasons, is not uncommon for some flower crops.

However, another solution seems much more attractive, both economically and scientifically, namely management of photosynthetic assimilates so that even at low light levels more of what the plant manufactures can be allocated to the flower(s) [and developing fruits(s)]. We have cited sufficient evidence for daylength, CO_2, and chemically enhanced sink strength of inflorescence to believe that this can be translated into commercial practice. Management of roses, by selective cutting of flowers, is apparently a means of favoring flower development on shoots in the best competitive position for assimilate partitioning to the developing flower (Section VIII; Table 6). There may be other cultural practices, such as lowering night temperature, increasing water stress by reducing irrigation frequency, or using selected osmotica in the irrigation solution, that can improve flower initiation and development without excessive reduction in photosynthesis itself.

Still the most likely developments should come from chemical regulators that more or less selectively reduce stem elongation without seriously impeding apical meristematic function (e.g., growth retardants) or can more or less selectively enhance flower initiation and/or development (e.g., growth retardants, GAs, cytokinins, ethylene-releasing compounds). Specific examples will be cited in Section VI. The paper by Chapman and Sadjadi is an excellent example of the combined use of stem growth inhibitors and cytokinins to increase yield in *Vicia faba* through reallocation of assimilates.[981]

2. Photoperiod

a. Daylength Management

Daylength is an easily managed parameter in greenhouses and for this reason control of photoperiod is commonly found in production of chrysanthemums, carnations, poinsettia, azalea, and other photoperiodically sensitive flower crops.[251] The species cited above are vegetatively propagated and, hence, the juvenile phase is never a problem. However, minimum plant size for optimum response to photoperiodic treatments is important, and daylength management at this early stage of propagation is equally important to prevent premature flowering. For SDP, daylength is lengthened, often by low intensity light breaks near the middle of the night during early stages of growth, so that plants will grow vegetatively to some predetermined size, after which daylength is reduced to 8 to 9 hr by covering entire benches of greenhouses with black cloth.

Several studies have had as their goal the conservation of light energy required to prevent flowering under natural SD. In *Chrysanthemum* duration of supplementary lighting varies with season and latitude, presumably owing to the direct relationship between daytime irradiance and nightime irradiance required to prevent flowering.[982] Repeated light-dark cycles during the night break, suggested for chrysanthemum by Cathey and Borthwick,[983] conserve radiant energy since plants are illuminated for only 6 min every 30 min during the prescribed night break instead of the usual 4 hr of continuous light. Other cycles, say 1 min of light every 5 min, are necessary for other *Chrysanthemum* cvs.,[984] but the energy savings remain approximately 80%. Interval lighting is not identical to uninterrupted light.[985] In some species interval lighting is more effective, but in others continuous lighting may be required if plants are to be held vegetative. For chrysanthemum cvs. investigated, about 3 to 5 weeks of SD promote most rapid inflorescence initiation and development, yet it is not unusual to see "blackclothing" for many additional weeks until the inflorescence shows color.[986] This is an unnecessary practice that should be eliminated to reduce labor costs and improve flower quality.[986]

Recognition of the fact that the phytochrome system seems to be the primary photoreceptor for most daylength-controlled responses (Volume I, Chapter 4) has encouraged studies on light quality required (and the best light source) for promotion or inhibition of flowering. In *Chrysanthemum,* during the night interruption, a high R/FR ratio from fluorescent tubes is more effective than a relatively low ratio from incandescent lamps.[987,988]

Paulet reported that a 30-sec night interruption by a red laser beam will induce flowering in *Cichorium intybus,* a LDP,[989] and Paleg and Aspinall have shown that a laser beam emitting light at 633 nm is extraordinarily effective in inhibiting induction by a long night in *Pharbitis nil,* a SDP.[990] The narrow laser beam strikes the leaves (cotyledons) primarily, with perhaps only rare, accidental exposure of the shoot tip (plumular) tissue, confirming that at least one flower controlling phytochrome pool is in the leaves. However, there is an excellent study with *Pharbitis* showing sensitivity of the plumule to red light interruption.[991] More selective irradiance with lasers should be instrumental in distinguishing among physiologically active photoreceptors in leaves and in receptor buds. Eventually as prices of lasers are reduced some applications may be developed to give more precise control of the flowering response. This would seem to be a particularly valid application in cases where high irradiance is required primarily for "sink" activation and not for enhanced photosynthesis.

b. Loss of Daylength Requirement with Age

We have discussed acquisition of daylength and low temperature sensitivity with age in considerations of juvenility (Volume I, Chapter 7), but of equally great practical significance in some flower crops is the loss of daylength requirements for flower initiation and development. In SD "requiring" chrysanthemums, grown in LD, inflorescence initiation will occur on a shoot system after a certain number of leaves have been initiated. This number is strongly heritable and is not very great in some cvs. (Volume I, Chapter 7, Table 2), e.g., 'Bright Golden Anne' will initiate a terminal inflorescence on a shoot after only 18 leaves are formed.[152] Upon initiation of the terminal inflorescence, lateral buds will break leading to a complete alteration in the form of the plant and making eventual inflorescence (spray) development much less desirable for commerce. Flower growers must remove the terminal shoot bud (pinch) and give SD induction well prior to initiation of the maximum leaf number to insure proper inflorescence form at anthesis.

Practical chemical controls, to inhibit premature inflorescence initiation, have been investigated,[992,993] but a genetic solution is not out of the question if the genes for increasing maximum leaf number can be isolated and introduced into otherwise desirable cvs.

With increasing age almost all LD *Fuchsia* cvs. flower under SD when grown in containers in the greenhouse; hence, depending upon the plant size desired at sale and the requirement for a predictable bloom date, extra lighting may not be required.

c. Day Neutrality

The absence of daylength requirements for flowering, i.e., day neutrality, would facilitate exchange of genotypes across a wide latitudinal range and would perhaps lead to development of early maturing varieties. In soybeans it has been found only in a few cvs. from the earliest maturing groups.[994,995] Apparently maturation date in soybean is determined primarily by the critical nightlength for flower initiation (and also for pod development) and not to length of the juvenile phase. In rice apparent day neutrality (or very short critical night length) is found in cvs. with short and long juvenile phases. Since the inheritance of daylength controls for flower initiation and development in many species is multigenic and multiallelic,[948,971,996] the "answer" to day neutrality will have complex genetic solutions, probably affecting many physiological systems.

3. Low Temperature Requirements for Vernalization

In many instances in biennial crops the vegetative parts only are marketed (lettuce, celery, carrot, table and sugar beets, Brussels sprouts, etc.) and, hence, premature flowering, with inflorescence stalk elongation, is not desirable. Sowing dates in such cases must be selected so that early spring growth, at low temperatures, is during the juvenile phase when the plants are relatively insensitive to vernalization, and subsequent LD do not promote flower initiation. In general, the longer the juvenile period, the earlier the planting date.

Data for sugar beets grown in Scotland reveal the possibilities for selecting varieties in which seedlings are relatively less responsive to spring chilling. This extension of chilling insensitivity is probably due to some combination of an extended juvenile phase as well as increased requirements for chilling,[997] but the field experiments reported do not permit distinguishing which of the two factors is most important. Since the progeny from the nonbolting plants also do not bolt when they are planted early, there is clear indication that one has a manageable heritable trait to work with.[998]

Control of bolting and flowering in some biennial crops may come by elevating soil temperatures. In celery and Chinese cabbage, raising root zone temperature to 30°C negates the effects of vernalizing temperatures to the shoot system such that the plants remain vegetative even in LD conditions that follow.[999,1000] At present, soil heating is far from practical in most locations, but there is research in progress to use rejected thermal energy from power stations to heat water for drip irrigation systems. Should this prove feasible, then maintaining some biennial species vegetative by heating the root system may be feasible, too.

Since many biennial species grow best at cool but not cold temperatures (15 to 20°C), they are grown commercially in relatively restricted climatic zones and times of year. A small part of the Salinas Valley of California with year-round cool temperatures is quite literally the U.S. garden for winter vegetable crops — another example of phenoclimatology in practice. For seed production in biennial crops, early autumn sowing is a common practice, to insure adequate growth through the juvenile phase, in time for winter chilling, followed by flowering and seed production during the LD of the following spring and summer.

Practical examples for controlled low temperature treatment of flower crops are difficult to find. *Matthiola incana,* a biennial or perennial species requiring low temperatures for flower initiation and development, is grown outdoors in some Mediterranean climates. For some cvs. fractional vernalization on a daily basis in greenhouses is adequate to insure normal and reasonably rapid flower development. For example, temperatures near 10°C for more than 20 hr and 18°C for 4 hr give the most rapid flowering, but shorter exposures to 10°C are also satisfactory if delayed anthesis is acceptable.[1001,1002] The lesson here is that most low temperature-requiring greenhouse flower crops may be grown during the late fall to early spring interval, and thus, by bringing in outside air, or by reducing greenhouse heating if outdoor temperatures are very low, exploit the naturally low temperatures available.

According to Halevy, *Limonium sinuatum* may be the only biennial flower crop that is given a true vernalization treatment at an early seedling stage (8 weeks at 11°C); after this, plants are shifted to higher temperatures (22 to 27°C) for rapid inflorescence development to anthesis.[1003] However, in this case, too, normal but slower flower initiation and development will occur if plants are held continuously at temperatures somewhat below 18°C.

4. Water (and Osmotic) Stress

a. Inhibition of Reproductive Development: Response to Acute Stress

Water deficits occur frequently under field conditions; a brief (acute) episode may reduce inflorescence initiation and development such that severe yield depressions result. The impact upon yield depends on when the stress occurs during crop development as well as upon the species. In maize, wheat, oats, barley, etc., with terminal inflorescences and a single period for initiation of reproductive structures, stress at the time of initiation will severely reduce yield.[1004] As expected in these crops irrigation improves yield. Yield increases have warranted installation of irrigation systems on the fringe of the U.S. "corn belt", primarily to avoid stress at pollination or during ear initiation when effects upon yield are particularly severe. Stress applied earlier during the vegetative phases of growth or later at tasseling restricts vegetative growth, but has little effect ultimately on yield. This suggests that corn plants have more than adequate photosynthetic surface to fill the ears initiated.

Multieared corn is affected much less severely by moisture stress. The reason for this is that pollen is often in excess and ear initiation occurs over a longer interval; hence, stress for a limited period is not likely to reduce sites for grain filling. For the same reason, crops with axillary flowering (e.g., soybean) that initiate inflorescences over long periods are not as severely affected by stress. Stress applied to soybeans during the flowering period results in fewer pods in the lower part of the plants, but this loss is often compensated by greater pod production at a later stage in the upper parts. A rule of thumb is that once plants have reached one half the mature height at the time when stress occurs they are capable of producing a large number of pods after the plants resume growth (Figure 7). Stress during flowering in soybean causes increased flower abortion rather than depressed flower initiation.[1004]

b. Enhanced Reproductive Development: Response to Acute or Chronic Stress

Crops grown in saline (high osmotic) environments have depressed yields due to overall reduction in growth, yet reproductive development may be precocious and numbers of flowers may be increased. The same results are reported for crops under some degree of water deficiency, a normal occurrence even for irrigated crops grown in arid regions with high evapotranspirative demands.[1005-1009] A classic observation with *Citrus* is that flower bud differentiation commences within 9 days after relieving a prolonged summer and fall drought.[1010] Nir and co-workers found at least two critical stages for floral differentiation in lemon, the first following drought and the second, independent of drought, which is required for complete development of the terminal flower and subsequent axillary flowers.[1006] In tomato promotion or inhibition of abortion by water stress is dependent upon the flowering

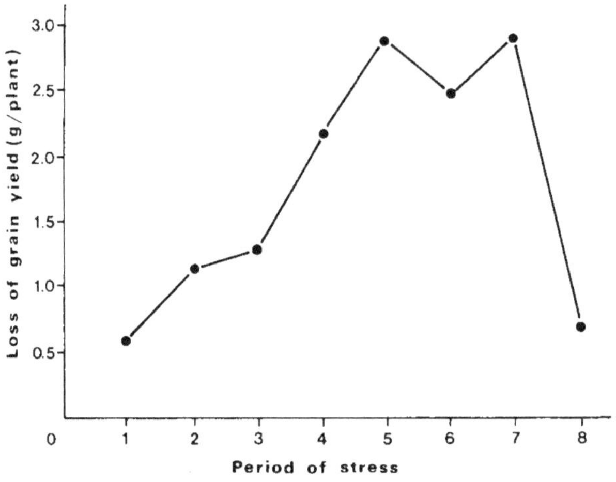

FIGURE 7. Relationship between period of water stress and reduction of yield in soybean *(Glycine max)*. Control yield = 6.49 g per plant. Periods 1 to 3: flowering; periods 4 to 8: bean filling. (Reproduced from Shaw, R. H. and Laing, D. R., *Plant Environment and Efficient Water Use*, Pierre, W. H., Kirkham, D., Pesek, J., and Shaw, R. H., Eds., Am. Soc. Agron., Madison, Wis., 1966, 73. By permission of the American Society of Agronomy.)

stage. Water stress before the flower buds have reached 4 mm may reduce abortion under low irradiance.[1011] The practical significance of these studies has been the search for a physiological measure of stress, to know when and for how long to use stress to optimize effects on reproductive development without overall yield depression (which must occur because of reduced photosynthesis).

Investigations of physiological changes in plants following water stress are numerous, but few have direct bearing on reproductive development. Allsopp suggests that water stress directly promotes flowering as a result of increased carbohydrate levels in stressed plants.[1012] Soluble solid concentrations are higher in water-stressed *Bougainvillea* plants, where water stress promotes inflorescence development.[1007] Although no causation has been established between soluble solids and inflorescence development, similar correlations have been noted with other environmental and chemical factors promoting flowering in *Bougainvillea* (Chapter 9). Many authors suggest that stress-induced increases in carbohydrate level may in part account for the observed increases in flowering (see Volume II, Chapter 8; this volume, Chapter 9). But stress alters so many systems and so many major chemical components in plant tissues[1013] that it is premature to say which may be involved in the flowering process. For this reason it seems unlikely that we will find naturally occurring compounds that substitute for drought, except by accident. And that accident has occurred. Growth retardants, discovered in screening tests for other applications, have been used experimentally in *Citrus* to replace water stress-induced promotion of flower initiation.[1005,1014] These compounds increase inflorescence development in *Bougainvillea* to levels beyond those seen even in salt and/or water-stressed plants,[231,1015] but no one suggests that growth retardants mimic water stress effects upon physiological systems.

5. *Nitrogen Nutrition*

There are research data showing that the amount and form of N applied affect flower initiation and sex expression in some species (Volume I, Chapter 2; Volume III, Chapter 3), but we do not know of examples where agricultural practices have been altered as a result of these findings. Although some of the effects upon flowering of ammonium vs.

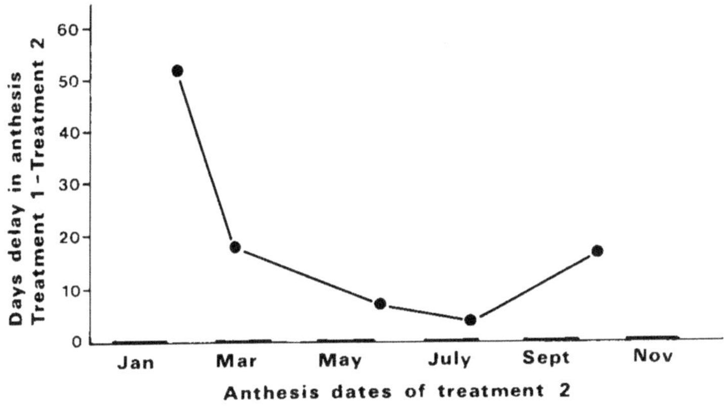

FIGURE 8. Effect of season and N, P, and Ca fertilization on delay of anthesis for greenhouse-grown tomatoes (*Lycopersicon esculentum*). Levels of fertilizers in treatment 1 = 4 N, 4 Ca, and 1 P and in treatment 2 = 1 N, 1 Ca, 4 P. (From Bunt, A. C., *Plant Soil*, 31, 97, 1969. With permission.)

nitrate nutrition are spectacular, as in apple trees,[1016,1017] they are also controversial in that some research stations have not been able to reproduce the results. In most greenhouses, nutrients are provided in a modified Hoagland's solution where there is little or no ammonium ion.[1018] In our greenhouse nutrient solutions, the ratio of nitrate to ammonium is about 9:1 and in the original formula only nitrate is used. The complex of responses to nitrate and ammonium ion nutrition of intact photosynthesizing plants is well described by Barker and Mills.[266] There are often large effects upon soil pH and availability of other mineral elements and consequently there cannot be direct and selective effects on the flowering process. Even though the mechanisms through which the results are obtained may remain unknown, the overall impression is that with more refined quantitative research N could be treated in much the same manner as the light integral or water stress with improved control of flowering time, flower number, or quality.

A study by Bunt on greenhouse tomato production, as a function of N, P, and Ca levels, shows how important mineral nutrition is in affecting the time to anthesis (Figure 8);[267] with high N and Ca (essentially 4 times that of the low level), up to a 50-day delay in anthesis is seen in the winter-cultivated crops. Note that the delay is negligible in the summer-grown greenhouse crops.

In general, we believe that varying nutrition as a function of season and/or as a function of soil temperature, particularly in greenhouses where the shoot system may be kept at relatively lower temperatures than that of the soil, may have important economic consequences for flower and vegetable crop production systems.

IV. SPECIAL PROBLEMS WITH WOODY PERENNIAL SPECIES

A long juvenile phase, characteristic of plants with a tree-like growth habit, and large annual variation in crop yield (alternate bearing) are two problems essentially unique to woody perennial species.

A. Juvenility

Delayed flowering in many tree species impedes breeding programs and, perhaps of more practical importance in these cases, the ease of vegetative propagation through rootings of stem cuttings is inversely related to age; quite often cuttings from mature plants will not root, whereas cuttings from seedlings will root relatively easily. Hence, there is a very practical interest in rejuvenation. Defining terms to deal with the variety of subproblems within the large framework of juvenility has been difficult; to some extent discussion of the subject here is hindered by a lack of precise terms and quantitative measurements of the juvenile phase for most tree species.

1. Reducing the Juvenile Phase

Minimum node number to flower would seem to be a prerequisite measure before one can realistically compare advantages accruing to one or another technique for shortening the juvenile phase, but there are difficulties with the node number concept (Volume I, Chapter 7).

Breeding programs for apples and pears have provided the basic data showing that the length of the juvenile period is a heritable trait (Volume I, Chapter 7).[1019-1022] Results of these studies show that the length of the juvenile phase is inversely correlated with early seedling vigor (growth rate). Unfortunately, the data do not show whether the seedlings flower at an earlier node number, i.e., whether vigor includes accelerated rates of leaf initiation. Another approach to reducing the length of the juvenile period was used by Visser.[1021] He found that Malling IX and Malling XVI apple rootstocks shorten the juvenile phase of apple seedlings irrespective of growth rate of the scion. *Malus hupehensis* and *M. sikkimensis* rootstocks extend the juvenile phase and both rootstocks reduce scion growth rate, as measured by trunk diameter. Again, node to first flower was not recorded. Without this information we must consider the possibility that the rate of leaf initiation is accelerated in seedlings grafted on Malling IX or Malling XVI rootstocks and retarded in the scions on *M. hupehensis* or *M. sikkimensis*.

There is no indication that chemical treatments are as yet generally useful in terminating the juvenile phase, although CEPA treatments force earlier flowering in mango and apple,[1023,1024] and some GAs markedly shorten the time to inflorescence formation in some coniferous species.[756]

There is little question that leaf number or area (Volume I, Chapter 7) or some other aspect of plant size is important in determining the juvenile periods, i.e., the vigor factor which is rarely defined quantitatively and confuses many physiologists. The best recommendation for the plant breeder working with trees is to force plants into as rapid growth as possible until flowering occurs.[217,1025,1026] Greenhouse conditions are an excellent aid for forcing rapid growth but, since field studies are required to evaluate economic performance of a cv., seedlings must be transplanted to the field.

Once the desired fruit quality and growth habit have been developed, grafting of adult wood onto selected rootstocks or rooting of adult cuttings is practical; the time-to-flowering of the new propagule is not determined by juvenile traits but by factors that determine the duration of the "transition period".[1027] The transition period in woody species may be equivalent to "induction" in herbaceous species. It will not receive special attention here because we have dealt with this aspect of flowering under several other headings (water stress, light, growth retardants, etc).

2. Extending the Juvenile Phase — Rejuvenation

Propagation of plants by rooting of stem cuttings is without question the easiest method of developing and maintaining uniform genotypes (clones). For this reason much of the horticultural industry relies on cutting propagation and accounts for research interest on rooting of stem cuttings of valuable selections. Many adult trees are notoriously difficult to root from cuttings, whereas seedling shoots, or shoots from the base of old trees, root readily. Since attempts to rejuvenate trees by chemical means have not been generally successful,[1028,1029] more classic horticultural approaches are used. Forcing sprouting from the lower trunk by continuous pruning often provides an abundance of sprouts that are relatively easy

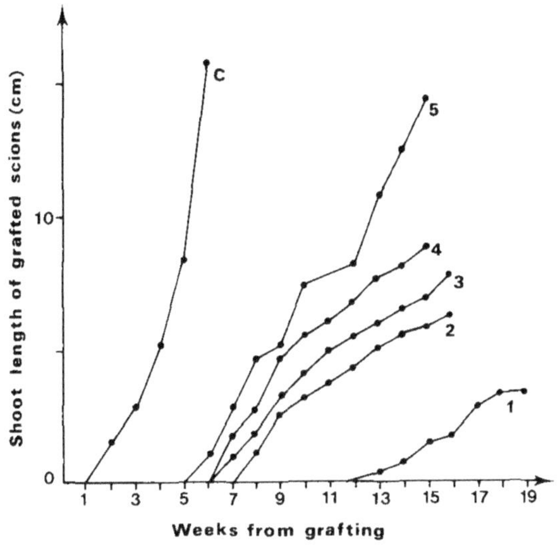

FIGURE 9. Growth curves of shoots issued from grafted buds collected on a 95-year-old tree of *Sequoia sempervirens*. 1 to 5 = 1st to 5th generation respectively. Each generation represents another graft generation; after bud has grown and produced a shoot, another bud is taken and grafted onto another 8-month seedling. C = control, an 8-month bud grafted on an 8-month stock. Terminal or axillary buds are used as scions. They are inserted in the terminal position of the decapitated seedlings (rootstocks). (From Franclet, A., *Ann. AFOCEL 1980*, 11, 1981. With permission.)

to root.[1028,1030] Franclet describes recent progress in rejuvenation by: (1) serial subculturing in vitro of adult lateral buds and (2) serial grafting of adult buds onto juvenile rootstocks followed by in vitro propagation of the scion buds so that clones from rooted cuttings can be developed.[1031] This has been accomplished with tissues taken from the largest *Sequoia sempervirens* known (Figure 9)[1032,1033] as well as from Douglas fir.[1034] During the rejuvenation process the growth rate of the scion-derived tissue gradually increases to that of juvenile tissue (Figure 9) and flower initiation is reduced or delayed considerably; thus, the scions are rejuvenated with respect to a number of morphological and physiological features that characterize the juvenile state. Using grafts to juvenile seedlings, rootable cuttings of adult, difficult-to-root *Cupressus sempervirens* and *Eucalyptus grandis* have been obtained.[1031,1035]

B. Alternate (Biennial) Bearing

Year-to-year cyclic fluctuations in fruit production, called alternate or biennial bearing, in perennial tree and vine crops have been thoroughly reviewed by Davis, Jonkers, and Monselise and Goldschmidt.[1036-1038] Fruit development in many of these species occurs nearly simultaneously with flower initiation for the crop of next year; thus the concept of nutrient competition between developing fruit and nearby meristematic sites for flower initiation receives much weight in the literature. Even when the period of flower initiation is separated in time from that of major increases in fruit growth, as in olive and some *Citrus* species, emphasis is given to nutritional factors. Although some good correlations between storage carbohydrates and flower initiation have been uncovered,[1036,1039,1040] there are nevertheless clear indications of hormonal influences in all cases. There are pronounced influences of the developing fruit, particularly seeded fruit, on the bearing and nonbearing spurs. Grochowska found that starch content in spurs bearing seeded fruit is much lower than in nonbearing spurs.[1041] This observation can be correlated with that of Chan and Cain for apple and of Gil and co-workers for pear that parthenocarpic cvs. are not as severely prone to biennial bearing.[1042-1044] Apparently something(s) diffusing from developing seed inhibit(s) floral initiation in spurs, perhaps through its effect upon starch deposition and/or utilization. GA-induced inhibition of flower initiation in many deciduous fruit trees is well known (Volume II, Chapter 6, Section IV.A); hence, this class of regulators received special attention. There are now several studies showing diffusion of GA-like substances from seeded fruit.[1044,1045]

GAs alone may not account for the role of developing seed in inhibiting flower initiation, but GAs have been used experimentally to reduce flower initiation when fruit load is light.

Growth retardants, compounds that appear to inhibit GA biosynthesis or reduce GA activity, have been used commercially to increase flower initiation when fruit load is heavy. Most successful treatments have been with the growth retardant daminozide.[908,1038,1046] Early fruit thinning with a variety of auxin-like compounds or CEPA or 2,4-dinitro orthocresol is still the most practical and commonly used treatment in the heavy-bearing year. However, these treatments are also fraught with problems, particularly that of overthinning. The optimum fruit carrying capacity of a tree is often impossible to judge at the time of fruit set, and even thereafter; moreover it often depends on environmental parameters that cannot be controlled. Hence, recommendations based on empirical evidence one season may not serve well in another year. Clearly this is a field of research where more fundamental knowledge will lead to important improvements in horticultural practice. The reviews by Dennis for fruit set and by Monselise and Goldschmidt are particularly instructive in analyzing the various experimental approaches that are likely to resolve the alternate bearing problem.[907,1038]

V. MODIFICATION OF SEX EXPRESSION

Hybrid seed and pickling cucumber production in *Cucumis sativus* are greatly facilitated by the use of CEPA and GAs. As reviewed in Chapter 7, CEPA promotes early female flower formation. GAs, on the other hand, promote male flower formation, even in gynoecious lines. CEPA treatments result in significant yield increases and at advanced dates. The vines are smaller and the crop is more uniform, warranting a once-over machine harvest.

The value of small, early-yielding vines encouraged development of gynoecious lines that produce primarily female flowers under most cultural conditions. Seed production for these lines, however, is a problem because they have insufficient male flowers with viable pollen. This is where GA-induced male flower formation is of value. Commercial seed producers treat one out of three rows of gynoecious cvs. with GAs at the first true leaf stage; these treated rows produce pollen for the unsprayed rows permitting seed production in the normally parthenocarpic fruit.

Although many other species — particularly among the Cucurbitaceae — respond to chemical (and environmental) modification of sex expression, we know of no other important commercial applications. Since, however, hybridization is favored by producing male sterile lines (self pollination is prevented), we expect that geneticists will continually find use for sex-modifying treatments.

VI. CHEMICAL CONTROLS OF FLOWERING

Throughout these volumes chemical controls of flowering have been integrated with other parameters. In this section we try to pick up those chemicals that have greatest practical value. What we find is a broad miscellany without logical grouping. The reason for this apparent disarray is that we still lack good hypotheses accounting for the action of these compounds. Surely all of the compounds affect many physiological systems and not merely the flowering process.

A. Growth Retardants

Growth retardants increase or advance flower bud initiation not only in fruit trees (Volume II, Chapter 6), but also in some flower crops such as azalea,[1047,1048] bougainvillea,[1015] camellia,[1049] and geranium.[1050] Daminozide is used routinely as a foliar spray in the production of potted azaleas for both increased flower bud formation and size control; CCC is used in the production of potted geraniums from seed. Daminozide is widely used in pot chrysanthemum culture, as is CCC in the production of poinsettias; such treatments reduce stem elongation without seriously delaying or reducing inflorescence development. Owing to great success achieved with these compounds new, more effective growth retardants are likely to be made available for research and eventual application to ornamental crops, but we are not as sanguine concerning food crops owing to problems in registration. CCC is used to increase yield and prevent lodging in wheat in some countries, but is not registered for use on any food crop in the U.S. and Canada. This problem of safety of chemicals limits optimism for future developments by agricultural chemical companies who have obvious commercial interests in extending their application to food crops.

B. Ethylene/Ethylene-Releasing Compounds

Ethylene, acetylene (released from calcium carbide sprays), and NAA, which increases ethylene production in many species, cause early and uniform flower initiation in pineapple plants.[1051,1052] Uniform flower initiation is one of the essential steps in obtaining uniform fruit ripening which facilitates once-over harvesting. Moreover, timing can be controlled for the convenience of the grower. CEPA is now the compound of choice in some commercial pineapple production. Flower induction in ornamental bromeliads represents a small use of ethylene-releasing compounds.[1053] As discussed above, ethylene-releasing compounds are used in modifying sex expression in Cucurbitaceae, where their application results in increased yield and more uniform crop. A potential use for ethylene-releasing compounds may be to prevent premature budding in chrysanthemum.[992,993] Much more research on ethylene influences on flowering should commence with the use of inhibitors of ethylene biosynthesis (AVG)[647] and ethylene action (silver-containing compounds, such as STS).[647,649,650] A naturally occurring ethylene precursor, ACC, is now widely available for research (see Volume II, Chapter 7 and review by Lieberman[647]). These compounds will permit more precise investigations of timing and site of ethylene-related responses. Recall that high CO_2 enhances flower development in some species where it has little or no effect upon photosynthesis rates (or dry weight gain). In such cases CO_2 may be competing with ethylene and thereby reduce its physiological activity, perhaps in the apical meristematic or developing flower tissues.

C. Gibberellins

Although GAs have great potential for modifying reproductive development (Volume II, Chapter 6 and Volume III, Chapter 7), few commercial applications have been found. In seed production of head lettuce GA_3 applications insure uniform inflorescence initiation and stalk elongation when applied subsequent to chilling and before leaf primordia have tightly enfolded the apical meristem, i.e., at the three- to four-leaf stage of vegetative growth. Seed production is increased and harvesting is facilitated by such treatments.[1054] As mentioned above, GAs are also used for commercial production of seeds in gynoecious cucumber lines.

In California, inflorescence development in artichokes is advanced to harvestable size by GA_3 treatment during the winter months when yields are otherwise low. Young buds (or whole plants) are treated at least 3 months in advance of the planned winter harvest.[1055] Total yields are not increased by these treatments; one is concerned only with faster inflorescence development.

D. Triiodobenzoic Acid

TIBA often increases flower production and reduces stem elongation in soybeans.[1056,1057] Although TIBA formulations have been sold for field applications, particularly to reduce lodging, we are not aware that they are in wide use in soybean production.

E. Cytokinins

To our knowledge cytokinins are not used commercially to modify reproductive development. This is surprising in view of all the "good" things that happen when tomatoes, chrysanthemums, or bougainvillea receive appropriate treatments (Chapter 7). Also, at least one synthetic cytokinin, BA, is widely available, generally quite active, and hence should find its way into agricultural practice.

VII. GIRDLING

Although not widely practiced, not even by tree breeders, ringing, girdling, and bark inversion accelerate reproductive development of branches and even of entire trees. Girdled citrus trees flower approximately 2 years earlier than nongirdled trees.[1058] Sax's studies on bark inversion and the use of graft unions to hasten flowering in apples and pears are classic examples of this genre of experiment that dates back more than 150 years.[1059,1060] Explanations to account for the increased flowering rely heavily on some form of a nutrient diversion hypothesis. Knight said in 1820 that ringing intercepted the "descending current" that "accumulates above the decorticated space when it is...carried upward, to be expended in an increased production of blossoms."[1061] However, much more than photosynthetic assimilate flow is altered by interrupting phloem connections to the root system. Root-system function itself should be altered as assimilate supply is limited and, hence, root-derived materials will arrive to the shoot (via the xylem) in reduced quantities.

VIII. TRAINING SYSTEMS AND PRUNING

Espalier and other training systems that bring branches to more horizontal positions encourage earlier and greater flower initiation. In mature larch trees the sagging horizontal branches flower 1 to 2 years earlier and with many more flowers than the upright ones. Tying down upright branches encourages the same increase in reproductive development. The combined effect of tying down and girdling is much greater than additive, hence one suspects that at least two different physiological systems are affected.[1062] Gravitational influences on redistribution of exogenously applied cytokinin and auxin suggest that altered hormonal relationships play a large role in the growth and flowering responses of laterally oriented branches. Unfortunately, we find no chemical treatments that substitute for espalier or other training systems. Growth retardants come closest to creating the desired effect.

Management of greenhouse roses for cut flower production depends strongly on pruning practices.[1063,1064] Axillary bud break from the lowermost portion of plants is highly desirable because shoots derived from these buds ("bottom breaks") are more vigorous, make significant photosynthetic assimilate contributions to the root system, and in general insure a longer, more productive life span for the plant. Studies directed toward encouraging bottom breaks have added several significant findings of practical importance in rose culture. Bottom breaks generally suffer a greater degree of inflorescence abortion than shoots derived from the upper axillary buds.[1064] Furthermore, flowers are initiated and develop more rapidly from upper than lower buds (Table 6).[551,552] Since bottom breaks are vigorous and generally of large diameter, but give rise to many blind shoots, they should be cut back to be used as a parent shoot for production of marketable flowers. Climate control, particularly greenhouse night temperature, is also an important factor in formation of shoots from upper buds following pruning[413,1065] and for extending production of roses over a longer season and for reducing energy consumption. The complexity of rose plant management goes beyond the discussion presented, but the essence of the argument is clear: there is positional information in axillary buds of roses, as in tobacco (Volume II, Chapter 1 and Volume III, Chapter 6), that influences inflorescence initiation and development. Hence, the flower cutting procedure in removing one crop has very important consequences for the succeeding crop.[552]

CONCLUSIONS

Applications of fundamental investigations of reproductive development have found their way into agronomic, horticultural, and silvicultural practice. After a review of each topic covered in these volumes, one is left with the impression that plant physiologists working in the area of reproduction will find even greater applications for their findings. This conclusion seems warranted because of relatively recent emphasis on the role of assimilate partitioning and its management in the flowering process. Some emerging chemical, environmental, and even cultural controls of assimilate partitioning seem particularly promising. In Woolhouse's recent essay on hormonal controls of senescence and reproduction, resource allocation is always central to his arguments.[1066] We find this true in most of the reviews examined in agricultural journals, whether the subject is alternate bearing, light interception, yield improvement, shoot/root ratios, etc. The impression is that concepts concerning resource allocation can be fitted to almost any agricultural problem and vice versa.

Collaboration of physiologists with geneticists and agronomists or horticulturists has produced sometimes astonishing success in engineering crops to fit many climatic regions. To date, all advances have been made using conventional approaches to breeding with little or no reference to molecular phenomena. In this work the importance of contributions by morphologists — skilled at defining developmental stages — working with physiologists and geneticists, cannot be underestimated.

ADDENDUM

INTRODUCTION

Information about inflorescence and flower development has accumulated during the publication of this volume. To update our work as much as possible, we present in this section a few selected references which were overlooked in the initial review period or appeared in 1983 or early 1984. The proceedings of a symposisum on "Light and the Flowering Process" (Littlehampton, England in September 1983) will be published shortly,[1067] and a review by the Durands on sex differentiation[1068] should be referred to directly, because they cannot be excerpted adequately in this addendum.

CHAPTER 1 — MORPHOLOGY OF INFLORESCENCE AND FLOWER DEVELOPMENT

The importance of the size of primordia at initiation in relation to phyllotactic and other changes occurring during morphogenesis of the reproductive structures was stressed in Section III.E. This is further exemplified in several recent papers. Rosenberg and Bonnett show that stamen primordia in flowers of two cytoplasmic male-sterile cvs. of tobacco are smaller and have a different shape very early after emergence than corresponding primordia in a male-fertile cv.[1069] Similarly, in flowers of *Pelargonium* × *hortorum,* 10 stamen primordia are initiated but only 7 will develop into normal fertile stamens. The other 3, which are conspicuously smaller at the primordial stage, will develop into sterile filamentous staminodia.[1070] In corn, spikelets are grouped in pairs arranged in rows along the inflorescence axis. In the male inflorescence (tassel) there is a pedicelled and a sessile spikelet in each pair, and it was found by Cheng and colleagues that the pedicelled spikelet arises from a larger primordium than the sessile spikelet.[1071]

Not all initial inequality results in different mature forms, however, since the dissimilar primordia of the two spikelets of a pair in the female inflorescence (cob) of corn develop at equal rates and give rise to two almost identical spikelets.[1071]

Lord and co-workers have pursued their comparative descriptions of the ontogeny of chasmogamous and cleistogamous flowers in various species. Although primordia of the two floral types are indistinguishable at inception and during organogenesis until anther differentiation in *Lamium amplexicaule* and *Collomia grandiflora,*[556,1072] the primordia of the chasmogamous flowers in *Viola odorata* are already larger than the primordia of the cleistogamous flowers at inception.[1073] This early difference in *Viola* is presumably related to the fact that the two types of flowers are more divergent at maturity than those of *Lamium* and *Collomia.*

Interestingly, anther tissue differentiation seems to trigger some changes in the growth rates of various floral parts in *Collomia.*[1072] Indeed, a decrease in growth of anther length as well as an increase in growth of calyx circumference relative to growth in calyx length coincide with microspore mitosis in both chasmogamous and cleistogamous flowers. Also, microspore mitosis is coincident with an increase in growth in corolla length relative to growth in calyx length in chasmogamous flowers.

Various examples of fusion of floral organs were given in Section III.F. According to Moore and Walker, the mechanisms involved in these fusions are basically identical to those occurring during autografting of nonwounded callus tissues in vitro.[1074]

The problem of the genesis of forms and its relationships with the polarity of cell enlargement and division was discussed in Section V.B. Green has recently investigated the patterns of cell files in the epidermis and the directionality of cellulose microfibril reinforcements in the outer epidermal walls of vegetative and reproductive meristems.[1075] In vegetative meristems of several species with whorled or spiral phyllotaxis he found that the dome surface is highly structured, showing various sectors which correlate well with the production of "hoop" reinforced areas for leaves and internodes on the dome. The surface of the reproductive meristem of *Catharanthus* (= *Vinca*) *major*, at a very early stage of flower morphogenesis, is totally different, since it exhibits no cell file and cellulose alignment pattern at all. Whether this condition is specific of *Catharanthus* or typical of all reproductive meristems and when does it arise during the floral transition are questions that should be solved before the significance of this exciting observation can be gauged.

CHAPTER 3 — CONTROL BY NUTRITION

The influence of nitrogen upon inflorescence development in *Helianthus annus* has been investigated in a recent study which shows that N supply does not affect the duration of floral development (time between floret initiation and anthesis), but the rate of floret production per day and hence final floret number varies directly with available nitrogen.[1076] Relief of N stress at different stages of plant development suggests that floret number is a function of the plant's content of reduced N at the beginning of floret production.

Still, in *Helianthus annuus*, boron deficiency was found to result in inflorescence malformations:[1077] bracts and ray florets develop in the central region of the capitulum (where only disc florets are normally seen). Such abnormalities could be the products of splits which occur in the receptacle surface of B-deficient plants. Wounding the receptacle was shown to induce similar abnormalities (see below, addendum to Chapter 6).

CHAPTER 4 — CONTROL BY LIGHT

One new study suggests that male sex expression is enhanced by low irradiance. It shows that in three of four hybrids of gynoecious × monoecious cucumbers, increasing plant density reduces pistillate flowers, whereas in two of these hybrids increases in staminate flower production are associated with increasing density.[1078] Plant density has little effect on the numbers of pistillate and staminate flowers on gynoecious inbreds of cucumber, indicating, once again, that there is a strong environment × genotype interaction.

Some interesting papers concerning the effect of plant density and light intensity upon sex expression were overlooked earlier. In varied species, including spinach[1079] and *Atriplex confertifolia*,[1080] increased male flower production is associated with increased plant density (low light). Oil palm (*Elaeis guineensis*) also produces more female inflorescences when at high light intensity.[1081]

CHAPTER 5 — CONTROL BY TEMPERATURE AND WATER STRESS

Investigations reported in Section I.A.1. of this chapter show that the average daily temperature is the major factor determining the rate of flower development in tomato and bulbous iris. In chrysanthemum, however, Lepage and associates found that, although there is a close relationship between the daily mean temperature and the number of days to visible inflorescence bud, further development to anthesis is more dependent on day temperature than on night temperature.[1082] In contrast, Mohapatra and co-workers found that the rate of early development of the wheat spike is unaffected by temperature. Time interval from double-ridge to stamen initiation is the same at both 20 and 30°C.[1083] However, the final number of spikelets initiated is considerably lower at the high temperature than at the low, corroborating observations reported in Section I.B.

As in other species (see Section I.C of this chapter), temperature affects the number of floral parts in tomato flowers.[1084] A low-temperature treatment significantly increases the number of petals, stamens, and carpels. Low temperature also promotes the development of normal stamens with viable pollen in the stamenless-2 mutant of tomato.[1085]

In rice, flowering is very sensitive to water deficit. Panicles, 3 to 5 days after emergence from the protective flagleaf sheath, are the most susceptible to conditions which reduce the water status and lead to sterility.[1086] These newly emerged panicles apparently lack active mechanisms to regulate or inhibit water loss since they transpire at rates that are independent of their water potential.[1086]

Water stress has also been found to reduce the total number of flowers and percentage of chasmogamous flowers in *Collomia grandiflora*.[1087]

CHAPTER 6 — THE BASIC ROLE OF CORRELATIVE INFLUENCES

Surgical manipulations of the reproductive structure described in Section II.B have shown that flower meristems pass through a sequence of distinct phases that govern the nature of successively initiated floral organs. These phases seem to precede the actual emergence of the corresponding primordia. Wounding experiments by Palmer and Marc indicate that physiological phases governing the nature of the structures to be produced also occur during the develoment of the receptacle of *Helianthus annuus*.[1077] Wounding the capitulum at an early stage from receptacle dome formation through appearance of the first row of disc florets results in the initiation of involucral bracts and ray and disc florets in the wounded area. Wounding at later stages does not affect differentiation. This observation indicates that the fate of the cells in the central region of the receptacle remains optional during early stages of reproductive structure differentiation and is fixed when the first row of disc florets is initiated, i.e., before the emergence of these structures in the center of the capitulum. These results also suggest that involucral bracts are true floral organs and not products of the meristem in transition from leaf production to the formation of the capitulum. Interestingly, Palmer and Marc observed that outwardly developing disc florets, induced by wounding, spread across the receptacle surface until they encounter the inwardly moving front of naturally formed disc florets, showing that neither the natural nor the wound-induced florets can inhibit floret production at a distance.[1077]

CHAPTER 7 — EXOGENOUS SUBSTANCES

Promotive concentrations of auxin vary with the stage of reproductive development in tobacco. In explants consisting of the epidermis and several layers of subepidermal cortex collected from flower stalks, a low auxin concentration promotes flower bud initiation, but further development is supported by a higher auxin level.[1088] Recall that high auxin levels inhibit initiation in several species (see Section II.A of this chapter). In tulip, where auxin is involved in the control of stalk elongation, Gabryszewska and Saniewski observed that only basipetal transport of the regulator is responsible for this phenomenon.[1089] When stalk explants isolated from bulbs are cultured in normal (basal end down) or inverted (basal end up) positions, IAA in the medium stimulates the growth of the latter only.

The role of ethylene in the control of male sterility and sex expression has been further investigated. In cereal crops, the time of CEPA application relative to the stage of pollen production is of crucial importance. Lethal effects are induced from early stages of sporogenous cell formation to the vacuolate microspore stage.[1090] In *Cucurbita pepo*, CEPA causes suppression of male flowers and earlier production of female flowers. This increased female tendency results from the abortion of the male flower buds in which sex was already determined at the time of the treatment and from the preferential development into female flowers of buds which still were at the hermaphroditic stage.[1091] When plants are sprayed with AVG no female flowers are produced; male flower production is unaffected.[1092] The interesting fact here is that even when CEPA is applied to AVG-treated plants, there is still complete inhibition of female flower production. Similarly, AVG-treated plants subsequently exposed to 4000 ppm ethylene for 2 days never produce female flowers. Although spraying AVG-treated plants with ACC does not reverse the inhibition, application of ACC via a cut petiole for a 72-hr period following AVG application causes female flowers to form.[1092] Hume and Lovell suggest that increased femaleness induced in squash by CEPA or ethylene would be due to increased levels of ACC in response to these treatment.[1093] ACC and not ethylene would thus be the factor controlling female flower production in this case.

Ruckenbauer and Kirby reported that in barley kinetin sprayed on the leaves at the second or third-leaf stage increases the number of spikelet primordia as well as shoot apex development and final ear size.[1094] Cytokinins also enhance the production of normal spikelets on cultured tassels of *Zea mays*.[1095] Their effectiveness is dependent on their structure. The results of Polowick and Greyson clearly demonstrate that the concentration of the cytokinin in the medium profoundly affects the development of the tassel. At optimum concentrations, development is normal, while at supraoptimal concentrations abnormalities are induced in the spikelets.[1095] In rose, application of BA completely prevents the abortion of the flower developing on the third axilary branch (from the top) of decapitated shoots.[1096] This branch, usually enclosed within the canopy, is subjected to low light conditions and to apical control. A promotive effect of cytokinins in rose was postulated in Section V.A.

The feminizing effect of cytokinins in *Mercurialis annua* has been demonstrated once more.[1097] Tumor tissues induced in either female or male *Mercurialis* by a virulent strain of *Agrobacterium tumefaciens* were found to contain higher cytokinin levels than healthy tissues. These tumors are able to shift plant sex expression from male to female. Thus, a tumor close to a developing inflorescence of a male plant may induce the development of female flowers in this inflorescence. Also, when young male inflorescences are grown in vitro in the immediate proximity of female tumor tissues, they produce normal female flowers.

ABA appears to play some role in the control of cleistogamy in *Collomia grandiflora*, where it reduces the total number of flowers and the percentage of chasmogamous flowers.[1087] Cleistogamous flowers are produced in positions in the inflorescence where chasmogamous flowers typically occur.

A new class of chemicals, phthalimide derivatives, has been found to promote maleness in cucumber. They induce male flowers in gynoecious cvs. and prolong the male phase in monoecious cvs.[1098] The mode of action of phthalimides is not known, but they were shown to mimic many responses typically elicited by GA-like substances.[1099]

The review by the Durands[1098] enlarges upon our coverage of exogenous growth regulator effects and includes the following valuable generalization concerning the role of substances in sex expression: "Sex conversion is thus the repression of the normal program induced by the presence of the sex regulator genes for the sex currently expressed, and the experimental induction of the opposite program, incompatible with these regulator genes."

CHAPTER 8 — LEVELS, DISTRIBUTION, AND METABOLISM OF ENDOGENOUS SUBSTANCES

Studies reported in Section III.A of this chapter indicate that there is increased ethylene production in tulip at the time of rest completion. Results of Nell and associates also imply a role of ethylene in development of azalea buds. GA_3 or cold storage at 5°C for 6 weeks results in increased ethylene production in flower buds 4 days after the end of these treatment which overcome dormancy and produce flowering plants of equal quality.[1100] The increased ethylene production precedes any noticeable elongation of the flower buds. Ethylene is not detectable in untreated plants.

We have repeatedly emphasized the complexity of the control of sex expression and flower development by growth regulators and in interpreting results from studies of endogenous levels with those of exogenous applications. The work by Takahashi and associates provides a nice illustration of this complexity.[1101] Using two different cucumber cvs., one (Higan-fushinari) producing more female flowers in LD and the other (Matsunomidori) where femaleness is increased in SD, the Japanese workers found that in both lines endogenous GA levels are higher and ethylene is evolution less in SD than in LD. Female sex expression of these two cvs. is, however, increased by exogenously applied ethylene and decreased in response to GAs. These results raise once again the problem of the physiological meaning of growth regulator levels expressed per unit weight of tissue extracted.

We previously reported that the cytokinin concentration in xylem sap of *Phaseolus* P47 is higher in SD conditions, which stimulate flower development to anthesis, than in LD, which cause flower bud abscission (see Section V.A of this chapter). Unfortunately, as the volumes of exudate collected per unit time in these two photoperiodic treatments were unknown, cytokinin fluxes in these conditions could not be estimated. D. G. and C. B. Morgan now have reported that, although the volume of sap exuded from plants is always greater in LD, the rate of cytokinin production per plant is significantly higher in SD before anthesis.[1102] This result is in agreement with the common observation that cytokinin levels are higher in conditions which promote reproductive development to anthesis.

A 7-day dark-treatment which induces flower bud abortion in *Iris* results in an increased ABA level in these buds. Zeatin injection in the reproductive structure stimulates flower development and suppresses the accumulation of ABA, suggesting that these growth regulators are involved in the control of flower development.[1103]

Some studies suggest that polyamines and their biosynthetic enzymes have a role in growth and development of reproductive structures. *Nicotiana tabacum* mutants with altered polyamine metabolism exhibit abnormal floral development. One line produces flowers with a second row of petals in place of anthers.[1104] Another line, with elevated polyamine levels, produces flowers with anthers in the place of ovules.[1105] Arginine decarboxylase (ADC), ornithine decarboxylase (ODC), and *S*-adenosyl-L-methionine carboxylase (all polyamine biosynthetic enzymes) are higher before than at anthesis in floral buds of *Phaseolus vulgaris*.[1106] In mature flowers, female parts have higher ADC and ODC activities than other parts on a fresh weight basis. On a protein basis there are no differences among different flower parts. The main polyamine in flower buds is putrescine, which falls in titer after anthesis. Spermidine and spermine are minor components. Pistils have the higher content of all these three polyamines.[1106]

CHAPTER 9 — PHOTOSYNTHESIS, ASSIMILATE SUPPLY, AND UTILIZATION

Some new papers clearly indicate that assimilate distribution within the plant strongly affects reproductive development.

At anthesis, the dry weight of the terminal inflorescence of chrysanthemum is progressively reduced if the later axillary inflorescence buds are removed from branched plants; the weight reduction is equal to the weight of the axillary inflorescences removed.[1107] Disbudding does not reduce the total dry weight of the plants, suggesting that total assimilate production is not affected and that axillary shoots compete strongly with the terminal inflorescence for assimilates.

Application of CCC to tomato plants during early inflorescence development reduces the incidence of inflorescence abortion at high temperature and under low light conditions (Chapter 7, Section IV.A.2). Nourai and Harris showed that CCC increases the dry weight of the inflorescence and decreases the dry weight of the remainder of the shoot system.[1108]

Thus, the growth retardant reduces competition by vegetative parts of the shoot for metabolites. Russel and Morris showed that the developing inflorescence of tomato receives assimilates mainly from leaves in the two orthostichies adjacent to the radial position of the inflorescence on the vertical axis of the plant; these include leaves which are major contributors of carbon to the root system and the shoot apex.[1109] Thus, under conditions of reduced photoassimilate availability, the inflorescence is likely to suffer by competition from these major vegetative sinks, and any treatment which selectively reduces vegetative growth potential is likely to increase the availability of assimilates for inflorescence development.[1109]

Assimilate accumulation in the flower bud of *Iris* is reduced after a 7-day dark period which increases ABA level.[1103] Zeatin injection in the flower bud causes increased assimilate transport to the reproductive structure concomitant with a reduction in ABA content.[1103]

CHAPTER 11 — APPLIED ASPECTS OF FLORAL INITIATION AND DEVELOPMENT

Evaporative cooling was shown again to improve bud break when applied during rest,[1110] whereas it delays flowering when applied after rest completion.[1111,1112] The effects of delaying blossoming by evaporative cooling after rest completion on frost resistance are variable in apple.[11,1112] Buds from watered trees increase their water content and therefore become less hardy than those from unwatered trees. Hamer concluded that delaying blossoming by this method does not necessarily confer the expected benefits in terms of frost resistance, although it may be of value in delaying the harvest date.[1111]

Kronenberg[1113] has described a more complex model for 8 apple cvs., accounting for flower development and blossoming date as a function of temperature sums (post-chilling); the key feature added in this new model is that the base temperature for effective development changes with the progression from winter to spring. Another complexity found, if the model were to be predictive, is that different base temperatures would have to be assigned to different locations and different cvs. Thus, strong gene-environment interactions were found.

In Chapter 1, Section IV.B of this volume, we described several examples of the tendency towards synchronous development of reproductive structures; we suggested in Section IV.E.1 that abortion of parts of an inflorescence might be caused by insufficient synchronization of floret development within an inflorescence. This concept, important in yield formation, is nicely illustrated in 2 papers. In standard wheat, it is known that the fastest growing spikelet primordia occur in the middle and not at the base or at the top of the prospective spike (Chapter 1, Fig. 3A). This results in a developing ear with a spindle-shaped appearance. High-yielding cvs. (e.g., derivatives of the Norin-cv.) are characterized by higher numbers of developing spikelets than in standard wheats and in these high-yielding cvs. Fisher noticed that the developing ear is almost cylindrical due to the fact that a greater number of spikelets are developing synchronously along the spike.[114] In corn, Harris and co-workers show that the abortion of second ears in one-eared lines and third ears in two-eared (prolific) lines appears to be conditional upon lack of synchronized silking with the nearest upper developing ear.[115] In both cereals, synchronous development is associated with higher numbers of reproductive structures developing to maturity, and consequently higher yields, and non-synchronous development with lower yields. These results indicate that one of the main regulatory factors in the initiation and/or development of reproductive structures in cereals is that of temporal coordination among the structures. This coordination might not be based simply on a competition for nutrients, but also on an exchange of regulatory chemicals, either promotors or inhibitors.[1115,1116]

REFERENCES

1. **Sattler, R.,** *Organogenesis of Flowers. A Photographic Text-Atlas,* University of Toronto Press, Toronto, Canada, 1973.
2. **Moncur, M. W.,** *Floral Initiation in Field Crops. An Atlas of Scanning Electron Micrographs,* CSIRO, Canberra, 1981.
3. **Jones, H. A. and Emsweller, S. L.,** *Hilgardia,* 10, 415, 1936.
4. **Hartsema, A. M.,** *Encyclopedia of Plant Physiology,* Vol. 16, Ruhland, W., Ed., Springer-Verlag, Berlin, 1961, 123.
5. **Shoub, J. and de Hertogh, A. A.,** *J. Am. Soc. Hortic. Sci.,* 100, 32, 1975.
6. **Bernier, G.,** *Caryologia,* 15, 303, 1962.
7. **Besnard-Wibaut, C.,** *Cellular and Molecular Aspects of Floral Induction,* Bernier, G., Ed., Longman, London, 1970, 37.
8. **Orr, A. R.,** *Am. J. Bot.,* 65, 466, 1978.
9. **Diomaiuto-Bonnand, J.,** *C. R. Acad. Sci.,* 282, 1511, 1976.
10. **Bernier, G.,** *Mém. Acad. R. Belg. Cl. Sci.,* 16(1), 1, 1964.
11. **Sachs, R. M.,** *Annu. Rev. Plant Physiol.,* 16, 73, 1965.
12. **Vaughan, J. G.,** *J. Linn. Soc. Lond. Bot.,* 55, 279, 1955.
13. **Nougarède, A., Bronchart, R., Bernier, G., and Rondet, P.,** *Rev. Gén. Bot.,* 71, 205, 1964.
14. **Saint-Côme, R.,** *Rev. Gén. Bot.,* 73, 241, 1966.
15. **Bonnett, O. T.,** *J. Agric. Res.,* 53, 445, 1936.
16. **Barnard, C.,** *Aust. J. Bot.,* 3, 1, 1955.
17. **Rawson, H. M.,** *Aust. J. Biol. Sci.,* 23, 1, 1970.
18. **Rawson, H. M.,** *Aust. J. Agric. Res.,* 22, 537, 1971.
19. **Kirby, E. J. M.,** *J. Agric. Sci.,* 82, 437, 1974.
20. **Barnard, C.,** *Aust. J. Bot.,* 5, 1, 1957.
21. **Evans, M. W. and Grover, F. O.,** *J. Agric. Res.,* 61, 481, 1940.
22. **Bonnett, O. T.,** *J. Agric. Res.,* 54, 927, 1937.
23. **Artschwager, E., Brandes, E. W., and Starrett, R. C** *J. Agric. Res.,* 39, 1, 1929.
24. **Bonnett, O. T.,** *J. Agric. Res.,* 51, 451, 1935.
25. **Bonnett, O. T.,** *J. Agric. Res.,* 60, 25, 1940.
26. **Sharman, B. C.,** *Bot. Gaz. (Chicago),* 121, 192, 1960.
27. **Bruinsma, J. and Swart, J.,** *Acta Bot. Neerl.,* 11, 411, 1962.
28. **Nicholls, P. B. and May, L. H.,** *Aust. J. Biol. Sci.,* 16, 561, 1963.
29. **Popham, R. A.,** *Meristems and Differentiation,* Publ. No. 16, Brookhaven Symposia in Biology, 1964, 138.
30. **Lance, A.,** *Ann. Sci. Nat. Bot. Biol. Vég. 11e Sér.,* 18, 91, 1957.
31. **Cockshull, K. E.,** *Crop Processes in Controlled Environments,* Rees, A. R., Cockshull, K. E., Hand, D. W., and Hurd, R. G., Eds., Academic Press, London, 1972, 235.
32. **Jacqmard, A.,** *Bull. Cl. Sci. Acad. R. Belg.,* 50, 174, 1964.
33. **Molder, M. and Owens, J. N.,** *Can. J. Bot.,* 51, 535, 1973.
34. **Farr, C. H.,** *Bot. Gaz. (Chicago),* 59, 136, 1915.
35. **Léonard, M., Kinet, J. M., Bodson, M., Havelange, A., Jacqmard, A., and Bernier, G.,** *Plant Physiol.,* 67, 1245, 1981.
36. **Borthwick, H. A., Phillips, M., and Robbins, W. W.,** *Am. J. Bot.,* 18, 784, 1931.
37. **Magin, N.,** *Rev. Cytol. Biol. Vég.,* 21, 401, 1960.
38. **Diomaiuto-Bonnand, J.,** *Rev. Cytol. Biol. Vég.,* 29, 1, 1966.
39. **Favard, A.,** *C. R. Acad. Sci.,* 250, 1534, 1960.
40. **Ecole, D.,** *Rev. Cytol. Biol. Vég.,* 37, 127, 1974.
41. **Rickett, H. W.,** *Bot. Rev.,* 10, 187, 1944.
42. **Phelouzat, R.,** *Rev. Cytol. Biol. Vég.,* 26, 101, 1963.
43. **Durand, B.,** *Nat. Monspeliensia Bot.,* 9, 21, 1957.
44. **Tucker, S. C.,** *Adv. Plant Morphol.,* 1972, 359.
45. **Boke, N. H.,** *Am. J. Bot.,* 53, 534, 1966.
46. **Ross, R.,** *Am. J. Bot.,* 69, 369, 1982.
47. **Cheung, M. and Sattler, R.,** *Can. J. Bot.,* 45, 1609, 1967.
48. **Tucker, S. C.,** *Am. J. Bot.,* 62, 993, 1975.
49. **Soetiarto, S. R. and Ball, E.,** *Can. J. Bot.,* 47, 133, 1969.
50. **Lance, A. and Rondet, P.,** *C. R. Acad. Sci.,* 247, 123, 1958.
51. **Boke, N. H.,** *Am. J. Bot.,* 36, 535, 1949.
52. **Hicks, G. S.,** *Can. J. Bot.,* 51, 1611, 1973.

53. **Satina, S. and Blakeslee, A. F.,** *Am. J. Bot.*, 28, 862, 1941.
54. **Satina, S. and Blakeslee, A. F.,** *Am. J. Bot.*, 30, 453, 1943.
55. **Satina, S.,** *Am. J. Bot.*, 31, 493, 1944.
56. **Dermen, H. and Stewart, R. N.,** *Am. J. Bot.*, 60, 283, 1973.
57. **Dermen, H.,** *Am. J. Bot.*, 34, 32, 1947.
58. **Tucker, S. C.,** *Am. J. Bot.*, 68, 1017, 1981.
59. **Brulfert, J.,** *Rev. Gén. Bot.*, 72, 641, 1965.
60. **Sundberg, M. D.,** *Am. J. Bot.*, 69, 380, 1982.
60a. **Brett, J. F. and Posluszny, U.,** *Can. J. Bot.*, 60, 2133, 1982.
61. **Garrod, J. F. and Harris, G. P.,** *Ann. Bot. (London)*, 38, 1025, 1974.
62. **Lindenbaum, S., Ginzburg, C., and Halevy, A. H.,** *Ann. Bot. (London)*, 39, 219, 1975.
63. **Matzke, E. B.,** *Am. J. Bot.*, 25, 465, 1938.
64. **Bersillon, G.,** *Ann. Sci. Nat. Bot. Biol. Vég., 11e Sér.* 16, 225, 1955.
65. **Tucker, S. C.,** *Am. J. Bot.*, 47, 266, 1960.
66. **Blake, J.,** *Ann. Bot. (London)*, 26, 95, 1962.
67. **Meicenheimer, R. D.,** *Am. J. Bot.*, 66, 557, 1979.
68. **Cooper, D. C.,** *Bot. Gaz. (Chicago)*, 83, 399, 1927.
69. **Sawhney, V. K. and Greyson, R. I.,** *Can. J. Bot.*, 50, 1493, 1972.
70. **Lyndon, R. F.,** *Can. J. Bot.*, 57, 233, 1979.
71. **Meyer, V. G.,** *Bot. Rev.*, 32, 165, 1966.
72. **Erickson, R. O. and Meicenheimer, R. D.,** *Am. J. Bot.*, 64, 981, 1977.
73. **Williams, R. F.,** *The Shoot Apex and Leaf Growth*, Cambridge University Press, New York, 1975.
74. **Bernier, G. and Kinet, J. M.,** unpublished data, 1981.
75. **Moore, P. H.,** *Can. J. Bot.*, 49, 677, 1971.
76. **Lyndon, R. F.,** *Ann. Bot. (London)*, 42, 1343, 1978.
77. **Lyndon, R. F.,** *Ann. Bot. (London)*, 42, 1349, 1978.
78. **Meicenheimer, R. D.,** *Am. J. Bot.*, 69, 1108, 1982.
79. **Jacobs, W. P. and Raghavan, V.,** *Phytomorphology*, 12, 144, 1962.
80. **Langer, R. H. M. and Bussell, W. T.,** *Ann. Bot. (London)*, 28, 163, 1964.
81. **Tucker, S. C.,** *Am. J. Bot.*, 48, 60, 1961.
82. **Cusick, F.,** *Trends in Plant Morphogenesis*, Cutter, E. G., Ed., Longman, London, 1966, 170.
83. **Baum, H.,** *Österreich. Bot. Z.*, 95, 86, 1948.
84. **Baum, H.,** *Österreich. Akad. Wiss. Math.-Naturwiss. Kl. Abt. I*, 157, 17, 1948.
85. **Kaplan, D. R.,** *Am. J. Bot.*, 55, 406, 1968.
86. **Nishino, E.,** *Bot. Mag.*, 95, 1, 1982.
87. **Boke, N. H.,** *Am. J. Bot.*, 35, 413, 1948.
88. **Nishino, E.,** *Bot. Mag.*, 89, 191, 1976.
89. **Nishino, E.,** *Bot. Mag.*, 91, 263, 1978.
90. **Singh, V.,** *Phytomorphology*, 29, 239, 1979.
91. **Daniel, E. and Sattler, R.,** *Phytomorphology*, 28, 151, 1978.
92. **Kaplan, D. R.,** *Am. J. Bot.*, 55, 933, 1968.
93. **Lamond, M. and Vieth, J.,** *Can. J. Bot.*, 50, 1633, 1972.
94. **Trapp, A.,** *Bot. Stud.*, 5, 1, 1956.
95. **Tepfer, S. S.,** *Univ. Calif. Publ. Bot.*, 25, 513, 1953.
96. **Walker, D. B.,** *Am. J. Bot.*, 62, 457, 1975.
97. **Walker, D. B.,** *Protoplasma*, 86, 43, 1975.
98. **Boeke, J. H.,** *Acta Bot. Neerl.*, 20, 570, 1971.
99. **Brooks, R. M.,** *Hilgardia*, 13, 249, 1940.
100. **Jackson, G.,** *Am. J. Bot.*, 21, 453, 1934.
101. **Judson, J. E.,** *Am. J. Bot.*, 16, 69, 1929.
102. **Kaplan, D. R.,** *Am. J. Bot.*, 54, 1274, 1967.
103. **McCoy, R. W.,** *Am. J. Bot.*, 27, 600, 1940.
104. **Heslop-Harrison, J.,** *Meristems and Differentiation*, Publ. No. 16, Brookhaven Symposia in Biology, 1964, 109.
105. **Heimlich, L. F.,** *Am. J. Bot.*, 14, 227, 1927.
106. **Lazarte, J. E. and Palser, B. F.,** *Am. J. Bot.*, 66, 753, 1979.
107. **de Mason, D. A., Stolte, K. W., and Tisserat, B.,** *Can. J. Bot.*, 60, 1439, 1982.
108. **Galun, E. and Porath, D.,** *Cellular and Molecular Aspects of Floral Induction*, Bernier, G., Ed., Longman, London, 1970, 431.
109. **Galun, E., Jung, Y., and Lang, A.,** *Dev. Biol.*, 6, 370, 1963.
110. **Rodrigues Pereira, A. S.,** *Planta*, 80, 349, 1968.

111. **Heslop-Harrison, J.,** *Biol. Rev.*, 32, 38, 1957.
112. **Berghoef, J. and Bruinsma, J.,** *Z. Pflanzenphysiol.*, 93, 303, 1979.
113. **Barnard, C.,** *Aust. J. Bot.*, 5, 115, 1957.
114. **Smith, D. L.,** *Ann. Bot. (London)*, 30, 475, 1966.
115. **Nitsch, J. P., Kurtz, E. B., Jr., Liverman, J. L., and Went, F. W.,** *Am. J. Bot.*, 39, 32, 1952.
116. **Frankel, R. and Galun, E.,** *Pollination Mechanisms, Reproduction and Plant Breeding*, Springer-Verlag, Berlin, 1977.
117. **Lloyd, D. G. and Webb, C. J.,** *Bot. Rev.*, 43, 177, 1977.
118. **Lord, E. M.,** *Bot. Rev.*, 47, 421, 1981.
119. **Lord, E.,** *Bot. Gaz. (Chicago)*, 140, 39, 1979.
120. **Tennakone, K., Dayatilaka, R. K. D., and Ariyaratne, S.,** *Ann. Bot. (London)*, 50, 397, 1982.
121. **Ornduff, R.,** *J. South Afr. Bot.*, 40, 169, 1974.
122. **Tucker, S. C.,** *Am. J. Bot.*, 69, 743, 1982.
123. **Lance, A. and Rondet, P.,** *C. R. Acad. Sci.*, 249, 745, 1959.
124. **Lawalrée, A.,** *Cellule*, 52, 215, 1948.
125. **Simon, L.,** L'*Impatiens balsamina* L. Cultivar "Buisson fleuri". Morphogenèse et Ontogénèse Expérimentales, Ph.D. thesis, University of Nantes, France, 1973.
126. **Tucker, S. C.,** *Am. J. Bot.*, 69, 1389, 1982.
127. **Tucker, S. C. and Gifford, E. M., Jr.,** *Am. J. Bot.*, 53, 433, 1966.
128. **Philipson, W. R.,** *Ann. Bot. (London)*, 12, 147, 1948.
129. **Greyson, R. I. and Sawhney, V. K.,** *Bot. Gaz. (Chicago)*, 133, 184, 1972.
130. **Lyndon, R. F.,** *Ann. Bot. (London)*, 43, 539, 1979.
131. **Bennett, M. D., Finch, R. A., Smith, J. B., and Rao, M. K.,** *Proc. R. Soc. Lond.*, B138, 301, 1973.
132. **Buttrose, M. S., Grant, W. J. R., and Sedgley, M.,** *Aust. J. Bot.*, 29, 385, 1981.
133. **Blumenfeld, A.,** *Plant Physiol.*, 55, 265, 1975.
134. **Erickson, R. O.,** *Am. J. Bot.*, 35, 729, 1948.
135. **Williams, R. F.,** *Aust. J. Biol. Sci.*, 19, 949, 1966.
136. **Jeffcoat, B. and Cockshull, K. E.,** *J. Exp. Bot.*, 23, 722, 1972.
137. **Tompsett, P. B.,** *Ann. Bot. (London)*, 42, 889, 1978.
138. **Wareing, P. F.,** *The Physiology of Forest Trees*, Thimann, K. V., Critchfield, W. B., and Zimmerman, M. H., Eds., Ronald Press, New York, 1958, 643.
139. **Naylor, F. L.,** *Bot. Gaz. (Chicago)*, 103, 146, 1941.
140. **Heslop-Harrison, J. and Heslop-Harrison, Y.,** *The Induction of Flowering. Some Case Histories*, Evans, L. T., Ed., Macmillan, Melbourne, 1969, 205.
141. **Auderset, G. and Greppin, H.,** *Saussurea*, 7, 73, 1976.
142. **Chao, M. D.,** *Plant Physiol.*, 22, 393, 1947.
143. **Sachs, R. M.,** *Am. J. Bot.*, 55, 62, 1968.
144. **Greyson, R. I. and Tepfer, S. S.,** *Am. J. Bot.*, 53, 485, 1966.
145. **Nicholls, P. B.,** *Ann. Bot. (London)*, 38, 827, 1974.

145a. **Bugnon, F. and Bessis, R.,** *Biologie de la Vigne. Acquisitions Récentes et Problèmes Actuels*, Masson, Paris, 1968.

146. **Posluszny, U. and Sattler, R.,** *Am. J. Bot.*, 61, 209, 1974.
147. **Hackett, W. P. and Sachs, R. M.,** *Proc. Am. Soc. Hortic. Sci.*, 88, 606, 1966.
148. **Hubbell, D. S.,** *J. Agric. Res.*, 48, 91, 1934.
149. **Horridge, J. S. and Cockshull, K. E.,** *Sci. Hortic.*, 2, 273, 1974.
150. **Nell, T. A. and Rasmussen, H. P.,** *J. Am. Soc. Hortic. Sci.*, 104, 18, 1979.
151. **Popham, R. A. and Chan, A. P.,** *Am. J. Bot.*, 39, 329, 1952.
152. **Cockshull, K. E.,** *J. Hortic. Sci.*, 51, 441, 1976.
153. **Schwabe, W. W.,** *J. Exp. Bot.*, 2, 223, 1951.
154. **Kirk, S. C., Morrow, I. B., and Jacobs, W. P.,** *Phytomorphology*, 17, 410, 1967.
155. **Kinet, J. M.,** unpublished data, 1978.
156. **Resende, F.,** *Encyclopedia of Plant Physiology*, Vol. 18, Ruhland, W., Ed., Springer-Verlag, Berlin, 1967, 257.
157. **Kirby, E. J. M. and Rymer, J. L.,** *Ann. Bot. (London)*, 38, 565, 1974.
158. **Thomas, J. F. and Raper, C. D., Jr.,** *Crop Sci.*, 16, 667, 1976.
159. **Simon, L.,** *C. R. Acad. Sci.*, 278, 45, 1974.
160. **Steeves, T. A.,** *Trends in Plant Morphogenesis*, Cutter, E. G., Ed., Longmans, London, 1966, 200.
161. **Rees, A. R.,** *Bot. Rev.*, 32, 1, 1966.
162. **Rees, A. R.,** *The Growth of Bulbs. Applied Aspects of the Physiology of Ornamental Bulbous Crop Plants*, Academic Press, London, 1972.
163. **de Hertogh, A.,** *Sci. Hortic.*, 2, 313, 1974.
164. **Doorenbos, J.,** *Meded. Landbouwhogesch. Wageningen*, 53(1), 1, 1953.

165. **Samish, R. M.,** *Annu. Rev. Plant Physiol.*, 5, 183, 1954.
166. **Vegis, A.,** *Annu. Rev. Plant Physiol.*, 15, 185, 1964.
167. **Carpenter, E. D. and Watson, D. P.,** *Proc. Am. Soc. Hortic. Sci.*, 86, 806, 1965.
168. **Millbocker, D. C. and Wright, R. D.,** *J. Am. Soc. Hortic. Sci.*, 104, 76, 1979.
169. **Lona, F.,** *Eingeschaften und Wirkungen der Gibberelline*, Knapp, R., Ed., Springer-Verlag, Berlin, 1962, 73.
170. **Lona, F.,** *Planta*, 82, 145, 1968.
171. **Chandler, W. H. and Tufts, W. P.,** *Proc. Am. Soc. Hortic. Sci.*, 30, 180, 1933.
172. **Mess, M. G.,** *Port. Acta Biol. Sér. A*, 5, 25, 1958.
173. **Lona, F.,** *Frutticoltura*, 33, 35, 1961.
174. **Corgan, J. N. and Martin, G. C.,** *HortScience*, 6, 405, 1971.
175. **Sunderland, N.,** *J. Exp. Bot.*, 12, 446, 1961.
176. **Bernier, G., Kinet, J. M., and Bronchart, R.,** *Physiol. Vég.*, 5, 311, 1967.
177. **Taillandier, J.,** *Z. Pflanzenphysiol.*, 87, 395, 1978.
178. **Green, P. B. and Brooks, K. E.,** *Am. J. Bot.*, 65, 13, 1978.
179. **Green, P. B.,** *Annu. Rev. Plant Physiol.*, 31, 51, 1980.
180. **Green, P. B. and Lang, J. M.,** *Planta*, 151, 413, 1981.
181. **Emons, A. M. C.,** *Protoplasma*, 113, 85, 1982.
182. **Dermen, H.,** *Cytologia*, 34, 541, 1969.
183. **Brown, C. L. and Sax, K.,** *Am. J. Bot.*, 49, 683, 1962.
184. **Yeoman, M. M. and Brown, R.,** *Ann. Bot. (London)*, 35, 1101, 1971.
185. **Lintilhac, P. M.,** *Am. J. Bot.*, 61, 135, 1974.
186. **Lintilhac, P. M.,** *Am. J. Bot.*, 61, 230, 1974.
187. **Lintilhac, P. M. and Vesecky, T. B.,** *Am. J. Bot.*, 67, 1477, 1980.
188. **Lyndon, R. F.,** *Ann. Bot. (London)*, 49, 281, 1982.
189. **Considine, J. A., Knox, R. B., and Frankel, O. H.,** *Ann. Bot. (London)*, 50, 647, 1982.
190. **Marushige, K. and Marushige, Y.,** *Plant Cell Physiol.*, 3, 319, 1962.
191. **Barber, J. T. and Steward, F. C.,** *Dev. Biol.*, 17, 326, 1968.
192. **Kahlem, G.,** *C. R. Acad. Sci.*, 270, 1314, 1970.
193. **Piérard, D., Jacqmard, A., and Bernier, G.,** *C. R. Acad. Sci.*, 289, 761, 1979.
194. **Raff, J. W., Hutchinson, J. F., Knox, R. B., and Clarke, A. E.,** *Differentiation*, 12, 179, 1979.
195. **Durand, B., Durand, R., Kahlem, G., Champault, A., and Delaigue, M.,** *Rev. Cytol. Biol. Vég.*, 37, 249, 1974.
196. **Kahlem, G.,** *Dev. Biol.*, 50, 58, 1976.
197. **Kahlem, G.,** *Z. Pflanzenphysiol.*, 76, 80, 1975.
198. **Jaiswal, V. S. and Kumar, A.,** *Z. Pflanzenphysiol.*, 102, 299, 1981.
199. **Bazin, M., Chabin, A., and Durand, R.,** *Dev. Biol.*, 44, 288, 1975.
200. **Cutter, E. G.,** *Plant Anatomy, Experiment and Interpretation. Part 2. Organs*, Edward Arnold, London, 1971.
201. **Harris, G. P. and Scott, M. A.,** *Ann. Bot. (London)*, 33, 143, 1969.
202. **Kinet, J. M., Hurdebise, D., Parmentier, A., and Stainier, R.,** *J. Am. Soc. Hortic. Sci.*, 103, 724, 1978.
203. **Léonard, M., Kinet, J. M., Bodson, M., and Bernier, G.,** *Physiol. Plant.*, 57, 85, 1983.
204. **Heslop-Harrison, J.,** *Plant Physiology. A Treatise, Vol. VIC: Physiology of Development: From Seeds to Sexuality*, Steward, F. C., Ed., Academic Press, New York, 1972, 133.
205. **Greyson, R. I. and Raman, K.,** *Am. J. Bot.*, 62, 531, 1975.
206. **Sawhney, V. K. and Greyson, R. I.,** *Am. J. Bot.*, 60, 514, 1973.
207. **Galun, E., Jung, Y., and Lang, A.,** *Nature (London)*, 194, 596, 1962.
208. **Tepfer, S. S., Greyson, R. I., Craig, W. R., and Hindman, J. L.,** *Am. J. Bot.*, 50, 1035, 1963.
209. **Tepfer, S. S., Karpoff, A. J., and Greyson, R. I.,** *Am. J. Bot.*, 53, 148, 1966.
210. **Blake, J.,** *Nature (London)*, 211, 990, 1966.
211. **Blake, J.,** *J. Exp. Bot.*, 20, 113, 1969.
212. **Brulfert, J. and Fontaine, D.,** *Biol. Plant.*, 9, 439, 1967.
213. **Fontaine, D.,** *Mém. Soc. Bot. Fr.*, 297, 1970.
214. **Porath, D. and Galun, E.,** *Ann. Bot. (London)*, 31, 283, 1967.
215. **Hicks, G. S. and Sussex, I. M.,** *Can. J. Bot.*, 48, 133, 1970.
216. **Raman, K. and Greyson, R. I.,** *Am. J. Bot.*, 65, 180, 1978.
217. **Zimmerman, R. H.,** *J. Am. Soc. Hortic. Sci.*, 96, 404, 1971.
218. **Gates, C. T., Bouma, D., and Groenewegen, H.,** *Aust. J. Agric. Res.*, 12, 1050, 1961.
219. **Alexander, D. McE.,** *J. Hortic. Sci.*, 41, 85, 1966.
220. **Mullins, M. G.,** *Nature (London)*, 209, 419, 1966.

221. **Couvillon, G. A., King, G. A., Moore, C. M., and Bush, P.,** *HortScience,* 10, 78, 1975.
222. **Friedlander, M., Atsmon, D., and Galun, E.,** *Plant Cell Physiol.,* 18, 1343, 1977.
223. **Beyer, J. J.,** *Meded. Landbouwhogesch. Wageningen,* 46(5), 3, 1942.
224. **Sharman, B. C.,** *New Phytol.,* 46, 20, 1947.
225. **Gott, M. B., Gregory, F. G., and Purvis, O. N.,** *Ann. Bot. (London),* 19, 87, 1955.
226. **Waddington, S. R., Cartwright, P. M., and Wall, P. C.,** *Ann. Bot. (London),* 51, 119, 1983.
227. **Damptey, H. B. and Aspinall, D.,** *Ann. Bot. (London),* 40, 23, 1976.
228. **Nerson, H., Sibony, M., and Pinthus, M. J.,** *Ann. Bot. (London),* 45, 203, 1980.
229. **Aspinall, D. and Paleg, L. G.,** *Bot. Gaz. (Chicago),* 124, 429, 1963.
230. **Friend, D. J. C., Fisher, J. E., and Helson, V. A.,** *Can. J. Bot.,* 41, 1663, 1963.
231. **Hackett, W. P. and Sachs, R. M.,** *Proc. Am. Soc. Hortic. Sci.,* 92, 615, 1968.
232. **Moe, R. and Kristoffersen, T.,** *Acta Hortic.,* 14, 157, 1969.
233. **Cockshull, K. E. and Hughes, A. P.,** *J. Hortic. Sci.,* 47, 113, 1972.
234. **Graves, C. J. and Sutcliffe, J. F.,** *Ann. Bot. (London),* 38, 729, 1974.
235. **Ojehomon, O. O., Zehni, M. S., and Morgan, D. G.,** *Ann. Bot. (London),* 37, 871, 1973.
236. **Shillo, R. and Halevy, A. H.,** *Sci. Hortic.,* 4, 79, 1976.
237. **Shillo, R. and Halevy, A. H.,** *Sci. Hortic.,* 15, 187, 1981.
238. **Srinivasan, C. and Mullins, M. G.,** *Am. J. Enol. Vitic.,* 32, 47, 1981.
239. **Bodson, M.,** *J. Am. Soc. Hortic. Sci.,* 108, 382, 1983.
240. **Cumming, B. G.,** *The Induction of Flowering. Some Case Histories,* Evans, L. T., Ed., Macmillan, Melbourne, 1969, 156.
241. **Koller, D., Kigel, J., and Ovadia, S.,** *Planta,* 136, 13, 1977.
242. **Evans, L. T.,** *New Phytol.,* 59, 163, 1960.
243. **Reid, J. B.,** *Z. Pflanzenphysiol.,* 93, 297, 1979.
244. **Berghoef, J. and Bruinsma, J.,** *Phytomorphology,* 30, 231, 1980.
245. **Shifriss, O. and Galun, E.,** *Proc. Am. Soc. Hortic. Sci.,* 67, 479, 1956.
246. **Rood, S. B., Pharis, R. P., and Major, D. J.,** *Plant Physiol.,* 66, 793, 1980.
247. **Chapman, P. J.,** *N.Y. State Hortic. Soc. Proc.,* 111, 146, 1966.
248. **Ballard, J. K., Proebsting, E. L., Jr., and Tukey, R. B.,** *Wash. State Univ. Ext. Circ.,* 369, 1971.
249. **de Munk, W. J.,** *Neth. J. Plant Pathol.,* 79, 41, 1973.
250. **Moe, R., de Hertogh, A. A., and Dilley, D. R.,** *Meld. Norg. Landbrukshoegsk.,* 57 (45), 1, 1978.
251. **Larson, R. A.,** *Introduction to Floriculture,* Academic Press, New York, 1980.
252. **Halevy, A. H. and Mayak, S.,** *Hortic. Rev.,* 1, 204, 1979.
253. **Halevy, A. H. and Mayak, S.,** *Hortic. Rev.,* 3, 59, 1981.
254. **Nell, T. A. and Larson, R. A.,** *J. Hortic. Sci.,* 49, 323, 1974.
255. **Kraus, E. J. and Kraybill, H. R.,** *Ore. Agric. Exp. Stn. Bull.,* 149, 1, 1918.
256. **Nightingale, G. T., Schermerhorn, L. G., and Robbins, W. R.,** *N.J. Agric. Exp. Stn. Bull.,* 461, 1 1928.
257. **Howlett, F. S.,** *Ann. Bot. (London),* 50, 767, 1936.
258. **Murneek, A. E.,** *Mo. Agric. Exp. Stn. Res. Bull.,* 268, 1, 1937.
259. **Borthwick, H. A. and Parker, M. W.,** *Bot. Gaz. (Chicago),* 101, 341, 1939.
260. **Polster, H.,** *Beitr. Biol. Pflanz.,* 25, 228, 1938.
261. **Wittwer, S. H. and Teubner, F. G.,** *Am. J. Bot.,* 44, 125, 1957.
262. **Eguchi, T., Matsumura, T., and Ashizawa, M.,** *Proc. Am. Soc. Hortic. Sci.,* 72, 343, 1958.
263. **Fisher, K. J.,** *J. Hortic. Sci.,* 44, 407, 1969.
264. **Abdalla, A. A. and Verkerk, K.,** *Neth. J. Agric. Sci.,* 18, 111, 1970.
265. **Haynes, R. J. and Goh, K. M.,** *Biol. Rev.,* 53, 465, 1978.
266. **Barker, A. V. and Mills, H. A.,** *Hortic. Rev.,* 2, 395, 1980.
267. **Bunt, A. C.,** *Plant Soil,* 31, 97, 1969.
268. **Ryan, E., Smillie, G. W., and McAleese, D. M.,** *Ir. J. Agric. Res.,* 11, 295, 1972.
269. **Gardner, V. R.,** *Mo. Agric. Exp. Stn. Res. Bull.,* 57, 1, 1923.
270. **Schaffner, J. H.,** *Ohio J. Sci.,* 25, 172, 1925.
271. **Schaffner, J. H.,** *Bull. Torr. Bot. Club,* 52, 35, 1925.
272. **Darrow, G. M.,** *J. Agric. Res.,* 34, 393, 1927.
273. **Murneek, A. E.,** *Mo. Agric. Exp. Stn. Res. Bull.,* 106, 1, 1927.
274. **Tiedjens, V. A.,** *J. Agric. Res.,* 36, 721, 1928.
275. **Tibeau, M. E.,** *Plant Physiol.,* 11, 731, 1936.
276. **Hall, W. C.,** *Plant Physiol.,* 24, 753, 1949.
277. **Talley, P. J.,** *Plant Physiol.,* 9, 731, 1934.
278. **Bouillenne, R., Bouillenne, M., and Ghenne, L.,** *C. R. Soc. Biol.,* 114, 189, 1933.
279. **Bouillenne, M. and Bouillenne, R.,** *Bull. Cl. Sci. Acad. R. Belg.,* 21, 642, 1935.

280. **Stanfield, J. F.,** *Am. J. Bot.*, 24, 710, 1937.
281. **Neidle, E. K.,** *Bot. Gaz. (Chicago)*, 100, 607, 1939.
282. **El Hinnawy, E. I.,** *Meded. Landbouwhogesch. Wageningen*, 56(9), 1, 1956.
283. **Scully, N. J., Parker, M. W., and Borthwick, H. A.,** *Bot. Gaz. (Chicago)*, 107, 218, 1945.
284. **Shanks, J. B. and Link, C. B.,** *Proc. Am. Soc. Hortic. Sci.*, 58, 329, 1951.
285. **Shanks, J. B., Link, C. B., and Preston, W. H., Jr.,** *Proc. Am. Soc. Hortic. Sci.*, 65, 441, 1955.
286. **Langer, R. H. M. and Liew, F. K. Y.,** *Aust. J. Agric. Res.*, 24, 647, 1973.
287. **Langer, R. H. M. and Hanif, M.,** *Ann. Bot. (London)*, 37, 743, 1973.
288. **Blake, J. and Harris, G. P.,** *Ann. Bot. (London)*, 24, 247, 1960.
289. **Naylor, A. W.,** *Bot. Gaz. (Chicago)*, 103, 342, 1941.
290. **de Munk, W. J., Hoogeterp, P., and Slootweg, G.,** *Acta Hortic.*, 109, 81, 1980.
291. **Eastwood, T.,** *Proc. Am. Soc. Hortic. Sci.*, 59, 531, 1952.
292. **Brantley, B. B. and Warren, G. F.,** *Plant Physiol.*, 35, 741, 1960.
293. **Stout, A. B.,** *Am. J. Bot.*, 10, 57, 1923.
294. **Howlett, F. S.,** *J. Agric. Res.*, 58, 79, 1939.
295. **Bilderback, D. E.,** *Am. J. Bot.*, 59, 525, 1972.
296. **Peterson, C. M.,** *Am. J. Bot.*, 60, 381, 1973.
297. **de Jong, A. W., Smit, A. L., and Bruinsma, J.,** *Z. Pflanzenphysiol.*, 72, 227, 1974.
298. **Berghoef, J. and Bruinsma, J.,** *Z. Pflanzenphysiol.*, 93, 345, 1979.
299. **Seeley, J. G.,** *Proc. Am. Soc. Hortic. Sci.*, 56, 466, 1950.
300. **Besford, R. T. and Maw, G. A.,** *Plant Soil*, 42, 395, 1975.
301. **Jenkins, J. M., Jr. and Stuart, N. W.,** *Proc. Am. Soc. Hortic. Sci.*, 54, 477, 1949.
302. **Wakhloo, J. L.,** *J. Exp. Bot.*, 26, 425, 1975.
303. **Wakhloo, J. L.,** *Flora*, 155, 486, 1965.
304. **Seeley, J. C.,** *Proc. Am. Soc. Hortic. Sci.*, 56, 439, 1950.
305. **Menary, R. C. and van Staden, J.,** *Aust. J. Plant Physiol.*, 3, 201, 1976.
306. **Hill, H., Davis, M. B., and Johanson, F. B.,** *Sci. Agric.*, 15, 110, 1934.
307. **Howell, R. W.,** *Plant Physiol.*, 29, 477, 1954.
308. **Adams, P., Graves, C. J., and Winsor, G. W.,** *J. Sci. Food Agric.*, 26, 1899, 1975.
309. **Adams, P., Graves, C. J., and Winsor, G. W.,** *J. Hortic. Sci.*, 55, 89, 1980.
310. **Smith, H. J., McIlrath, W. J., and Bogorad, L.,** *Bot. Gaz. (Chicago)*, 118, 174, 1956.
311. **Löhnis, M. P.,** *Meded. Landbouwhogesch. Wageningen*, 44(3), 3, 1940.
312. **Graham, R. D.,** *Nature (London)*, 254, 514, 1975.
313. **Graham, R. D.,** *J. Exp. Bot.*, 27, 717, 1976.
314. **Agarwala, S. C., Chatterjee, C., Sharma, P. N., Sharma, C. P., and Nautiyal, N.,** *Can. J. Bot.*, 57, 1946, 1979.
315. **Dell, B.,** *Ann. Bot. (London)*, 48, 599, 1981.
316. **Garner, W. W. and Allard, H. A.,** *J. Agric. Res.*, 18, 553, 1920.
317. **Garner, W. W. and Allard, H. A.,** *J. Agric. Res.*, 23, 871, 1923.
318. **Teeri, J. A.,** *Ecology*, 57, 374, 1976.
319. **Eguchi, T.,** *Proc. Imp. Acad. Jpn.*, 13, 332, 1937.
320. **Nitsch, J. P.,** *Encyclopedia of Plant Physiology*, Vol. 15 (Part 1), Ruhland, W., Ed., Springer-Verlag, Berlin, 1965, 1537.
321. **Vince-Prue, D.,** *Photoperiodism in Plants*, McGraw-Hill, London, 1975.
322. **Hamner, K. C.,** *The Induction of Flowering. Some Case Histories*, Evans, L. T., Ed., Macmillan, Melbourne, 1969, 62.
323. **Struckmeyer, B. E.,** *Bot. Gaz. (Chicago)*, 103, 182, 1941.
324. **Nanda, K. K. and Krishnamoorthy, H. N.,** *Planta*, 72, 338, 1967.
325. **Cook, R. E.,** *J. Exp. Bot.*, 27, 541, 1976.
326. **Sato, T. and Oda, Y.,** *Plant Cell Physiol.*, 18, 1041, 1977.
327. **Lush, W. M., Evans, L. T., and Wien, H. C.,** *Field Crops Res.*, 3, 173, 1980.
328. **Joffe, A.,** *South Afr. J. Agric. Sci.*, 1, 435, 1958.
329. **Crawford, R. M. M.,** *Ann. Bot. (London)*, 25, 78, 1961.
330. **Cleland, C. F. and Briggs, W. R.,** *Plant Physiol.*, 42, 1553, 1967.
331. **Nwoke, F. I. O.,** *J. Exp. Bot.*, 31, 1657, 1980.
332. **Salisbury, F. B.,** *The Flowering Process*, Pergamon Press, Oxford, 1963.
333. **Langton, F. A.,** *Sci. Hortic.*, 7, 277, 1977.
334. **Paleg, L. G. and Aspinall, D.,** *Bot. Gaz. (Chicago)*, 125, 149, 1964.
335. **Cottrell, J. E., Dale, J. E., and Jeffcoat, B.,** *Ann. Bot. (London)*, 50, 57, 1982.
336. **Williams, R. F. and Williams, C. N.,** *Aust. J. Biol. Sci.*, 21, 835, 1968.
337. **Evans, L. T., Wardlaw, I. F., and Fischer, R. A.,** *Crop Physiology. Some Case Histories*, Evans, L. T., Ed., Cambridge University Press, New York, 1975, 101.

338. **Abou Dahab, A. M.,** *Meded. Landouwhogesch. Wageningen,* 67(13), 1, 1967.
339. **Harris, G. P. and Ashford, M.,** *J. Hortic. Sci.,* 41, 397, 1966.
340. **Schneider, E. F.,** *Can. J. Plant Sci.,* 47, 112, 1967.
341. **Sachs, R. M. and Bretz, C. F.,** *Proc. Am. Soc. Hortic. Sci.,* 80, 581, 1962.
342. **Marc, J. and Palmer, J. H.,** *Field Crops Res.,* 4, 155, 1981.
343. **Piringer, A. A., Downs, R. J., and Borthwick, H. A.,** *Am. J. Bot.,* 50, 86, 1963.
344. **Austin, J. P.,** *Am. J. Bot.,* 28, 244, 1941.
345. **Doorenbos, J.,** *Euphytica,* 8, 69, 1959.
346. **Hughes, A. P. and Cockshull, K. E.,** *Ann. Bot. (London),* 29, 131, 1965.
347. **Schwartz, A. and Koller, D.,** *Bot. Gaz. (Chicago),* 136, 41, 1975.
348. **Heide, O. M.,** *Meld. Norg. Landbrukshoegsk.,* 59(14), 1, 1980.
349. **Moss, G. I. and Heslop-Harrison, J.,** *Ann. Bot. (London),* 32, 833, 1968.
350. **Koller, D., Kigel, J., and Ovadia, S.,** *Planta,* 137, 133, 1977.
351. **Seidlová-Blumová, F.,** *Biol. Plant,* 3, 156, 1961.
352. **Chaussat, R.,** *C. R. Acad. Sci.,* 270, 86, 1970.
353. **Bondon, F.,** *Phytotronique et Prospective Horticole. Phytotronique II,* Chouard, P. and de Bilderling, N., Eds., Gauthier-Villars, Paris, 1972, 135.
354. **Bagnard, C., Bernier, G., and Arnal, C.,** *Physiol. Vég.,* 10, 237, 1972.
355. **Bagnard, C.,** *Can. J. Bot.,* 58, 2335, 1980.
356. **Murneek, A. E.,** *Bot. Gaz. (Chicago),* 102, 269, 1940.
357. **Biddulph, O.,** *Bot. Gaz. (Chicago),* 97, 139, 1935.
358. **Greulach, V. A.,** *Bot. Gaz. (Chicago),* 103, 698, 1942.
359. **Madsen, G. C.,** *Bot. Gaz. (Chicago),* 109, 120, 1947.
360. **Chouard, P.,** *Bull. Soc. Bot. Fr.,* 109, 219, 1962.
361. **Philipson, W. R.,** *New Phytol.,* 33, 359, 1934.
362. **Evans, L. T.,** *Grasses and Grasslands,* Barnard, C., Ed., Macmillan, London, 1964, 126.
363. **Fisher, J. E.,** *Bot. Gaz. (Chicago),* 133, 78, 1972.
364. **Galinat, W. C. and Naylor, A. W.,** *Am. J. Bot.,* 38, 38, 1951.
365. **Wycherley, P. R.,** *Ann. Bot. (London),* 18, 119, 1954.
366. **Youngner, V. B.,** *Am. J. Bot.,* 47, 753, 1960.
367. **Krishnamoorthy, H. N. and Nanda, K. K.,** *Planta,* 80, 43, 1968.
368. **Debraux, G. and Simon, L.,** *Can. J. Bot.,* 47, 1791, 1969.
369. **Lecocq, M.,** *Mém. Soc. Bot. Fr.,* 209, 1969.
370. **Bagnard, C.,** Etude Expérimentale de la Réversion d'Inflorescence chez *Sinapis alba* L., Ph.D. thesis, University of Dijon, France, 1978.
371. **Tournois, J.,** *C. R. Acad. Sci.,* 155, 297, 1912.
372. **Tournois, J.,** *Ann. Sci. Nat. Bot. Biol. Vég.* 9e Sér., 19, 49, 1914.
373. **Heslop-Harrison, J.,** *Proc. Linn. Soc. Lond.,* 172 Session, 108, 1961.
374. **Mann, L. K.,** *Bot. Gaz. (Chicago),* 103, 780, 1942.
375. **Jones, K. L.,** *Am. J. Bot.,* 34, 371, 1947.
376. **Tothill, J. C. and Knox, R. B.,** *Aust. J. Agric. Res.,* 19, 869, 1968.
377. **Rudich, J., Baker, L. R., Scott, J. W., and Sell, H. M.,** *J. Am. Soc. Hortic. Sci.,* 101, 48, 1976.
378. **Cantliffe, D. J.,** *J. Am. Soc. Hortic. Sci.,* 106, 133, 1981.
379. **Fukushima, E., Matsuo, E., and Fujieda, K.,** *J. Fac. Agric. Kyushu Univ.,* 14, 349, 1968.
380. **Loy, J. B.,** *J. Am. Soc. Hortic. Sci.,* 96, 641, 1971.
381. **Thompson, A. E.,** *Mem. Cornell Agric. Exp. Stn.,* 336, 1955.
382. **Canham, A. E.,** *Acta Hortic.,* 14, 109, 1969.
383. **Heide, O. M.,** *Z. Pflanzenphysiol.,* 61, 279, 1969.
384. **van Veen, J. W. H.,** *Acta Hortic.,* 14, 39, 1969.
385. **Cockshull, K. E. and Hughes, A. P.,** *Ann. Bot. (London),* 35, 915, 1971.
386. **Takahashi, H., Saito, T., and Suge, H.,** *Plant Cell Physiol.,* 23, 1, 1982.
387. **Nielsen, C. S.,** *Bot. Gaz. (Chicago),* 104, 99, 1942.
388. **Batch, J. J. and Morgan, D. G.,** *Nature (London),* 250, 165, 1974.
389. **Heslop-Harrison, J. and Heslop-Harrison, Y.,** *Nature (London),* 181, 100, 1958.
390. **Heslop-Harrison, J.,** *Ann. Bot. (London),* 23, 345, 1959.
391. **Chouard, P.,** *C. R. Acad. Sci.,* 226, 1831, 1948.
392. **Evans, L. T.,** *Nature (London),* 178, 1301, 1956.
393. **Heslop-Harrison, J.,** *Phytomorphology,* 11, 378, 1961.
394. **Lord, E. M.,** *Ann. Bot. (London),* 44, 757, 1980.
395. **Lord, E. M.,** *Ann. Bot. (London),* 49, 261, 1982.
396. **Nitsch, J. P.,** *Proc. Am. Soc. Hortic. Sci.,* 70, 526, 1957.

397. **Wareing, P. F.,** *Annu. Rev. Plant Physiol.*, 7, 191, 1956.
398. **Sawhney, S., Sawhney, N., and Nanda, K. K.,** *Plant Cell Physiol.*, 19, 647, 1978.
399. **Morgan, D. G. and Zehni, M. S.,** *Ann. Bot. (London)*, 46, 37, 1980.
400. **Layzell, D. B. and Horton, R. F.,** *Can. J. Bot.*, 56, 1844, 1978.
401. **Wienk, J. F.,** *Meded. Landbouwhogesch. Wageningen*, 63(3), 1, 1963.
402. **Bagnard, C.,** *Can. J. Bot.*, submitted.
403. **Parmentier, A. and Kinet, J. M.,** *Rev. Agric.*, 33, 133, 1980.
404. **Cathey, H. M.,** *The Induction of Flowering. Some Case Histories*, Evans, L. T., Ed., Macmillam, Melbourne, 1969, 268.
405. **Deitzer, G. F., Hayes, R. G., and Jabben, M.,** *Plant Physiol.*, 69, 597, 1982.
406. **Sachs, R. M. and Hackett, W. P.,** *Acta Hortic.*, 68, 29, 1977.
407. **Friend, D. J. C., Helson, V. A., and Fisher, J. E.,** *Can. J. Plant Sci.*, 41, 418, 1961.
408. **Zehni, M. S., Saad, F. S., and Morgan, D. G.,** *Nature (London)*, 227, 628, 1970.
409. **Moe, R.,** *J. Am. Soc. Hortic. Sci.*, 97, 796, 1972.
410. **Kinet, J. M.,** *Sci. Hortic.*, 6, 15, 1977.
411. **Hughes, A. P. and Cockshull, K. E.,** *Ann. Bot. (London)*, 35, 927, 1971.
412. **Post, K. and Howland, J. E.,** *Proc. Am. Soc. Hortic. Sci.*, 47, 446, 1946.
413. **Zieslin, N. and Halevy, A. H.,** *Sci. Hortic.*, 3, 383, 1975.
414. **Nell, T. A. and Rasmussen, H. P.,** *J. Am. Soc. Hortic. Sci.*, 104, 21, 1979.
415. **May, P. and Antcliff, A. J.,** *J. Hortic. Sci.*, 38, 85, 1963.
416. **Buttrose, M. S.,** *Vitis*, 9, 121, 1970.
417. **Calvert, A.,** *J. Hortic. Sci.*, 34, 154, 1959.
418. **Calvert, A.,** *J. Hortic. Sci.*, 39, 182, 1964.
419. **Cooper, A. J.,** *J. Hortic. Sci.*, 39, 92, 1964.
420. **Sachs, R. M., Hackett, W. P., Tse, A. T. Y., Gallaher, C. M., Bruening, G., Salerno, D., Scott, I., and Graesser, R.,** unpublished data, 1968 to 1983.
421. **Shillo, R. and Halevy, A. H.,** *Sci. Hortic.*, 4, 131, 1976.
422. **Shillo, R. and Halevy, A. H.,** *Sci. Hortic.*, 4, 139, 1976.
423. **Shillo, R., Valis, G., and Halevy, A. H.,** *Sci. Hortic.*, 14, 367, 1981.
424. **Fortanier, E. J. and Zevenbergen, A.,** *Neth. J. Agric. Sci.*, 21, 145, 1973.
425. **Mastalerz, J. W.,** *Proc. Am. Soc. Hortic. Sci.*, 87, 502, 1965.
426. **Kamerbeek, G. A. and Durieux, A. J. B.,** *Acta Hortic.*, 23, 71, 1971.
427. **Durieux, A. J. B., Kamerbeek, G. A., and van Meeteren, U.,** *Sci. Hortic.*, 18, 287, 1983.
428. **Litlere, B. and Strømme, E.,** *Acta Hortic.*, 51, 285, 1975.
429. **Calvert, A.,** *J. Hortic. Sci.*, 44, 117, 1969.
430. **Kinet, J. M.,** *Rev. Agric.*, 35, 1767, 1982.
431. **Kinet, J. M. and Léonard, M.,** *Acta Hortic.*, 134, 117, 1983.
432. **Kudryavtsev, V. A.,** *Fiziol. Rast.*, 11, 409, 1964.
433. **Kinet, J. M., Léonard, M., and Linotte, C.,** unpublished data, 1978 to 1983.
434. **Mor, Y. and Halevy, A. H.,** *Plant Physiol.*, 66, 990, 1980.
435. **Mor, Y., Halevy, A. H., and Porath, D.,** *Plant Physiol.*, 66, 996, 1980.
436. **Mae, T. and Vonk, C. R.,** *Acta Bot. Neerl.*, 23, 321, 1974.
437. **May, P.,** *Aust. J. Biol. Sci.*, 18, 463, 1965.
438. **Cockshull, K. E., Sachs, R. M., and Tjosvold, S. A.,** unpublished data, 1979 to 1983.
439. **Sachs, R. M. and Hackett, W. P.,** *HortScience*, 4, 103, 1969.
440. **Resende, F. and Viana, M. J.,** *Port. Acta Biol. Sér. A*, 2, 211, 1948.
441. **Shibata, O. and Takimoto, A.,** *Plant Cell Physiol.*, 16, 513, 1975.
442. **Sato, T. and Oda, Y.,** *Plant Cell Physiol.*, 19, 537, 1978.
443. **Teeri, J. A. and Tonsor, S. J.,** *Can. J. Bot.*, 59, 388, 1981.
444. **Friend, D. J. C.,** *Can. J. Bot.*, 43, 345, 1965.
445. **Kooistra, E.,** *Euphytica*, 16, 1, 1967.
446. **Gregg, K. B.,** *Selbyana*, 1, 101, 1975.
447. **Resende, F.,** *Port. Acta Biol. Sér. A*, 2, 365, 1949.
448. **Smeets, L.,** *Sci. Hortic.*, 4, 255, 1976.
449. **Smeets, L.,** *Sci. Hortic.*, 12, 343, 1980.
450. **Rawson, H. M. and Bagga, A. K.,** *Aust. J. Plant Physiol.*, 6, 391, 1979.
451. **Calvert, A.,** *J. Hortic. Sci.*, 39, 194, 1964.
452. **Lake, J. V.,** *J. Hortic. Sci.*, 42, 1, 1967.
453. **Armitage, A. M., Carlson, W. H., and Flore, J. A.,** *J. Am. Soc. Hortic. Sci.*, 106, 643, 1981.
454. **Moss, G. I.,** *J. Hortic. Sci.*, 44, 311, 1969.
455. **Dosser, A. L. and Larson, R. A.,** *J. Am. Soc. Hortic. Sci.*, 106, 247, 1981.

456. **Morgan, J. V. and Clarke, E. J.,** *Acta Hortic.*, 51, 77, 1975.
457. **Berninger, E.,** *Sci. Hortic.*, 10, 271, 1979.
458. **Abdalla, A. A. and Verkerk, K.,** *Neth. J. Agric. Sci.*, 16, 71, 1968.
459. **Moss, G. I.,** *Aust. J. Agric. Res.*, 27, 399, 1976.
460. **Moe, R.,** *Physiol. Plant.*, 24, 291, 1971.
461. **Shillo, R. and Halevy, A. H.,** *Isr. J. Agric. Res.*, 13, 141, 1963.
462. **Shillo, R. and Halevy, A. H.,** *Sci. Hortic.*, 4, 157, 1976.
463. **Warrington, I. J., Dunstone, R. L., and Green, L. M.,** *Aust. J. Agric. Res.*, 28, 11, 1977.
464. **Bagga, A. K. and Rawson, H. M.,** *Aust. J. Plant Physiol.*, 4, 877, 1977.
465. **Lewis, D.,** *J. Hortic. Sci.*, 28, 207, 1953.
466. **Hurd, R. G. and Cooper, A. J.,** *J. Hortic. Sci.*, 42, 181, 1967.
467. **Hurd, R. G. and Cooper, A. J.,** *J. Hortic. Sci.*, 45, 19, 1970.
468. **Blaauw, A. H., Luyten, I., and Hartsema, A. M.,** *Proc. K. Ned. Akad. Wet.*, 35, 485, 1932.
469. **Halevy, A. H. and Zieslin, N.,** *Acta Hortic.*, 14, 149, 1969.
470. **Beisland, A. and Kristoffersen, T.,** *Acta Hortic.*, 14, 97, 1969.
471. **Kohl, H. C., Jr.,** *Proc. Am. Soc. Hortic. Sci.*, 77, 540, 1961.
472. **Atherton, J. G. and Harris, G. P.,** *Sci. Hortic.*, 13, 371, 1980.
473. **Galun, E.,** *Experientia*, 12, 218, 1956.
474. **Brooking, I. R.,** *Aust. J. Plant Physiol.*, 3, 589, 1976.
475. **Satake, T. and Hayase, H.,** *Proc. Crop Sci. Soc. Jpn.*, 39, 468, 1970.
476. **Satake, T. and Hayase, H.,** *Proc. Crop Sci. Soc. Jpn.*, 43, 36, 1974.
477. **Braak, J. P.,** *Euphytica*, 17, 311, 1968.
478. **Randolph, L. F.,** *Proc. Natl. Acad. Sci. U.S.A.*, 18, 222, 1932.
479. **Sax, K.,** *Am. J. Bot.*, 24, 218, 1937.
480. **Iwahori, S.,** *J. Jpn. Soc. Hortic. Sci.*, 34, 33, 1965.
481. **Saini, H. S. and Aspinall, D.,** *Ann. Bot. (London)*, 49, 835, 1982.
482. **Cathey, H. M.,** *Proc. Am. Soc. Hortic. Sci.*, 69, 485, 1957.
483. **Cockshull, K. E.,** *Ann. Bot. (London)*, 44, 451, 1979.
484. **van Bragt, J. and Zijlstra, F. A.,** *Z. Pflanzenphysiol.*, 64, 139, 1971.
485. **van Bragt, J. and van Ast, K. J.,** *Sci. Hortic.*, 4, 117, 1976.
486. **Richardson, E. A., Seeley, S. D., and Walker, D. R.,** *HortScience*, 9, 331, 1974.
487. **Freeman, M. W. and Martin, G. C.,** *J. Am. Soc. Hortic. Sci.*, 106, 333, 1981.
488. **Gilreath, P. R. and Buchanan, D. W.,** *J. Am. Soc. Hortic. Sci.*, 106, 426, 1981.
489. **Pettersen, H. and Kristoffersen, T.,** *Acta Hortic.*, 14, 27, 1969.
490. **Kawata, J.,** *Acta Hortic.*, 47, 201, 1975.
491. **Moe, R. and Wickstrøm, A.,** *Sci. Hortic.*, 10, 187, 1979.
492. **Rees, A. R. and Turquand, E. D.,** *Exp. Hortic.*, 20, 49, 1969.
493. **Weinberger, J. H.,** *Proc. Am. Soc. Hortic. Sci.*, 56, 122, 1950.
494. **Darrow, G. M.,** *Proc. Am. Soc. Hortic. Sci.*, 41, 189, 1942.
495. **Rigby, B. and Dana, M. N.,** *J. Am. Soc. Hortic. Sci.*, 97, 145, 1972.
496. **Chandler, W. H., Kimball, M. H., Philp, G. L., Tufts, W. P., and Weldon, G. P.,** *Calif. Agric. Exp. Stn. Bull.*, 611, 1, 1937.
497. **Brown, D. S., Griggs, W. H., and Iwakiri, B. T.,** *Calif. Agric.*, 21, 10, 1967.
498. **Rees, A. R., Turquand, E. D., and Briggs, J. B.,** *Exp. Hortic.*, 23, 52, 1972.
499. **Eady, F. and Eaton, G. W.,** *Can. J. Plant Sci.*, 49, 637, 1969.
500. **Rees, A. R.,** *J. Hortic. Sci.*, 44, 27, 1969.
501. **Brooks, R. M. and Philp, G. L.,** *Proc. Am. Soc. Hortic. Sci.*, 39, 190, 1941.
502. **Brown, D. S.,** *Proc. Am. Soc. Hortic. Sci.*, 59, 111, 1952.
503. **Denny, F. E. and Stanton, E. N.,** *Am. J. Bot.*, 15, 337, 1928.
504. **Schneider, E. F.,** *J. Exp. Bot.*, 19, 817, 1968.
505. **Rees, A. R.,** *J. Hortic. Sci.*, 48, 149, 1973.
506. **Weinberger, J. H.,** *Proc. Am. Soc. Hortic. Sci.*, 63, 157, 1954.
507. **Erez, A. and Lavee, S.,** *J. Am. Soc. Hortic. Sci.*, 96, 711, 1971.
508. **Shanks, J. B. and Link, C. B.,** *Proc. Am. Soc. Hortic. Sci.*, 90, 353, 1967.
509. **Erez, A., Samish, R. M., and Lavee, S.,** *Physiol. Plant.*, 19, 650, 1966.
510. **Gilreath, P. R. and Buchanan, D. W.,** *J. Am. Soc. Hortic. Sci.*, 106, 321, 1981.
511. **Westwood, M. N. and Bjornstad, H. O.,** *J. Am. Soc. Hortic. Sci.*, 103, 142, 1978.
512. **Husain, I. and Aspinall, D.,** *Ann. Bot. (London)*, 34, 393, 1970.
513. **Zavadskaya, I. G. and Skazkin, F. D.,** *Dokl. Akad. Nauk SSSR*, 131, 692, 1960.
514. **Saini, H. S. and Aspinall, D.,** *Ann. Bot. (London)*, 48, 623, 1981.
515. **Moss, G. I. and Downey, L. A.,** *Crop Sci.*, 11, 368, 1971.
516. **Brown, W. V.,** *Bot. Gaz. (Chicago)*, 113, 438, 1952.

517. **Seifriz, W.,** *Am. J. Bot.*, 10, 32, 1923.
518. **Went, F. W.,** *The Experimental Control of Plant Growth*, Chronica Botanica, Waltham, Mass., 1957.
519. **Holdsworth, M.,** *J. W. Afr. Sci. Assoc.*, 7, 28, 1961.
520. **Pagacz, E. A.,** *Bull. Agric. Congo Belge Ruanda-Urundi*, 50, 1531, 1959.
521. **Alvim, P. de T.,** *Science*, 132, 354, 1960.
522. **Piringer, A. A. and Borthwick, H. A.,** *Turrialba*, 5, 72, 1955.
523. **Mess, M. G.,** *Port. Acta Biol. Sér. A*, 4, 342, 1957.
524. **Alvim, P. de T.,** *Turrialba*, 8, 64, 1958.
525. **Chailakhyan, M. Kh., Yanina, L. I., and Florova, I. A.,** *Fiziol. Rast.*, 17, 709, 1970.
526. **Viana, M. J.,** *Port. Acta Biol. Sér. A*, 15, 91, 1977/79.
527. **Chailakhyan, M. Kh.,** *Can. J. Bot.*, 39, 1817, 1961.
528. **Peterson, C. A. and Fletcher, R. A.,** *Can. J. Bot.*, 51, 1899, 1973.
529. **Smith, D. L.,** *Ann. Bot. (London)*, 33, 505, 1969.
530. **Alexander, D. McE. and Woodham, R. C.,** *J. Hortic. Sci.*, 38, 307, 1963.
531. **Mullins, M. G.,** *J. Exp. Bot.*, 18, 206, 1967.
532. **Chailakhyan, M. Kh. and Khryanin, V. N.,** *Planta*, 138, 185, 1978.
533. **Chailakhyan, M. Kh. and Khryanin, V. N.,** *Planta*, 142, 207, 1978.
534. **Phatak, S. C., Wittwer, S. H., and Teubner, F. G.,** *Proc. Am. Soc. Hortic. Sci.*, 88, 527, 1966.
535. **Kinet, J. M.,** *Agricontact*, 42, (X)I, 1975.
536. **Cooper, A. J. and Hurd, R. G.,** *J. Hortic. Sci.*, 43, 243, 1968.
537. **de Jong, A. W. and Bruinsma, J.,** *Z. Pflanzenphysiol.*, 72, 220, 1974.
538. **Zieslin, N. and Halevy, A. H.,** *Sci. Hortic*, 4, 73, 1976.
539. **Ramina, A., Hackett, W. P., and Sachs, R. M.,** *Plant Physiol.*, 64, 810, 1979.
540. **Mullins, M. G.,** *J. Exp. Bot.*, 19, 532, 1968.
541. **Tse, A. T. Y., Ramina, A., Hackett, W. P., and Sachs, R. M.,** *Plant Physiol.*, 54, 404, 1974.
542. **Kinet, J. M.,** *Sci. Hortic.*, 6, 27, 1977.
543. **Abdul, K. S. and Harris, G. P.,** *Ann. Bot. (London)*, 42, 1361, 1978.
544. **Galun, E.,** *Physiol. Plant.*, 12, 48, 1959.
545. **Nanda, K. K., Krishnamoorthy, H. N., and Lata, K.,** *Plant Cell Physiol.*, 10, 357, 1969.
546. **Kinet, J. M. and Léonard, M.,** *Abstr. 2nd Congr. Fed. Eur. Soc. Plant Physiol.*, Santiago de Compostel Spain, 1980, 436.
547. **Yeung, E. C. and Peterson, R. L.,** *Can. J. Bot.*, 50, 73, 1972.
548. **Mockaitis, J. M. and Kivilaan, A.,** *Nature (London)*, 202, 216, 1964.
549. **Anaïs, G.,** *Ann. Amélior. Plantes*, 21, 55, 1971.
550. **Raman, K., Walden, D. B., and Greyson, R. I.,** *Ann. Bot. (London)*, 45, 183, 1980.
551. **Byrne, T. G. and Doss, R. P.,** *J. Am. Soc. Hortic. Sci.*, 106, 98, 1981.
552. **Zieslin, N. and Byrne, T. G.,** *Sci. Hortic.*, 15, 179, 1981.
553. **Kirby, E. J. M. and Faris, D. G.,** *J. Exp. Bot.*, 21, 787, 1970.
554. **Cottrell, J., Dale, J. E., and Jeffcoat, B.,** *Plant Sci. Lett.*, 22, 161, 1981.
555. **Berghoef, J. and Bruinsma, J.,** *Z. Pflanzenphysiol.*, 94, 407, 1979.
556. **Lord, E. M.,** *Bot. Gaz. (Chicago)*, 143, 63, 1982.
557. **Heslop-Harrison, J. and Heslop-Harrison, Y.,** *Portug. Acta Biol. Sér. A*, 5, 79, 1958.
558. **Resende, F.,** *Portug. Acta Biol. Sér. A*, Vol. Goldschmidt, R. B., 1949/1951, 729.
559. **Resende, F. and Viana, M. J.,** *Portug. Acta Biol. Sér. A*, 6, 77, 1959.
560. **Catarino, F. M.,** *Portug. Acta Biol., Sér. A*, 8, 267, 1964.
561. **Goldschmidt, E. E. and Huberman, M.,** *J. Exp. Bot.*, 25, 534, 1974.
562. **Plack, A.,** *Nature (London)*, 180, 1218, 1957.
563. **Pool, R. M.,** *Am. J. Enol. Vitic.*, 26, 43, 1975.
564. **Berghoef, J. and Bruinsma, J.,** *Z. Pflanzenphysiol.*, 93, 377, 1979.
565. **Tizio, R.,** *C. R. Acad. Sci.*, 289, 401, 1979.
566. **Polowick, P. L. and Greyson, R. I.,** *Plant Sci. Lett.*, 26, 139, 1982.
567. **Bilderback, D. E., Karpoff, A. J., and Tepfer, S. S.,** *Am. J. Bot.*, 55, 1042, 1968.
568. **Hicks, G. S., Browne, R., and Sand, S. A.,** *Can. J. Bot.*, 59, 1665, 1981.
569. **Ganapathy, P. S.,** *Biol. Plant.*, 11, 165, 1969.
570. **Coleman, R. E. and Nickell, L. G.,** *Nature (London)*, 201, 941, 1964.
571. **Mohan Ram, H. Y. and Wadhi, M.,** *Ann. Bot. (London)*, 32, 825, 1968.
572. **Rivière, S.,** *C. R. Acad. Sci.*, 267, 1439, 1968.
573. **Konar, R. N. and Nataraja, K.,** *Acta Bot. Neerl.*, 18, 680, 1969.
574. **Cline, M. G.,** *Plant Sci. Lett.*, 6, 181, 1976.
575. **Bilderback, D. E.,** *Am. J. Bot.*, 58, 203, 1971.
576. **Peterson, C. M.,** *Am. J. Bot.*, 61, 693, 1974.

577. **Haccius, B., Bhandari, N. N., and Hausner, G.,** *J. Exp. Bot.*, 25, 695, 1974.
578. **Hicks, G. S. and McHughen, A.,** *Planta*, 121, 193, 1974.
579. **Hicks, G. S. and McHughen, A.,** *Plant Sci. Lett.*, 8, 141, 1977.
580. **McHughen, A.,** *J. Exp. Bot.*, 33, 162, 1982.
581. **McHughen, A.,** *Ann. Bot. (London)*, 41, 1073, 1977.
582. **Hicks, G. S.,** *Can. J. Bot.*, 53, 77, 1975.
583. **Hicks, G. S.,** *Plant Sci. Lett.*, 17, 81, 1979.
584. **Cusick, F.,** *Trans. R. Soc. Edinburgh Sect. B*, 63, 153, 1956.
585. **Soetiarto, S. R. and Ball, E.,** *Can. J. Bot.*, 47, 1067, 1969.
586. **Hicks, G. S. and Sussex, I. M.,** *Bot. Gaz. (Chicago)*, 132, 350, 1971.
587. **Jensen, L. C. W.,** *Can. J. Bot.*, 49, 487, 1971.
588. **Hicks, G. S.,** *Can. J. Bot.*, 50, 2396, 1972.
589. **Hicks, G. S.,** *Bot. Gaz. (Chicago)*, 134, 154, 1973.
590. **Hicks, G. S. and Bell, J.,** *Can. J. Bot.*, 53, 231, 1975.
591. **McHughen, A.,** *Ann. Bot. (London)*, 46, 633, 1980.
592. **McHughen, A.,** *Bot. Gaz. (Chicago)*, 141, 389, 1980.
593. **Meins, F., Jr. and Binns, A. N.,** *The Clonal Basis of Development*, Subtelny, S. and Sussex, I. M., Eds., Academic Press, New York, 1978, 185.
594. **Obšil, K.,** *Planta*, 29, 468, 1939.
595. **Lang, A.,** *Encyclopedia of Plant Physiology*, Vol. 14, Ruhland, W., Ed., Springer-Verlag, Berlin, 1961, 909.
596. **de Zeeuw, D.,** *Proc. K. Ned. Akad. Wet. Ser. C*, 59, 535, 1956.
597. **Salisbury, F. B.,** *Plant Physiol.*, 30, 327, 1955.
598. **Khatoon, S., Seidlová, F., and Krekule, J.,** *Biol. Plant.*, 15, 361, 1973.
599. **Margara, J.,** *C. R. Acad. Sci.*, 277, 497, 1973.
600. **Goh, C. J.,** *Ann. Bot. (London)*, 41, 763, 1977.
601. **Seidlová, F. and Khatoon, S.,** *Ann. Bot. (London)*, 40, 37, 1976.
602. **Söding, H.,** *Jahrb. Wiss. Bot.*, 65, 611, 1926.
603. **Kaldewey, H.,** *Planta*, 49, 300, 1957.
604. **op den Kelder, P., Benschop, M., and de Hertogh, A. A.,** *J. Am. Soc. Hortic. Sci.*, 96, 603, 1971.
605. **Hanks, G. R. and Rees, A. R.,** *New Phytol.*, 75, 605, 1975.
606. **Hanks, G. R. and Rees, A. R.,** *New Phytol.*, 78, 579, 1977.
607. **Menhenett, R.,** *New Phytol.*, 84, 271, 1980.
608. **Saniewski, M. and de Munk, W. J.,** *Sci. Hortic.*, 15, 363, 1981.
609. **Debraux, G. and Astié, M.,** *Photochem. Photobiol.*, 5, 457, 1966.
610. **Astié, M.,** *Ann. Sci. Nat. Bot. Biol. Vég., 12e Sér.*, 3, 619, 1962.
611. **Sharman, B. C.,** *Ann. Bot. (London)*, 42, 145, 1978.
612. **Laibach, F. and Kribben, F. I.,** *Ber. Dtsch. Bot. Ges.*, 62, 53, 1950.
613. **Laibach, F. and Kribben, F. I.,** *Beitr. Biol. Pflanz.*, 28, 64, 1950.
614. **Heslop-Harrison, J.,** *Colloq. Int. C.N.R.S.*, 123, 649, 1964.
615. **Wittwer, S. H. and Hillyer, I. G.,** *Science*, 120, 893, 1954.
616. **Galun, E.,** *Phyton (Buenos Aires)*, 13, 1, 1959.
617. **Ito, H. and Saito, T.,** *J. Jpn. Soc. Hortic. Sci.*, 25, 101, 1956.
618. **Ito, H. and Saito, T.,** *Tohoku, J. Agric. Res.*, 11, 287, 1960.
619. **Bose, T. K. and Nitsch, J. P.,** *Physiol. Plant.*, 23, 1206, 1970.
620. **Bhandari, M. C. and Sen, D. N.,** *Biochem. Physiol. Pflanz.*, 164, 450, 1973.
621. **Smith, D. L.,** *Ann. Bot. (London)*, 31, 19, 1967.
622. **Heslop-Harrison, J.,** *Physiol. Plant.*, 9, 588, 1956.
623. **Galoch, E.,** *Acta Soc. Bot. Pol.*, 47, 153, 1978.
624. **Chailakhyan, M. Kh. and Khryanin, V. N.,** *Planta*, 138, 181, 1978.
625. **Sawhney, V. K. and Greyson, R. I.,** *Can. J. Bot.*, 51, 2473, 1973.
626. **Durand, B.,** *Bull. Soc. Fr. Physiol. Vég.*, 13, 195, 1967.
627. **Champault, A.,** *C. R. Acad. Sci.*, 269, 1948, 1969.
628. **Champault, A.,** *Bull. Soc. Bot. Fr.*, 120, 87, 1973.
629. **Tsukamoto, Y.,** *J. Hortic. Assoc. Jpn.*, 26, 192, 1957.
630. **de Jong, A. W. and Bruinsma, J.,** *Z. Pflanzenphysiol.*, 73, 142, 1974.
631. **Hügel, B.,** *Z. Pflanzenphysiol.*, 80, 283, 1976.
632. **Ćulafić, Lj. and Nešković, M.,** *Physiol. Plant.*, 48, 588, 1980.
633. **de Jong, A. W. and Bruinsma, J.,** *Z. Pflanzenphysiol.*, 73, 152, 1974.
634. **Zimmerman, P. W., Crocker, W., and Hitchcock, A. E.,** *Proc. Am. Soc. Hortic. Sci.*, 28, 53, 1931.
635. **Zimmerman, P. W., Hitchcock, A. E., and Crocker, W.,** *Contrib. Boyce Thompson Inst.*, 3, 459, 1931.
636. **Hitchcock, A. E., Crocker, W., and Zimmerman, P. W.,** *Contrib. Boyce Thompson Inst.*, 4, 155, 1932.

637. **de Munk, W. J. and Hoogeterp, P.,** *Acta Hortic.*, 47, 149, 1975.
638. **de Munk, W. J. and Gijzenberg, J.,** *Sci. Hortic.*, 7, 255, 1977.
639. **de Proft, M., van Meeteren, U., de Greef, J. A., and Kamerbeek, G. A.,** *Abstr. Symp. Aspects and Prospects of Plant Growth Regulators*, Wageningen, Netherlands, 1980, 17.
640. **Cameron, A. C. and Reid, M. S.,** *HortScience*, 16, 761, 1981.
641. **Zieslin, N. and Halevy, A. H.,** *Physiol. Plant.*, 37, 326, 1976.
642. **Sanderson, K. C., Martin, W. C., Jr., Marcus, K. A., and Goslin, W. E.,** *HortScience*, 10, 611, 1975.
643. **Phatak, S. C., Jaworski, C. A., and Liptay, A.,** *HortScience*, 16, 181, 1981.
644. **Simmonds, J. A. and Cumming, B. G.,** *Sci. Hortic.*, 6, 71, 1977.
645. **Simmonds, J. A. and Cumming, B. G.,** *Sci. Hortic.*, 8, 57, 1978.
646. **Dicks, J. W. and Rees, A. R.,** *Sci. Hortic.*, 1, 133, 1973.
647. **Lieberman, M.,** *Annu. Rev. Plant Physiol.*, 30, 353, 1979.
648. **Lau, O. L. and Yang, S. F.,** *Plant Physiol.*, 58, 114, 1976.
649. **Beyer, E. M., Jr.,** *Plant Physiol.*, 58, 268, 1976.
650. **Beyer, E. M., Jr.,** *Plant Physiol.*, 63, 169, 1979.
651. **van Meeteren, U. and de Proft, M.,** *Physiol. Plant.*, 56, 236, 1982.
652. **Dekazos, E. D.,** *Proc. Fla. State Hortic. Soc.*, 92, 248, 1979.
653. **Dekazos, E. D.,** *HortScience*, 16, 520, 1981.
654. **Minima, F. G.,** *C. R. Dokl. Acad. Sci. URSS*, 21, 298, 1938.
655. **Minima, E. G. and Tylkina, L. G.,** *C. R. Dokl. Acad. Sci. URSS*, 55, 165, 1947.
656. **Heslop-Harrison, J. and Heslop-Harrison, Y.,** *Proc. R. Soc. Edinburgh Sect. B*, 66, 424, 1957.
657. **Heslop-Harrison, J. and Heslop-Harrison, Y.,** *New Phytol.*, 56, 352, 1957.
658. **McMurray, A. L. and Miller, C. H.,** *Science*, 162, 1397, 1968.
659. **Robinson, R. W., Shannon, S., and de la Guardia, M. D.,** *BioScience*, 19, 141, 1969.
660. **Iwahori, S., Lyons, J. M., and Smith, O. E.,** *Plant Physiol.*, 46, 412, 1970.
661. **Matsubara, S.,** *Sci. Rep. Fac. Agric. Okayama Univ.*, 49, 15, 1977.
662. **Augustine, J. J., Baker, L. R., and Sell, H. M.,** *J. Am. Soc. Hortic. Sci.*, 98, 197, 1973.
663. **Friedlander, M., Atsmon, D., and Galun, E.,** *Plant Cell Physiol.*, 18, 261, 1977.
664. **Karchi, Z.,** *J. Am. Soc. Hortic. Sci.*, 95, 515, 1970.
665. **Rudich, J., Halevy, A. H., and Kedar, N.,** *Planta*, 86, 69, 1969.
666. **Loy, J. B., Natti, T. A., Zack, C. D., and Fritts, S.,** *J. Am. Soc. Hortic. Sci.*, 104, 100, 1979.
667. **Robinson, R. W., Whitaker, Th. W., and Bohn, G. W.,** *Euphytica*, 19, 180, 1970.
668. **Splittstoesser, W. E.,** *Physiol. Plant.*, 23, 762, 1970.
669. **Saimbhi, M. S. and Thakur, M. R.,** *Physiol. Plant.*, 28, 383, 1973.
670. **Mohan Ram, H. Y. and Jaiswal, V. S.,** *Plant Growth Substances 1973*, Hirokawa, Tokyo, 1974, 987.
671. **Jaiswal, V. S. and Kumar, A.,** *J. Exp. Bot.*, 31, 497, 1980.
672. **Jindal, K. K. and Singh, R. N.,** *Z. Pflanzenphysiol.*, 78, 403, 1976.
673. **Kender, W. J. and Remaily, G.,** *HortScience*, 5, 491, 1970.
674. **Hughes, W. G., Bennett, M. D., Bodden, J. J., and Galanopoulou, S.,** *Ann. Appl. Biol.*, 76, 243, 1974.
675. **Law, J. and Stoskopf, N. C.,** *Can. J. Plant Sci.*, 53, 765, 1973.
676. **Karchi, Z. and Govers, A.,** *J. Am. Soc. Hortic. Sci.*, 97, 357, 1972.
677. **Krishnamoorthy, H. N. and Talukdar, A. R.,** *Z. Pflanzenphysiol.*, 79, 91, 1976.
678. **Lower, R. L., Miller, C. H., Baker, F. H., and McCombs, C. L.,** *HortScience*, 5, 434, 1970.
679. **Phatak, S. C. and Cantliffe, D. J.,** *HortScience*, 11, 27, 1976.
680. **Byers, R. E., Baker, L. R., Sell, H. M., Herner, R. C., and Dilley, D. R.,** *Proc. Natl. Acad. Sci. U.S.A.*, 69, 717, 1972.
681. **Shannon, S. and de la Guardia, M. D.,** *Nature (London)*, 223, 186, 1969.
682. **Rudich, J., Halevy, A. H., and Kedar, N.,** *Plant Physiol.*, 50, 585, 1972.
683. **Beyer, E., Jr.,** *HortScience*, 11, 195, 1976.
684. **Kalloo and Franken, S.,** *Gartenbauwissenschaft*, 43, 280, 1978.
685. **Atsmon, D. and Tabbak, C.,** *Plant Cell Physiol.*, 20, 1547, 1979.
686. **Den Nijs, A. P. M. and Visser, D. L.,** *Euphytica*, 29, 273, 1980.
687. **Owens, K. W., Tolla, G. E., and Peterson, C. E.,** *HortScience*, 15, 256, 1980.
688. **Risser, G. and Rode, J. C.,** *Ann. Amélior. Plantes*, 29, 349, 1979.
689. **Owens, K. W., Peterson, C. E., and Tolla, G. E.,** *HortScience*, 15, 654, 1980.
690. **Sarath, G. and Mohan Ram, H. Y.,** *Experientia*, 35, 333, 1979.
691. **Mohan Ram, H. Y. and Sett, R.,** *Proc. Ind. Acad. Sci.*, 88, 303, 1979.
692. **Mohan Ram, H. Y. and Sett, R.,** *Planta*, 149, 413, 1980.
693. **Menhenett, R.,** *Ann. Bot. (London)*, 43, 305, 1979.

694. **Harris, G. P., Jeffcoat, B., and Garrod, J. F.,** *Nature (London),* 223, 1071, 1969.
695. **Nicholls, P. B.,** *Aust. J. Plant Physiol.,* 5, 581, 1978.
696. **Greulach, V. A. and Haesloop, J. G.,** *Science,* 127, 646, 1958.
697. **de Hertogh, A. A. and Blakely, N.,** *J. Am. Soc. Hortic. Sci.,* 97, 320, 1972.
698. **Halevy, A. H. and Shoub, J.,** *J. Hortic. Sci.,* 39, 120, 1964.
699. **Motomura, Y. and Ito, H.,** *Tohoku J. Agric. Res.,* 24, 151, 1973.
700. **Sachs, R. M., Kofranek, A. M., De Bie, J., and Kubota, J.,** *Florist's Rev.,* 150, 85, 1972.
701. **Menhenett, R.,** *Ann. Appl. Biol.,* 87, 451, 1977.
702. **Dicks, J. W., Gilford, J. McD., and Rees, A. R.,** *Sci. Hortic.,* 2, 153, 1974.
703. **Halevy, A. H.,** *Proc. 18th Int. Hortic. Congr.,* Vol. 5, Tel-Aviv, 1972, 187.
704. **Menhenett, R.,** *Ann. Bot. (London),* 47, 359, 1981.
705. **Hackett, W. P. and Sachs, R. M.,** *Proc. Am. Soc. Hortic. Sci.,* 90, 361, 1967.
706. **Guardiola, J. L., Monerri, C., and Agusti, M.,** *Physiol. Plant.,* 55, 136, 1982.
707. **Abdul, K. S., Canham, A. E., and Harris, G. P.,** *Ann. Bot. (London),* 42, 617, 1978.
708. **Abdalla, A. A. and Verkerk, K.,** *Neth. J. Agric. Sci.,* 18, 105, 1970.
709. **Dicks, J. W.,** *Recent Developments in the Use of Plant Growth Retardants,* BPGRG Monogr. No. 4, 1980, 1.
710. **Kinet, J.M., Hurdebise, D., and Stainier, R.,** *Rev. Agric.,* 31, 413, 1978.
711. **Racoff, H.,** *Am. Camellia Yearbook,* 1964, 207.
712. **Rappaport, L.,** *Plant Physiol.,* 32, 440, 1957.
713. **Plack, A.,** *Nature (London),* 182, 610, 1958.
714. **Jeffcoat, B., Scott, M. A., and Harris, G. P.,** *Ann. Bot. (London),* 33, 515, 1969.
715. **Zieslin, N., Biran, I., and Halevy, A.H.,** *Plant Cell Physiol.,* 15, 341, 1974.
716. **Murakami, Y.,** *Plant Cell Physiol.,* 16, 337, 1975.
717. **Joiner, J. N., Washington, O., Johnson, C. R., and Nell, T. A.,** *Sci. Hortic.,* 18, 143, 1982.
718. **Heide, O. M.,** *Physiol. Plant.,* 22, 1001, 1969.
719. **van Bragt, J. and Dekker,P. H. M.,** *Neth. J. Agric. Sci.,* 21, 139, 1973.
720. **Raghuvanshi, S. S., Pathak, C. S., and Singh, R. R.,** *Bot. Gaz. (Chicago),* 142, 40, 1981.
721. **Lord, E. M. and Mayers, A. M.,** *Ann. Bot. (London),* 50, 301, 1982.
722. **Weijer, J.,** *Science,* 129, 896, 1959.
723. **Ohkawa, K.,** *J. Jpn. Soc. Hortic. Sci.,* 45, 289, 1976.
724. **Mohan Ram, H. Y. and Mehta, U.,** *J. Exp. Bot.,* 29, 653, 1978.
725. **Dubuc-Lebreux, M. A. and Vieth, J.,** *Bull. Soc. Bot. Fr.,* 123, 273, 1976.
726. **Kirby, E. J. M.,** *J. Exp. Bot.,* 22, 411, 1971.
727. **Hillman, W. S. and Smith, H. H.,** *J. Hered.,* 56, 2, 1965.
728. **Dewan, K.,** *Phytomorphology,* 30, 41, 1980.
729. **Nothmann, J. and Koller, D.,** *Hortic. Res.,* 13, 105, 1973.
730. **Brulfert, J.,** *C. R. Acad. Sci.,* 253, 517, 1961.
731. **Wittwer, S. H. and Bukovac, M. J.,** *Econ. Bot.,* 12, 213, 1958.
732. **Saito, T. and Ito, H.,** *J. Jpn. Soc. Hortic. Sci.,* 32, 278, 1963.
733. **Peterson, C. E. and Anhder, L. D.,** *Science,* 131, 1673, 1960.
734. **Wittwer, S. H. and Bukovac, M. J.,** *Naturwissenschaften,* 49, 305, 1962.
735. **Fuchs, E., Atsmon, D., and Halevy, A. H.,** *Plant Cell Physiol.,* 18, 1193, 1977.
736. **Rudich, J., Halevy, A. H., and Kedar, N.,** *J. Am. Soc. Hortic. Sci.,* 97, 369, 1972.
737. **Felippe, G. M.,** *Z. Pflanzenphysiol.,* 94, 79, 1979.
738. **Krishnamoorthy, H. N.,** *Plant Cell Physiol.,* 13, 381, 1972.
739. **Juneja, A.,** *Phytomorphology,* 21, 235, 1971.
740. **Ghosh, S. P. and Sen, S. P.,** *J. Hortic. Sci.,* 50, 91, 1975.
741. **Phatak, S. C., Wittwer, S. H., Honma, S., and Bukovac, M. J.,** *Nature (London),* 209, 635, 1966.
742. **Rana, R. S. and Jain, H. K.,** *Naturwissenschaften,* 55, 301, 1968.
743. **Kasembe, J. N. R.,** *Nature (London),* 215, 668, 1967.
744. **Lazarte, J. E. and Garrison, S. A.,** *J. Am. Soc. Hortic. Sci.,* 105, 691, 1980.
745. **Kubicki, B.,** *Genet. Pol.,* 10, 145, 1969.
746. **Kubicki, B.,** *Genet. Pol.,* 6, 153, 1965.
747. **Hicks, G. S. and Sand, S. A.,** *Plant Sci. Lett.,* 10, 257, 1977.
748. **Hansen, D. J., Bellman, S. K., and Sacher, R. M.,** *Crop Sci.,* 16, 371, 1976.
749. **von Witsch, H.,** *Planta,* 57, 357, 1961.
750. **Shifriss, O.,** *Science,* 133, 2061, 1961.
751. **Nunes, M. A.,** *Port. Acta Biol. Sér. A,* 8, 285, 1964.
752. **Sawhney, V. K.,** *Can. J. Bot.,* 59, 8, 1981.
753. **Eenink, A. H.,** *Euphytica,* 26, 31, 1977.

754. **van der Meer, Q. P. and van Bennekom, J. L.,** *Euphytica*, 22, 239, 1973.
755. **van der Meer, Q. P. and van Dam, R.,** *Euphytica*, 28, 717, 1979.
756. **Pharis, R. P. and Kuo, C. G.,** *Can. J. For. Res.*, 7, 299, 1977.
757. **Nickerson, N. H.,** *Am. J. Bot.*, 47, 809, 1960.
758. **Sawhney, V. K. and Greyson, R. I.,** *J. Am. Soc. Hortic. Sci.*, 97, 589, 1972.
759. **Bukovac, M. J. and Honma, S.,** *Proc. Am. Soc. Hortic. Sci.*, 91, 514, 1967.
760. **Sawhney, V. K. and Greyson R. I.,** *Can. J. Bot.*, 57, 2471, 1979.
761. **Mitchell, W. D. and Wittwer, S. H.,** *Science*, 136, 880, 1962.
762. **Ota, T.,** *J. Jpn. Soc. Hortic. Sci.*, 31, 329, 1962.
763. **Ghosh, M. S. and Bose, T. K.,** *Phyton (Buenos Aires)*, 27, 131, 1970.
764. **Halevy, A. H. and Rudich, J.,** *Physiol. Plant.*, 20, 1052, 1967.
765. **Khryanin, V. N. and Chailakhyan, M. Kh.,** *Fiziol. Rast.*, 26, 455, 1979.
766. **Rodriquez, B. P. and Lambeth, V. N.,** *J. Am. Soc. Hortic. Sci.*, 97, 90, 1972.
767. **Phillips, I. D. J.,** *J. Exp. Bot.*, 13, 213, 1962.
768. **van der Veen, R.,** *Acta Bot. Neerl.*, 17, 373, 1968.
769. **Stuart, N. W. and Cathey, H. M.,** *Proc. 15th Int. Hortic. Congr.*, Vol. 2, Nice, 1962, 391.
770. **Boodley, J. W. and Mastalerz, J. W.,** *Proc. Am. Soc. Hortic. Sci.*, 74, 681, 1959.
771. **Larson, R. A. and Sydnor, T. D.,** *J. Am. Soc. Hortic. Sci.*, 96, 786, 1971.
772. **Rudnicki, R. M., Nowak, J., and Saniewski, M.,** *Sci. Hortic.*, 4, 387, 1976.
773. **Shoub, J. and de Hertogh, A. A.,** *Sci. Hortic.*, 2, 55, 1974.
774. **Brian, P. W., Petty, J. H. P., and Richmond, P. T.,** *Nature (London)*, 184, 69, 1959.
775. **Corgan, J. N. and Widmoyer, F. B.,** *J. Am. Soc. Hortic. Sci.*, 96, 54, 1971.
776. **Jeffcoat, B.,** *J. Hortic. Sci.*, 52, 143, 1977.
777. **Bentley, B., Morgan, C. B., Morgan, D. G., and Saad, F. A.,** *Nature (London)*, 256, 121, 1975.
778. **Bennink, G. J. H.,** *Plant Growth Substances 1973*, Hirokawa, Tokyo, 1974, 974.
779. **Bennink, G. J. H. and de Vries, J. W. A.,** *Abstr. 2nd Symp. Plant Growth Regulators*, Sofia, 1975, 66.
780. **Goh, C. J.,** *New Phytol.*, 82, 375, 1979.
781. **Pharis, R. P.,** *Planta*, 105, 205, 1972.
782. **Kunneman, B. P. A. M.,** personal communication, 1980.
783. **Srinivasan, C. and Mullins, M. G.,** *Plant Physiol.*, 61, 127, 1978.
784. **Srinivasan, C. and Mullins, M. G.,** *Planta*, 145, 187, 1979.
785. **McHughen, A.,** *Can. J. Bot.*, 60, 845, 1982.
786. **Bagnard, C.,** unpublished data, 1982.
787. **Negi, S. S. and Olmo, H. P.,** *Science*, 152, 1624, 1966.
788. **Negi, S. S. and Olmo, H. P.,** *Vitis*, 10, 1, 1971.
789. **Iizuka, M. and Hashizume, T.,** *Jpn. J. Genet.*, 43, 393, 1968.
790. **Moore, J. N.,** *J. Am. Soc. Hortic. Sci.*, 95, 387, 1970.
791. **Durand, B.,** *C.R. Acad. Sci.*, 263, 1309, 1966.
792. **Durand, B.,** *C.R. Acad. Sci.*, 268, 2049, 1969.
793. **Takahashi, H., Suge, H., and Saito, T.,** *Plant Cell Physiol.*, 21, 525, 1980.
794. **Chailakhyan, M. Kh. and Khryanin, V. N.,** *Naturwissenschaften*, 67, 94, 1980.
795. **van Nigtevecht, G.,** *Genen Phaenen*, 13, 12, 1969.
796. **Doazan, J. P. and Cuellar, V.,** *Ann. Amélior. Plantes*, 20, 79, 1970.
797. **Erez, A., Lavee, S., and Samish, R. M.,** *J. Am. Soc. Hortic. Sci.*, 96, 519, 1971.
798. **Broome, O. C. and Zimmerman, R. H.,** *J. Am. Soc. Hortic. Sci.*, 101, 28, 1976.
799. **Angrish, R. and Nanda, K. K.,** *Z. Pflanzenphysiol.*, 106, 263, 1982.
800. **Lona, F.,** *Ateneo Parmense Acta Nat.*, 11, 537, 1975.
801. **Furuta, T. and Straiton, T. H., Jr.,** *Proc. Am. Soc. Hortic. Sci.*, 88, 591, 1966.
802. **Morgan, D. G.,** *Aspects and Prospects of Plant Growth Regulators*, BPGRG Monogr. No. 6, 1981, 151.
803. **Rudich, J. and Halevy, A. H.,** *Plant Cell Physiol.*, 15, 635, 1974.
804. **Mohan Ram, H. Y. and Jaiswal, V. S.,** *Planta*, 105, 263, 1972.
805. **Galoch, E.,** *Acta Physiol. Plant.*, 2, 3, 1980.
806. **Abdel-Gawad, H. A. and Ketellapper, H. J.,** *Plant Physiol.*, 44 (Suppl.), 15, 1969.
807. **Atsmon, D. and Galun, E.,** *Phytomorphology*, 10, 110, 1960.
808. **Rehm, S.,** *Nature (London)*, 170, 38, 1952.
809. **Sankhla, N.,** *Z. Pflanzenphysiol.*, 61, 350, 1969.
810. **Bisaria, A. K. and Bisaria, A. K., Jr.,** *J. Exp. Bot.*, 27, 337, 1976.
811. **Jain, V. K. and Mukherjee, D.,** *Phyton (Buenos Aires)*, 38, 89, 1980.
812. **Mohan Ram, H. Y. and Jaiswal, V. S.,** *Naturwissenschaften*, 58, 149, 1971.
813. **Robinson, R. W., Cantliffe, D. J., and Shannon, S.,** *Science*, 171, 1251, 1971.

814. **Bisaria, A. K.,** *Isr. J. Bot.*, 26, 209, 1977.
815. **Krishnamoorthy, H. N.,** *Z. Pflanzenphysiol.*, 65, 88, 1971.
816. **Bisaria, A. K.,** *Can. J. Bot.*, 55, 752, 1977.
817. **Naylor, A. W.,** *Proc. Natl. Acad. Sci. U.S.A.*, 36, 230, 1950.
818. **Löve, A. and Löve, D.,** *Ark. Bot.*, 32A, 1, 1945.
819. **Kopcewicz, J.,** *Z. Pflanzenphysiol.*, 65, 92, 1971.
820. **Gawienowski, A. M., Cheney, R. W., Jr., and Marsh, H. V.,** *Phytochemistry*, 10, 2033, 1971.
821. **Baszynski, T.,** *Naturwissenschaften*, 57, 48, 1970.
822. **Sironval, C.,** *C.R. Rech. IRSIA*, 18, 1, 1957.
823. **Trewavas, A. J.,** *Plant Cell Environ.*, 4, 203, 1981.
824. **Katunskij, V. M.,** *C.R. Dokl. Acad. Sci. URSS*, 3, 343, 1936.
825. **Borris, H. and Bussmann, K.,** *Jahrb. Wiss. Bot.*, 88, 519, 1939.
826. **Edelbluth, E. and Kaldewey, H.,** *Planta*, 131, 285, 1976.
827. **Cooke, A. R.,** *Plant Physiol.*, 29, 440, 1954.
828. **Gifford, E. M., Jr. and Nitsch, C.,** *Cellular and Molecular Aspects of Floral Induction*, Bernier, G., Ed., Longman, London, 1970, 304.
829. **Konishi, M.,** *Mem. Coll. Agric. Kyoto Univ.*, 75, 1, 1956.
830. **Bronchart, R.,** *Bull. Soc. R. Sci. Liège*, 26, 241, 1957.
831. **Gilson, M.,** *Bull. Cl. Sci. Acad. R. Belg.*, 43, 778, 1957.
832. **Teltscherová, L., Pavlová, L., and Pleskotová, D.,** *Biol. Plant.*, 19, 205, 1977.
833. **Kopcewicz, J., Centkowska, G., Kriesel, K., and Zatorska, Z.,** *Acta Soc. Bot. Pol.*, 48, 255, 1979.
834. **Zieslin, N. and Halevy, A. H.,** *Physiol. Plant.*, 37, 317, 1976.
835. **Moe, R.,** *Physiol. Plant.*, 24, 374, 1971.
836. **Conrad, K. and Mothes, K.,** *Naturwissenschaften*, 48, 26, 1961.
837. **Galoch, E.,** *Acta Physiol. Plant.*, 2, 31, 1980.
838. **Ćulafić, Lj. and Nešković, M.,** *Ark. Biol. Nauka Beograd*, 26, 119, 1974.
839. **Galun, E., Izhar, S., and Atsmon, D.,** *Plant Physiol.*, 40, 321, 1965.
840. **Sladký, Z.,** *Biol. Plant.*, 11, 208, 1969.
841. **Chromiński, A. and Kopcewicz, J.,** *Z. Pflanzenphysiol.*, 68, 184, 1972.
842. **Zieslin, N. and Halevy, A. H.,** *Physiol. Plant.*, 37, 331, 1976.
843. **Prince, T. A., Herner, R. C., and de Hertogh, A. A.,** *Sci. Hortic.*, 16, 77, 1982.
844. **Moe, R. and Wickstrøm, A.,** *Physiol. Plant.*, 28, 81, 1973.
845. **Moe, R., de Hertogh, A. A., and Dilley, D. R.,** *Meld. Norg. Landbrukshoegsk.*, 57(46), 1, 1978.
846. **Rudich, J., Halevy, A. H., and Kedar, N.,** *Plant Physiol.*, 49, 998, 1972.
847. **Fujita, Y. and Fujieda, K.,** *Plant Cell Physiol.*, 22, 667, 1981.
848. **Gregg, K. B.,** *Bot. Gaz. (Chicago)*, 143, 466, 1982.
849. **Sircar, P. K., Dey, B., Sanyal, T., Ganguly, S. N., and Sircar, S. M.,** *Phytochemistry*, 9, 735, 1970.
850. **Murakami, Y.,** *Plant Cell Physiol.*, 14, 91, 1973.
851. **Dathe, W. and Sembdner, G.,** *Biochem. Physiol. Pflanz.*, 175, 599, 1980.
852. **Zieslin, N., Madori, G., and Halevy, A. H.,** *J. Exp. Bot.*, 30, 15, 1979.
853. **Kaufman, P. B., Ghosheh, N. S., Nakosteen, L., Pharis, R. P., Durley, R. C., and Morf, W.,** *Plant Physiol.*, 58, 131, 1976.
854. **Parups, E. V.,** *Phyton (Buenos Aires)*, 39, 121, 1980.
855. **Suzuki, Y., Kurogochi, S., Murofushi, N., Ota, Y., and Takahashi, N.,** *Plant Cell Physiol.*, 22, 1085, 1981.
856. **Léonard, M. and Kinet, J. M.,** *Ann. Bot. (London)*, 50, 127, 1982.
857. **Raman, K. and Greyson, R. I.,** *Can. J. Bot.*, 55, 2115, 1977.
858. **Raman, K. and Greyson, R. I.,** *J. Exp. Bot.*, 29, 141, 1978.
859. **Saito, T. and Ito, H.,** *Tohoku J. Agric. Res.*, 14, 227, 1963.
860. **Atsmon, D., Lang, A., and Light, E. N.,** *Plant Physiol.*, 43, 806, 1968.
861. **Hayashi, F., Boerner, D. R., Peterson, C. E., and Sell, H. M.,** *Phytochemistry*, 10, 57, 1971.
862. **Hemphill, D. D., Jr., Baker, L. R., and Sell, H. M.,** *Euphytica*, 21, 285, 1972.
863. **Friedlander, M., Atsmon, D., and Galun, E.,** *Plant Cell Physiol.*, 18, 681, 1977.
864. **Kamieńska, A.,** *Roczn. Nauk Roln.*, 93A, 177, 1967.
865. **Sawhney, V. K.,** *J. Exp. Bot.*, 25, 1004, 1974.
866. **Leshem, Y. and Ophir, D.,** *Ann. Bot. (London)*, 41, 375, 1977.
867. **Neuffer, M. G., Jones, L., and Zuber, M. S.,** *The Mutants of Maize*, Crop Science Society of America, Madison, Wis., 1968.
868. **Phinney, B. O.,** *Plant Growth Regulation*, Klein, R. M., Ed., Iowa State University Press, Ames, 1961, 489.
869. **Browning, G.,** *J. Hortic. Sci.*, 48, 29, 1973.

870. **Sydnor, T. D. and Larson, R. A.,** *J. Am. Soc. Hortic. Sci.*, 100, 353, 1975.
871. **Aung, L. H. and de Hertogh, A. A.,** *Plant Cell Physiol.*, 8, 201, 1967.
872. **Aung, L. H. and de Hertogh, A. A.,** *Biochemistry and Physiology of Plant Growth Substances*, Wightman, F. and Setterfield, G., Eds., The Runge Press, Ottawa, Canada, 1968, 943.
873. **van Bragt, J.,** *Meded. Fac. Landbouwwetensch. Gent*, 36, 1301, 1971.
874. **Einert, A. E., Staby, G. L., and de Hertogh, A. A.,** *Can. J. Bot.*, 50, 909, 1972.
875. **Rakhimbaev, I. R., Syrtanova, G. A., and Solomina, V. F.,** *Fiziol. Rast.*, 25, 249, 1978.
876. **Aung, L. H., de Hertogh, A. A., and Staby, G.,** *Plant Physiol.*, 44, 403, 1969.
877. **Rodrigues Pereira, A. S.,** *Acta Bot. Neerl.*, 13, 302, 1964.
878. **Staby, G. L., de Hertogh, A. A., and Alpi, A.,** *J. Am. Soc. Hortic. Sci.*, 97, 189, 1972.
879. **van Staden, J., Zieslin, N., Spiegelstein, H., and Halevy, A. H.,** *Ann. Bot. (London)*, 47, 155, 1981.
880. **Davey, J. E. and van Staden, J.,** *Planta*, 130, 69, 1976.
881. **Sitton, D., Itai, C., and Kende, H.,** *Planta*, 73, 296, 1967.
882. **Davey, J. E. and van Staden, J.,** *Physiol. Plant.*, 43, 77, 1978.
883. **Beever, J. E. and Woolhouse, H. W.,** *Bull. R. Soc. N.Z.*, 12, 681, 1974.
884. **van Staden, J. and Dimalla, G. G.,** *Bot. Gaz. (Chicago)*, 141, 248, 1980.
885. **Vonk, C. R.,** *Acta Bot. Neerl.*, 23, 541, 1974.
886. **Vonk, C. R.,** *Physiol. Plant.*, 46, 235, 1979.
887. **Vonk, C. R. and Davelaar, E.,** *Physiol. Plant.*, 52, 101, 1981.
888. **Vonk, C. R.,** *Acta Bot. Neerl.*, 25, 153, 1976.
889. **Vonk, C. R.,** *Physiol. Plant.*, 44, 161, 1978.
890. **Even-Chen, Z., Sachs, R. M., and Hackett, W. P.,** *Plant Physiol.*, 64, 646, 1979.
891. **Borkowska, B. and Borkowski, J.,** *Bull. Acad. Pol. Sci.*, 23, 291, 1975.
892. **Galoch, E.,** *Acta Physiol. Plant.*, 2, 233, 1980.
893. **Kahlem, G., Champault, A., Louis, J. P., Bazin, M., Chabin, A., Delaigue, M., Dauphin, B., Durand, R., and Durand, B.,** *Physiol. Vég.*, 13, 763, 1975.
894. **Louis, J. P. and Durand, B.,** *Molec. Gen. Genet.*, 165, 309, 1978.
895. **Dauphin-Guerin, B., Teller, G., and Durand, B.,** *Planta*, 148, 124, 1980.
896. **Browning, G.,** *J. Hortic. Sci.*, 48, 297, 1973.
897. **Corgan, J. N. and Peyton, C.,** *J. Am. Soc. Hortic. Sci.*, 95, 770, 1970.
898. **Damptey, H. B., Coombe, B. G., and Aspinall, D.,** *Ann. Bot. (London)*, 42, 1447, 1978.
899. **Bowen, H. H. and Derickson, G. W.,** *HortScience*, 13, 694, 1978.
900. **Mielke, E. A. and Dennis, F. G., Jr.,** *J. Am. Soc. Hortic. Sci.*, 103, 446, 1978.
901. **Knights, B. A. and Smith, A. R.,** *Planta*, 133, 89, 1976.
902. **Knights, B. A. and Smith, A. R.,** *Planta*, 134, 115, 1977.
903. **Kopcewicz, J. and Chromiński, A.,** *Experientia*, 28, 603, 1972.
904. **Weaver, R. J. and McCune, S. B.,** *Hilgardia*, 29, 247, 1959.
905. **Sachs, R. M.,** *HortScience*, 12, 220, 1977.
906. **Evans, L. T. and Rawson, H. M.,** *Aust. J. Biol. Sci.*, 23, 245, 1970.
907. **Dennis, F. G., Jr.,** *Hortic. Rev.*, 1, 395, 1979.
908. **Williams, M. W.,** *Strategies of Plant Reproduction*, Meudt, W. J., Ed., BARC Symp. No. 6, Allanheld-Osmun, Totowa, 1982, 273.
909. **Winkler, A. J.,** *Hilgardia*, 4, 153, 1929.
910. **Fournioux, J. C. and Bessis, R.,** *Bull. Soc. Bot. Fr. Lett. Bot.*, 127, 23, 1980.
911. **van Meeteren, U.,** *Acta Hortic.*, 128, 37, 1981.
912. **Krenzer, E. G., Jr. and Moss, D. N.,** *Crop Sci.*, 15, 71, 1975.
913. **Gifford, R. M., Bremner, P. M., and Jones, D. B.,** *Aust. J. Agric. Res.*, 24, 297, 1973.
914. **Hughes, A. P. and Cockshull, K. E.,** *Ann. Bot. (London)*, 35, 899, 1971.
915. **Mattson, R. H. and Widmer, R. E.,** *J. Am. Soc. Hortic. Sci.*, 96, 484, 1971.
916. **Calvert, A. and Slack, G.,** *J. Hortic. Sci.*, 50, 61, 1975.
917. **Hughes, A. P. and Cockshull, K. E.,** *Ann. Bot. (London)*, 36, 533, 1972.
918. **Hand, D. W., Acock, B., and Cockshull, K. E.,** *Rep. Glasshouse Crops Res. Inst. 1973*, 51, 1974.
919. **Sawhney, R. and Cumming, B. G.,** *Can. J. Bot.*, 49, 2233, 1971.
920. **Cumming, B. G.,** *Can. J. Bot.*, 47, 1241, 1969.
921. **Ryle, G. J. A. and Powell, C. E.,** *Ann. Bot. (London)*, 36, 363, 1975.
922. **Hale, C. R. and Weaver, R. J.,** *Hilgardia*, 33, 89, 1962.
923. **Hussey, G.,** *J. Exp. Bot.*, 14, 326, 1963.
924. **Rodrigues Pereira, A. S.,** *Acta Bot. Neerl.*, 11, 97, 1962.
925. **Ho, L. C. and Rees, A. R.,** *New Phytol.*, 74, 421, 1975.
926. **Ho, L. C. and Rees, A. R.,** *New Phytol.*, 76, 59, 1976.
927. **Ho, L. C. and Rees, A. R.,** *New Phytol.*, 78, 65, 1977.

928. **Aung, L. H., Wright, R. D., and de Hertogh, A. A.,** *HortScience*, 11, 37, 1976.
929. **Ginzburg, C.,** *J. Exp. Bot.*, 25, 995, 1974.
930. **Robinson, M., Harav, I., Halevy, A. H., and Plaut, Z.,** *Ann. Bot. (London)*, 45, 113, 1980.
931. **Motomura, Y.,** *Tohoku J. Agric. Res.*, 33, 1, 1982.
932. **Shindy, W. W., Kliewer, W. M., and Weaver, R. J.,** *Plant Physiol.*, 51, 345, 1973.
933. **Tjosvold, S. A.,** Cytokinin and Gibberellin Induced Flowering in Relationship to Assimilate Transport in *Chrysanthemum morifolium* under Non-Inductive Conditions, M.S. thesis, University of California, Davis, 1981.
934. **Léonard, M.,** Etude du Rôle des Cytokinines, des Gibbérellines et de la Distribution des [^{14}C] Assimilats dans le Développement de l'Inflorescence de la Tomate (*Lycopersicon esculentum* Mill.), Ph.D. thesis, University of Liège, Belgium, 1981.
935. **Jacobs, W. P.,** *Plant Hormones and Plant Development*, Cambridge University Press, London, 1979.
936. **Harris, G. P. and Jeffcoat, B.,** *Ann. Bot. (London)*, 38, 77, 1974.
937. **Mohapatra, P. K., Aspinall, D., and Jenner, C. F.,** *Ann. Bot. (London)*, 49, 619, 1982.
938. **Russell, C. R. and Morris, D. A.,** *Ann. Bot. (London)*, 49, 89, 1982.
939. **Fawzi, A. F. A. and El-Fouly, M. M.,** *Physiol. Plant.*, 47, 245, 1979.
940. **Herold, A.,** *New Phytol.*, 86, 131, 1980.
941. **Pierard, D., Jacqmard, A., Bernier, G., and Salmon, J.,** *Planta*, 150, 397, 1980.
942. **Graham, C. F. and Wareing, P. F.,** *The Developmental Biology of Plants and Animals*, Blackwell Scientific, Oxford, 1976.
943. **Wardlaw, C. W.,** *Proc. R. Soc. Edinburgh Sect. B*, 66, 394, 1957.
943a. **Outlaw, W. H., Jr.,** *Annu. Rev. Plant Physiol.*, 31, 299, 1980.
944. **Kende, H. and Gardner, G.,** *Annu. Rev. Plant Physiol.*, 27, 267, 1976.
945. **Rubery, P. H.,** *Annu. Rev. Plant Physiol.*, 32, 569, 1981.
946. **Honma, S., Wittwer, S. H., and Phatak, S. C.,** *J. Hered.*, 54, 212, 1963.
947. **Vergara, B. S., Chang, T. T., and Lilis, R.,** *Int. Rice Res. Inst. Tech. Bull.*, 8, 1, 1969.
948. **Poehlman, J. M.,** *Breeding Field Crops*, Henry Holt & Co., New York, 1959.
949. **Major, D. J., Johnson, D. R., and Luedders, V. D.,** *Crop Sci.*, 15, 172, 1975.
950. **Major, D. J., Johnson, D. R., Tanner, J. W., and Anderson, I. C.,** *Crop Sci.*, 15, 174, 1975.
951. **Caddell, J. L. and Weibel, D. E.,** *Agron. J.*, 64, 473, 1972.
952. **Troyer, A. F.,** *Proc. 33rd Annu. Corn Sorghum Res. Conf.*, 1978, 1.
953. **Aldrich, S. R., Scott, W. O., and Leng, E. R.,** *Modern Corn Production*, 2nd ed., A & L Publications, Champaign, Ill., 1975.
954. **Mederski, H. J., Miller, M. E., and Weaver, C. R.,** *Agron. J.*, 65, 743, 1973.
955. **Proebsting, E. L., Jr. and Mills, H. H.,** *J. Am. Soc. Hortic. Sci.*, 103, 192, 1978.
956. **Proebsting, E. L., Jr. and Mills, H. H.,** *Proc. Am. Soc. Hortic. Sci.*, 78, 104, 1961.
957. **Monet, R. and Bastard, Y.,** *Ann. Amélior. Plantes*, 27, 717, 1977.
958. **Anderson, J. L., Ashcroft, G. L., Richardson, E. A., Alfaro, J. F., Griffin, R. E., Hanson, G. R., and Keller, J.,** *J. Am. Soc. Hortic. Sci.*, 100, 229, 1975.
959. **Bauer, M., Chaplin, C. E., Schneider, G. W., Barfield, B. J., and White, G. M.,** *J. Am. Soc. Hortic. Sci.*, 101, 452, 1976.
960. **Ashcroft, G. L., Richardson, E. A., and Seeley, S. D.,** *HortScience*, 12, 347, 1977.
961. **Richardson, E. A., Seeley, S. D., Walker, D. R., Anderson, J. L., and Ashcroft, G. L.,** *HortScience*, 10, 236, 1975.
962. **Aron, R. H., Sanders, C. G., Richardson, E. A., Seeley, S. D., and Walker, D. R.,** *HortScience*, 10, 559, 1975.
963. **Buchanan, D. W., Bartholic, J. F., and Biggs, R. H.,** *J. Am. Soc. Hortic. Sci.*, 102, 466, 1977.
964. **Hamer, P. J. C.,** *J. Hortic. Sci.*, 55, 157, 1980.
965. **Nuttonson, M. Y.,** *Vernalization and Photoperiodism*, Murneek, A. E. and Whyte, R. O., Eds., Chronica Botanica, Waltham, Mass., 1948, 129.
966. **Aitken, Y.,** *Flowering Time, Climate and Genotype*, Melbourne University Press, 1974.
967. **Pugsley, A. T.,** *Aust. J. Agric. Res.*, 22, 21, 1971.
968. **Klaimi, Y. Y. and Qualset, C. O.,** *Genetics*, 74, 139, 1973.
969. **Klaimi, Y. Y. and Qualset, C. O.,** *Genetics*, 76, 119, 1974.
970. **Pugsley, A. T.,** *Proc. 3rd Int. Genet. Symp.*, Canberra, 1968, 288.
971. **Qualset, C. O. and Vogt, H. E.,** *Proc. 3rd Int. Wheat Conf.*, Madrid, Spain, 1980, 166.
972. **Levy, J. and Peterson, M. L.,** *Crop Sci.*, 12, 487, 1972.
973. **Ormrod, D. P., Bunter, W. A., Jr., Finfrock, D. C., and Thysell, J. R.,** *Calif. Agric.*, 6-8 May, 1960.
974. **Fowells, H. A.,** *US Forest Service Agric. Handbook*, 271, 1965.
975. **Cain, J. C.,** *J. Am. Soc. Hortic. Sci.*, 96, 664, 1971.
976. **Jackson, D. I.,** *Aust. J. Biol. Sci.*, 22, 69, 1969.

977. **Buttrose, M. S.,** *Hortic. Abstr.*, 44, 319, 1974.
978. **Jackson, J. E.,** *Hortic. Rev.*, 2, 208, 1980.
979. **Bunt, A. C. and Powell, M. C.,** *Acta Hortic.*, 141, 1983.
980. **Cathey, H. M. and Campbell, L. E.,** *Hortic. Rev.*, 2, 491, 1980.
981. **Chapman, G. P. and Sadjadi, A. S.,** *Z. Pflanzenphysiol.*, 104, 265, 1981.
982. **Sachs, R. M., Kofranek, A. M., and Kubota, J.,** *HortScience*, 15, 609, 1980.
983. **Cathey, H. M. and Borthwick, H. A.,** *Proc. Am. Soc. Hortic. Sci.*, 78, 545, 1961.
984. **Kofranek, A. M.,** *Florist's Rev.*, 133, 23, 1963.
985. **Mastalerz, J. W.,** *The Greenhouse Environment*, John Wiley & Sons, New York, 1977.
986. **Kofranek, A. M. and Halevy, A. H.,** *HortScience*, 9, 543, 1974.
987. **Accati-Garibaldi, E., Kofranek, A. M., and Sachs, R. M.,** *Acta Hortic.*, 68, 51, 1977.
988. **Canham, A. E., Cockshull, K. E., and Hand, D. W.,** *Acta Hortic.*, 68, 63, 1977.
989. **Paulet, P.,** *Physiol. Vég.*, 17, 631, 1979.
990. **Paleg, L. G. and Aspinall, D.,** *Acta Hortic.*, 68, 69, 1977.
991. **Gressel, J., Zilberstein, A., Porath, D., and Arzee, T.,** *Photoreceptors and Plant Development*, de Greef, J., Ed., Antwerpen University Press, Belgium, 1980, 525.
992. **Cockshull, K. E. and Horridge, J. S.,** *J. Hortic. Sci.*, 53, 85, 1978.
993. **Cockshull, K. E., Horridge, J. S., and Langton, F. A.,** *J. Hortic. Sci.*, 54, 337, 1979.
994. **Criswell, J. G. and Hume, D. J.,** *Crop Sci.*, 12, 657, 1972.
995. **Polson, D. E.,** *Crop Sci.*, 12, 773, 1972.
996. **Johnson, H. W. and Bernard, R. L.,** *The Soybean*, Norman, A. G., Ed., Academic Press, New York, 1963, 1.
997. **Longden, P. C.,** *Aspects and Prospects of Plant Growth Regulators*, BPGRG Monogr. No. 6, 1981, 123.
998. **Whyte, R. O.,** *Crop Production and Environment*, Faber & Faber, London, 1946.
999. **Sachs, M. and Rylski, I.,** *Sci. Hortic.*, 12, 231, 1980.
1000. **Pressman, E. and Negbi, M.,** *J. Exp. Bot.*, 32, 821, 1981.
1001. **Post, K. and Bell, R. S.,** *Proc. Am. Soc. Hortic. Sci.*, 34, 630, 1936.
1002. **Kohl, H. C., Jr.,** *Proc. Am. Soc. Hortic. Sci.*, 72, 481, 1958.
1003. **Halevy, A. H.,** personal communication, 1983.
1004. **Shaw, R. H. and Laing, D. R.,** *Plant Environment and Efficient Water Use*, Pierre, W. H., Kirkham, D., Pesek, J., and Shaw, R. H., Eds., Am. Soc. Agron., Madison, Wis., 1966, 73.
1005. **Monselise, S. P. and Halevy, A. H.,** *Proc. Am. Soc. Hortic. Sci.*, 84, 141, 1964.
1006. **Nir, I., Goren, R., and Leshem, B.,** *J. Am. Soc. Hortic. Sci.*, 97, 774, 1972.
1007. **Sachs, R. M.,** *Accad. Econ-Agrar. Georgofili Firenze*, 22 (Ser. 7), 1, 1976.
1008. **Carbonneau, A. and Casteran, P.,** *Am. J. Enol. Vitic.*, 30, 3, 1979.
1009. **Grierson, W., Soule, J., and Kawada, K.,** *Hortic. Rev.*, 4, 247, 1982.
1010. **Abbott, C. E.,** *Am. J. Bot.*, 22, 476, 1935.
1011. **Atherton, J. G. and Othman, S.,** *Abstr. 21st Int. Hortic. Congr.*, Vol. 1, Hamburg, FRG, 1982, 1440.
1012. **Allsopp, A.,** *Encyclopedia of Plant Physiology*, Vol. 15 (Part 1), Ruhland, W., Ed., Springer-Verlag, Berlin, 1965, 504.
1013. **Paleg, L. G. and Aspinall, D.,** *The Physiology and Biochemistry of Drought Resistance in Plants*, Academic Press, New York, 1981.
1014. **Monselise, S. P. and Goren, R.,** *Proc. 1st Int. Citrus Symp.*, 3, 1969, 1105.
1015. **Hackett, W. P., Sachs, R. M., and De Bie, J.,** *Florist's Rev.*, 150, 56, 1972.
1016. **Grasmanis, V. O. and Leeper, G. W.,** *Aust. J. Biol. Sci.*, 20, 761, 1967.
1017. **Grasmanis, V. O. and Edwards, G. R.,** *Aust. J. Plant Physiol.*, 1, 99, 1974.
1018. **Hoagland, D. R. and Arnon, D. I.,** *Calif. Agric. Exp. Stn. Circ.*, 347, 4, 1950.
1019. **Visser, T. and de Vries, D. P.,** *Euphytica*, 19, 141, 1970.
1020. **Visser, T. and de Vries, D. P.,** *Euphytica*, 19, 293, 1970.
1021. **Visser, T.,** *J. Am. Soc. Hortic. Sci.*, 98, 26, 1973.
1022. **Zimmerman, R. H.,** *Acta Hortic.*, 56, 219, 1976.
1023. **Chacko, E. K., Kohli, R. R., Dore Swamy, R., and Randhawa, G. S.,** *Physiol. Plant.*, 32, 188, 1974.
1024. **Kender, W. J.,** *HortScience*, 9, 444, 1974.
1025. **Aldwinckle, H. S.,** *HortScience*, 10, 124, 1975.
1026. **Mika, A.,** *Acta Hortic.*, 56, 195, 1976.
1027. **Zimmerman, R. H.,** *Acta Hortic.*, 34, 139, 1973.
1028. **Schwabe, W. W.,** *Acta Hortic.*, 56, 45, 1976.
1029. **Rogler, C. E. and Hackett, W. P.,** *Physiol. Plant.*, 34, 141, 1975.
1030. **Libby, W. J. and Hood, J. V.,** *Acta Hortic.*, 56, 91, 1976.
1031. **Franclet, A.,** Proc. Colloq. Int. Culture "In Vitro" des Essences Forestières, Fontainebleau, France, 1981, 55.

1032. **Franclet, A.,** *Ann. AFOCEL 1980*, 11, 1981.
1033. **de la Goublaye de Nantois, H.,** Vieillissement et Rajeunissement chez le *Sequoia sempervirens* Endlicher en Relation avec la Propagation Végétative, 3rd Cycle thesis, University of Paris 6, France, 1981.
1034. **de la Goublaye de Nantois, T.,** Rajeunissement chez le Douglas *(Pseudotsuga menziesii)* en vue de la Propagation Végétative. Etude sur la Plagiotropie des Parties Aériennes et Racinaires, DEA University of Paris 6, France, 1980.
1035. **Paton, D. M., Willing, R. R., and Pryor, L. D.,** *Ann. Bot. (London)*, 47, 835, 1981.
1036. **Davis, L. D.,** *Proc. Am. Soc. Hortic. Sci.*, 70, 545, 1957.
1037. **Jonkers, H.,** *Sci. Hortic.*, 11, 303, 1979.
1038. **Monselise, S. P. and Goldschmidt, E. E.,** *Hortic. Rev.*, 4, 128, 1982.
1039. **Jones, W. W., Embleton, T. W., Steinacker, M. L., and Cree, C. B.,** *J. Am. Soc. Hortic. Sci.*, 95, 380, 1970.
1040. **Crane, J. C. and Nelson, M. M.,** *HortScience*, 6, 489, 1971.
1041. **Grochowska, M. J.,** *J. Hortic. Sci.*, 48, 347, 1973.
1042. **Chan, B. G. and Cain, J. C.,** *Proc. Am. Soc. Hortic. Sci.*, 91, 63, 1967.
1043. **Gil, G. F., Martin, G. C., and Griggs, W. H.,** *J. Am. Soc. Hortic. Sci.*, 97, 731, 1972.
1044. **Gil, G. F., Martin, G. C., and Griggs, W. H.,** *J. Am. Soc. Hortic. Sci.*, 98, 51, 1973.
1045. **Hoad, G. V.,** *Acta Hortic.*, 80, 93, 1978.
1046. **Weaver, R. J.,** *Plant Growth Substances in Agriculture*, W. H. Freeman Co., San Francisco, 1972.
1047. **Stuart, N. W.,** *Science*, 134, 50, 1961.
1048. **Cathey, H. M.,** *Proc. Am. Soc. Hortic. Sci.*, 86, 753, 1965.
1049. **Stuart, N. W.,** *Proc. 16th Int. Hortic. Congr.*, 5, 58, 1962.
1050. **White, J. W.,** *J. Am. Soc. Hortic. Sci.*, 95, 546, 1970.
1051. **Clark, H. E. and Kerns, K. R.,** *Science*, 95, 536, 1942.
1052. **Gowing, D. P.,** *Am. J. Bot.*, 43, 411, 1956.
1053. **Cathey, H. M. and Taylor, R. L.,** *Florist's Rev.*, 146, 38, 82, 1970.
1054. **Harrington, J. F.,** *Proc. Am. Soc. Hortic. Sci.*, 75, 476, 1960.
1055. **Snyder, M. J., Welch, N. C., and Rubatsky, V. E.,** *Abstr. West Reg. Meet. Am. Soc. Hortic. Sci.*, Berkeley, Calif., 1970.
1056. **Galston, A. W.,** *Am. J. Bot.*, 34, 356, 1947.
1057. **Ghorashy, S. R., Colville, W. L., and Ashworth, D. L.,** *Crop Sci.*, 9, 399, 1969.
1058. **Furr, J. R., Cooper, W. C., and Reece, P. C.,** *Am. J. Bot.*, 34, 1, 1947.
1059. **Sax, K.,** *J. Arnold Arboret.*, 35, 251, 1954.
1060. **Sax, K. and Johnson, A. G.,** *J. Arnold Arboret.*, 36, 110, 1955.
1061. **Knight, T. A.,** *Trans. Hortic. Soc. Lond.*, 4, 159, 1820.
1062. **Wareing, P. F.,** *Physiology of Tree Crops*, Luckwill, L. C. and Cutting, C. V., Eds., Academic Press, London, 1970, 1.
1063. **Zieslin, N. and Halevy, A. H.,** *Bot. Gaz. (Chicago)*, 137, 291, 1976.
1064. **Zieslin, N., Haaze, H., and Halevy, A. H.,** *Bot. Gaz. (Chicago)*, 137, 297, 1976.
1065. **Zieslin, N. and Mor, Y.,** *Sci. Hortic.*, 15, 67, 1981.
1066. **Woolhouse, H. W.,** *Strategies of Plant Reproduction*, Meudt, W. J., Ed., BARC Symp. Publ. No. 6, Allenheld-Osmun, Totowa, 1982, 201.
1067. **Vince-Prue, D., Thomas, B., and Cockshull, K., Eds.,** *Light and the Flowering Process*, Academic Press, London, 1984.
1068. **Durand, R. and Durand, B.,** *Physiol. Plant.*, 60, 267, 1984.
1069. **Rosenberg, S. M. and Bonnett, H. T.,** *Am. J. Bot.*, 70, 266, 1983.
1070. **Wetzstein, H. Y. and Armitage, A.,** *J. Am. Soc., Hortic. Sci.*, 108, 595, 1983.
1071. **Cheng, P. C., Greyson, R. I., and Walden, D. B.,** *Am. J. Bot.*, 70, 450, 1983.
1072. **Minter, T. C. and Lord, E. M.,** *Am. J. Bot.*, 70, 1499, 1983.
1073. **Mayers, A. M. and Lord, E. M.,** *Am. J. Bot.*, 70, 1556, 1983.
1074. **Moore, R. and Walker, D. B.,** *Protoplasma*, 115, 114, 1983.
1075. **Green, P. B.,** personal communication, 1984.
1076. **Steer, B. T. and Hocking, P. J.,** *Ann. Bot. (London)*, 52, 267, 1983.
1077. **Palmer, J. H. and Marc, J.,** *Plant Cell Physiol.*, 23, 1401, 1982.
1078. **Lower, R. L., Smith, O. S., and Ghaderi, A.,** *HortScience*, 18, 737, 1983.
1079. **Onyckwely, S. S. and Harper, J. L.,** *Nature (London)*, 282, 609, 1979.
1080. **Freeman, D. C. and Harper, K. T.,** *J. Range Manage.*, 33, 236, 1980.
1081. **Williams, C. N. and Thomas, R. L.,** *Ann. Bot. (London)*, 34, 957, 1970.
1082. **Lepage, I., De Jong, J., and Smeets, L.,** *Scientia Hortic.*, 22, 373, 1984.
1083. **Mohapatra, P. K., Aspinall, D., and Jenner, C. F.,** *Aust. J. Plant Physiol.*, 10, 1, 1983.
1084. **Sawhney, V. K.,** *Can. J. Bot.*, 61, 1258, 1983.

1085. **Sawhney, V. K.,** *J. Hered.*, 74, 51, 1983.
1086. **O'Toole, J. C., Hsiao, T. C., and Namuco, O. S.,** *Plant Sci. Lett.*, 33, 137, 1984.
1087. **Minter, T. C. and Lord, E. M.,** *Am. J. Bot.*, 70, 618, 1983.
1088. **van den ende, G., Croes, A. F., Kemp, A., and Barendse, G. W. M.,** *Physiol. Plant.*, 61, 114, 1984.
1089.
1090.
1091. **Hume, R. J. and Lovell, P. H.,** *Ann. Bot. (London)*, 52, 689, 1983.
1092. **Hume, R. J. and Lovell, P. H.,** *Physiol. Plant.*, 59, 324, 1983.
1093. **Hume, R. J. and Lovell, P. H.,** *Physiol. Plant.*, 58, 101, 1983.
1094. **Ruckenbauer, P. and Kirby, E. J. M.,** *J. Agric. Sci., Camb.*, 80, 211, 1973.
1095. **Polowick, P. L. and Greyson, R. I.,** *Can. J. Bot.*, 62, 830, 1984.
1096. **Mor, Y. and Halevy, A. H.,** *Physiol. Plant.*, 61, 119, 1984.
1097. **Guerin, B., Kahlem, G., Teller, G., and Durand, B.,** *Plant Physiol.*, 74, 139, 1984.
1098. **Xu, S. Y. and Bukovac, M. J.,** *J. Am. Soc. Hortic. Sci.*, 108, 278, 1983.
1099. **Suttle, J. C.,** *New Bull, BPGRG*, 6(2), 11, 1983.
1100. **Nell, T. A., Bodnaruk, W. H., Jr., Joiner, J. N., and Sheehan, T. J.,** *HortScience*, 18, 454, 1983.
1101. **Takahashi, H., Saito, T., and Suge, H.,** *Plant Cell Physiol.*, 24, 147, 1983.
1102. **Morgan, D. G. and Morgan, C. B.,** *Light and the Flowering Process*, Vince-Prue, D., Thomas, B., and Cockshull, K., Eds., Academic Press, London, 1984, in press.
1103. **Vonk, C. R. and Ribôt, S. A.,** *Plant Growth Regulation*, 1, 93, 1982.
1104. **Malmberg, R. L.,** *Cell*, 22, 603, 1980.
1105. **Malmberg, R. L. and McIndoo, J.,** *Nature (London)*, 305, 623, 1983.
1106. **Palavan, N. and Galston, A. W.,** *Physiol. Plant.*, 55, 438, 1982.
1107. **Cockshull, K. E.,** *J. Hortic. Sci.*, 57, 205, 1982.
1108. **Nourai, A. H. A. and Harris, G. P.,** *Scientia Hortic.*, 20, 341, 1983.
1109. **Russel, C. R. and Morris, D. A.,** *Ann. Bot. (London)*, 52, 357, 1983.
1110. **Erez, A. and Couvillon, G. A.,** *HortScience*, 18, 480, 1983.
1111. **Hamer, P. J. C.,** *J. Hortic. Sci.*, 56, 107, 1981.
1112. **Hamer, P. J. C.,** *J. Hortic. Sci.*, 58, 153, 1983.
1113. **Kronenberg, H. G.,** *Neth. J. Agric. Sci.*, 31, 259, 1983.
1114. **Fisher, J. E.,** *Can. J. Plant Sci.*, 53, 7, 1973.
1115. **Harris, R. E., Moll, R. H., and Stuber, C. W.,** *Crop Sci.*, 16, 843, 1976.
1116. **Michael, G.,** *Physiological Aspects of Crop Productivity*, Proc. 15th Colloq. Int. Potash Inst., Wageningen, 1980, 85.

GLOSSARY

Abaxial	The side of a lateral primordium or organ away from the axis
Abortion	Arrest of development of a structure: in some species, especially in bulbous plants, the terms abscission and blasting, which are mainly descriptive, are frequently used for abortion; abscission is the abortion of a structure which dries up and rapidly sheds; blasting is the abortion of a structure which shrivels, dries up, but usually does not shed; in certain studies, especially on rose, blindness is used for early abortion of the flower
Abscission	See Abortion
Adaxial	The side of lateral primordium or organ next to the axis
Androecious	Having primarily male flowers
Andromonoecious	Having male and hermaphroditic flowers on the same plant
Anthesis	Stage at which flower opens
Anticlinal	Perpendicular to the nearest surface
Apex	Apical part of the shoot including the apical meristem together with the immediate subjacent tissues and young leaf or flower primordia
Blasting	See Abortion
Blindness	See Abortion
Carpellody	The change of other floral organs into carpels (= pistillody)
Chasmogamy	Condition of flowers which reach anthesis normally
Cleistogamy	Condition of flowers which never reach anthesis and are self pollinated (opposite to chasmogamy)
Dioecious	Having unisexual flowers with the sexes on different plants
Disc florets	Central florets of the capitulum of some Compositae which are usually actinomorphic
Double flowers	Flowers in which the number of petals is much increased at the expense of other organs
Floral evocation	Processes in the apex required for irreversible commitment to initiate flower primordia
Florets	Individual flowers of Compositae and grasses; also other very small flowers that make up a very dense form of inflorescence
Gynoecious	Having primarily female flowers
Gynomonoecious	Having female and hermaphroditic flowers on the same plant
Hermaphroditic	(Of flowers) containing both functional stamens and ovary; (of plants) having hermaphroditic flowers
Heterostyly	The length of the style in relation to the other parts of the flower differs in different flowers of the same species
Meristem	That part of the shoot apex lying distal to the youngest leaf or flower primordium
Metamorphosis	The change of one floral organ into another
Monoecious	Having unisexual flowers, but both sexes on the same plant
Parastichy	An imaginary spiral linking up primordia (usually best recognized in transverse sections of apices) which differ by some constant in their order of initiation. In most angiosperms the primordia along such a spiral are in contact at their bases, hence the term "contact parastichy"; they are usually two sets of contact parastichies passing round the apex in opposite directions, characterized by numbers which are two consecutive terms from the Fibonacci series, e.g., (2 + 3), (3 + 5)[73,200]

Pedicel	The support of a single flower
Peduncle	The support of an inflorescence or partial inflorescence
Periclinal	Parallel to the nearest surface
Petalody	The change of other floral organs into petals
Pistillodium	A reduced infertile pistil
Pistillody	The change of other floral organs into pistil (= carpellody)
Plastochron	Time interval between the initiation of successive leaf primordia
Primordium	Refers to a nascent organ with height above its axil
Proliferation	Development of a new axis (reproductive or vegetative) inside a flower
Protandry	Stamens maturing before the gynoecium
Protogyny	Gynoecium maturing before the stamens
Ray florets	Marginal florets of the capitulum of some Compositae which are usually ligulate, i.e., zygomorphic
Reproductive axis	Any axis supporting a reproductive structure; this general term is used for stem, stalk, scape, peduncle, or pedicel
Reversion	Transition from a reproductive stage to a complete or partial vegetative condition
Scape	Leafless reproductive axis or peduncle arising from the ground
Sepalody	The change of other floral organs into sepals
Stalk	Any lengthened support of a structure
Staminodium	A reduced infertile stamen
Staminody	The change of other floral organs into stamens
Stem	The main axis
Trimonoecious	Having male, hermaphroditic, and female flowers on the same plant
Virescence	Development of one or several floral organs into leaf-like structures

ABBREVIATIONS

ABA	(±)-Abscisic acid
ACC	1-Amino-cyclopropane-1-carboxylic acid
ADOPB	1-Allyl-1-(3,7-dimethyloctyl)-piperidinium bromide
AMAB	Allyl trimethylammonium bromide
AMO - 1618	2′-Isopropyl-4′-(trimethylammonium chloride)-5′-methylphenyl-piperidine-1-carboxylate
Ancymidol	α-Cyclopropyl-α-(*p*-methoxyphenyl)-5-pyrimidine methyl alcohol
ATP	Adenosine triphosphate
AVG	L-2-Amino-4-(2-aminoethoxy)-*trans*-3-butenoic acid, aminoethoxyvinyl glycine
BA	6-Benzylaminopurine
BCB	Bromocholine bromide
B-nine	N-(Dimethylamino)succinamic acid, also termed B995, Alar, SADH
CCC	2-Chloroethyltrimethylammonium chloride (chlorocholine chloride)
CEPA	2-Chloroethylphosphonic acid
CR	Chilling requirement or cold requirement
CU	Chill unit
cv., cvs.	Cultivar, cultivars
2,4-D	2,4-Dichlorophenoxyacetic acid
Daminozide	See B-nine
DCMU	3(3,4-Dichlorophenyl)-1,1-dimethylurea
DNA	Deoxyribonucleic acid
DNP	Day-neutral plant
$E/m^2/s$	Einsteins per square meter per second, photon flux density of photosynthetically active radiation (400 to 700 nm) unit
eq	Equivalent
FR	Far-red light 700 to 770 nm (approximately)
GA(s)	Stands for gibberellin(s) in general; individual gibberellins are designated as GA_1, GA_2, etc.
GA-like substances	Substances occurring naturally and causing responses typical of known GAs, but not yet known chemically
GA_3	Gibberellic acid
GC-MS	Combined gas chromatography-mass spectrometry
GDH	Growing degree hour(s)
hr	Hour(s)
IAA	Indol-3-ylacetic acid
IBA	γ-(Indol-3-yl) butyric acid
J/m^2	Joule per square meter, radiant exposure unit
LD	Long day(s)
LDP	Long-day plant
lx	Lux, illuminance unit
MCEB	5-methyl-7-chloro-4-ethoxycarbonylmethoxy-2,1,3-benzothiadiazole
MH	Maleic hydrazide
min	Minute(s)
NAA	Naphth-1-ylacetic acid, α-naphthalene acetic acid
PBA	6-(Benzylamino)-9-(2-tetrahydropyranyl)-9*H*-purine
Pfr	Phytochrome in the far-red absorbing form
Phosfon(Ph)	2,4-Dichlorobenzyl-tributylphosphonium chloride
R	Red light (600 to 700 nm) approximately

RNA	Ribonucleic acid
RSA	Relative specific activity
SD	Short day(s)
SDP	Short-day plant
sec	Second(s)
SEM	Scanning electron microscopy
STS	Silverthiosulfate, $Ag(S_2O_3)_2^{3-}$
TIBA	2,3,5-Triiodobenzoic acid
W/m^2	Watt per square meter, irradiance unit
Wt	Weight

SPECIES INDEX

A

B

C

D

K

L

M

N

O

P

Q

R

S

T

V

W

X

Y

Z

SUBJECT INDEX

A

B

C

D

T

U

V

W

Z